GOOD SCIENCE

NSW Syllabus for the Australian Curriculum Stage 4

7+8

Rebecca **Cashmere**
Thomas **Cosgrove**
Emma **Craven**
Sarah **Edwards**
Haris **Harbas**
Niru **Kumar**
Vicki **Maggs**
Rhyce **Mahon**
Natalie **Stinson**

Series reviewers: Maria **de Lima** and Aaron **Elias**
Investigations reviewer: Margaret **Croucher**

Good Science Stage 4 NSW Syllabus for the Australian Curriculum

1st edition

Emma Craven

Rebecca Cashmere

Thomas Cosgrove

Sarah Edwards

Haris Harbas

Niru Kumar

Vicki Maggs

Natalie Stinson

Publisher: Catherine Charles-Brown

Content development: Patrick O'Duffy

Publishing director: Olive McRae

Project editor: Amy Nicholls-Diver

Copy editor: Sally Woollett

Proofreader: Jane Fitzpatrick

Indexer: Max McMaster

Illustrator: QBS Learning

Design manager: Jo Groud

Cover and text designer: Regine Abos

Production controller: Nicole Ackland

Digital production: Erin Dowling

Permissions researcher: Annthea Lewis

Typesetters: Cath Pirrett, Anne Stanhope, Leigh Ashforth

This edition published in 2021 by

Matilda Education Australia, an imprint
of Meanwhile Education Pty Ltd
Melbourne, Australia
T: 1300 277 235
E: customersupport@matildaed.com.au
www.matildaeducation.com.au

First edition published in 2020 by Macmillan Science and Education Australia Pty Ltd

Publication data

Author: Emma Craven, Rebecca Cashmere, Thomas Cosgrove, Sarah Edwards, Haris Harbas, Niru Kumar, Vicki Maggs, Natalie Stinson

Title: *Good Science Stage 4 NSW Syllabus for the Australian Curriculum*

ISBN: 9781420246124

A catalogue record for this book is available from the National Library of Australia

Printed in Malaysia by Vivar Printing
Sep-2024

MAKE EVERY LESSON A GOOD LESSON

CONTENTS

EARTH AND SPACE 72

CONTENTS

SYLLABUS CORRELATION GRID

	Stage 4 Outcomes		1 States of matter	2 Structure and properties of matter	3 Mixtures	4 Chemical change	5 Earth and the rock cycle	6 Earth, Sun and the Moon	7 Resources	8 Water as a resource	9 Forces	10 Fields	11 Energy	12 Energy as a resource	13 Classification	14 Cells	15 Body systems	16 Biotechnology	17 Ecosystems	Science skills	Investigations
Skills	SC4-4WS	A student identifies questions and problems that can be tested or researched and makes predictions based on scientific knowledge					✓				✓										✓
Skills	SC4-5WS	A student collaboratively and individually produces a plan to investigate questions and problems			✓																✓
Skills	SC4-6WS	A student follows a sequence of instructions to safely undertake a range of investigation types, collaboratively and individually																		✓	✓
Skills	SC4-7WS	A student processes and analyses data from a first-hand investigation and secondary sources to identify trends, patterns and relationships, and draw conclusions	✓					✓													✓
Skills	SC4-8WS	A student selects and uses appropriate strategies, understanding and skills to produce creative and plausible solutions to identified problems	✓	✓	✓	✓	✓	✓	✓	✓	✓	✓	✓	✓	✓	✓	✓	✓	✓	✓	✓
Skills	SC4-9WS	A student presents science ideas, findings and information to a given audience using appropriate scientific language, text types and representations	✓	✓	✓	✓	✓	✓	✓	✓	✓	✓	✓	✓	✓	✓	✓	✓	✓	✓	✓
Physical World	SC4-10PW	A student describes the action of unbalanced forces in everyday situations									✓	✓									
Physical World	SC4-11PW	A student discusses how scientific understanding and technological developments have contributed to finding solutions to problems involving energy transfers and transformations									✓	✓		✓							
Physical World	PW1	Change to an object's motion is caused by unbalanced forces acting on the object (ACSSU117).									✓										
Physical World	PW2	The action of forces that act at a distance may be observed and related to everyday situations.										✓									

	Stage 4 Outcomes		1 States of matter	2 Structure and properties of matter	3 Mixtures	4 Chemical change	5 Earth and the rock cycle	6 Earth, Sun and the Moon	7 Resources	8 Water as a resource	9 Forces	10 Fields	11 Energy	12 Energy as a resource	13 Classification	14 Cells	15 Body systems	16 Biotechnology	17 Ecosystems	Science skills	Investigations
	PW3	Energy appears in different forms including movement (kinetic energy), heat and potential energy, and causes change within systems (ACSSU155).											✓								
	PW4	Science and technology contribute to finding solutions to a range of contemporary issues; these solutions may impact on other areas of society and involve ethical considerations (ACSHE120, ACSHE135).												✓							
Earth and Space	SC4-12ES	A student describes the dynamic nature of models, theories and laws in developing scientific understanding of the Earth and solar system					✓	✓													
Earth and Space	SC4-13ES	A student explains how advances in scientific understanding of processes that occur within and on the Earth, influence the choices people make about resource use and management							✓	✓											
Earth and Space	ES1	Sedimentary, igneous and metamorphic rocks contain minerals and are formed by processes that occur within Earth over a variety of timescales. (ACSSU153)					✓														
Earth and Space	ES2	Scientific knowledge changes as new evidence becomes available. Some technological developments and scientific discoveries have significantly changed people's understanding of the solar system.						✓													
Earth and Space	ES3	Scientific knowledge influences the choices people make in regard to the use and management of the Earth's resources.							✓												
Earth and Space	ES4	Science understanding influences the development of practices in areas of human activity such as industry, agriculture and marine and terrestrial resource management. (ACSHE121, ACSHE136)								✓											

SYLLABUS CORRELATION GRID

	Stage 4 Outcomes		1 States of matter	2 Structure and properties of matter	3 Mixtures	4 Chemical change	5 Earth and the rock cycle	6 Earth, Sun and the Moon	7 Resources	8 Water as a resource	9 Forces	10 Fields	11 Energy	12 Energy as a resource	13 Classification	14 Cells	15 Body systems	16 Biotechnology	17 Ecosystems	Science skills	Investigations
Living World	SC4-14LW	A student relates the structure and function of living things to their classification, survival and reproduction													✓	✓	✓				
Living World	SC4-15LW	A student explains how new biological evidence changes people's understanding of the world														✓		✓	✓		
Living World	LW1	There are differences within and between groups of organisms; classification helps organise this diversity (ACSSU111)													✓						
Living World	LW2	Cells are the basic units of living things and have specialised structures and functions (ACSSU149)														✓					
Living World	LW3	Multicellular organisms contain systems of organs that carry out specialised functions that enable them to survive and reproduce (ACSSU150)															✓				
Living World	LW4	Scientific knowledge changes as new evidence becomes available, and some scientific discoveries have significantly changed people's understanding of the world. (ACSHE119, ACSHE134)																✓			
Living World	LW5	Science and technology contribute to finding solutions to conserving and managing sustainable ecosystems.																	✓		

Strand	Stage 4 Outcomes		1 States of matter	2 Structure and properties of matter	3 Mixtures	4 Chemical change	5 Earth and the rock cycle	6 Earth, Sun and the Moon	7 Resources	8 Water as a resource	9 Forces	10 Fields	11 Energy	12 Energy as a resource	13 Classification	14 Cells	15 Body systems	16 Biotechnology	17 Ecosystems	Science skills	Investigations
Chemical World	SC4-16CW	A student describes the observed properties and behaviour of matter, using scientific models and theories about the motion and arrangement of particles	✓	✓		✓															
Chemical World	SC4-17CW	A student explains how scientific understanding of, and discoveries about the properties of elements, compounds and mixtures relate to their uses in everyday life		✓	✓																
Chemical World	CW1	The properties of the different states of matter can be explained in terms of the motion and arrangement of particles. (ACSSU151)	✓			✓															
Chemical World	CW2	Scientific knowledge and developments in technology have changed our understanding of the structure and properties of matter.		✓																	
Chemical World	CW3	Mixtures, including solutions, contain a combination of pure substances that can be separated using a range of techniques. (ACSSU113)		✓	✓																
Chemical World	CW4	In a chemical change, new substances are formed, which may have specific properties related to their uses in everyday life.				✓															

1 STATES OF MATTER

Everything is made of matter. Solid, liquid and gas are the three main states of matter that we observe on Earth. A fourth state of matter, called plasma, occurs in lightning and inside the Sun. Each state of matter can be described by the way particles behave. In many everyday activities – such as melting chocolate, freezing ice cubes and evaporating water when we sweat – we are changing states of matter.

LEARNING LINKS

What do you already know about states of matter?

How are solid, liquid and gas substances different to each other?

Melting, freezing and evaporation are reversible changes. How do you know?

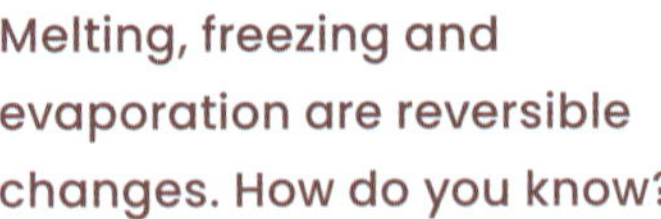

How do objects around you change when you add or remove heat?

2 SEE-KNOW-WONDER

List three things you can **see**, three things you **know** and three things you **wonder** about this image.

3 CRITICAL + CREATIVE THINKING

What if ... all the solids in the world became liquid overnight, and all the liquids became solid?

Alternatives: List all the ways that you could evaporate something without using fire.

Construction: How could you demonstrate the properties of solids, liquids and gases using a cup, a piece of string and 10 paperclips?

4 THE MOST!

Which of the states of matter do you think makes up most of the universe? You might be surprised to hear that it is plasma. Plasma makes up about 99% of known matter!

Plasma is found in amazing things like flames, stars and lightning. It requires a lot of energy to form. Lightning forms when electrical charges in the atmosphere create so much energy the air changes from gas to plasma.

1.1 THE PARTICLE MODEL

At the end of this lesson I will be able to:

- **describe** the behaviour of matter in terms of particles that are continuously moving and interacting.

KEY TERMS 

compress
squash into a smaller space

matter
particles that make up all physical substances; they have mass and take up space

particles
a very small amount of matter

particle model
a model used to describe the properties of solids, liquids and gases

volume
the space taken up by something

LITERACY LINK 

Create a flowchart that shows the relationship between solids, liquids and gases.

NUMERACY LINK 

Oxygen fills a room that is 6 m long, 5 m wide and 3 m high. What is the volume of the room?

Formula:
Volume = length x width x height

Figure 1.1 Water is the only substance that is found naturally as a solid, liquid and gas on Earth.

The **particle model** is a way of describing all the **matter** on Earth. This model states that all matter is made up of tiny **particles**, and that these particles are constantly moving. It can explain why matter behaves in certain ways. It can predict how matter will be affected by changing conditions such as pressure and temperature.

How can this model be used to describe the properties of solids, liquids and gases?

1 The particles in a solid are packed close together

Solids are materials such as metal or plastic. The particles in solids are packed closely together like bricks in a wall. The particles aren't still, they are constantly vibrating.

A solid has a fixed shape and a fixed **volume**. Because the particles in a solid are very strongly attracted to each other, the solid keeps its shape and doesn't spread out or flow. A solid cannot be easily **compressed**, because there's no room between particles for them to squeeze closer together.

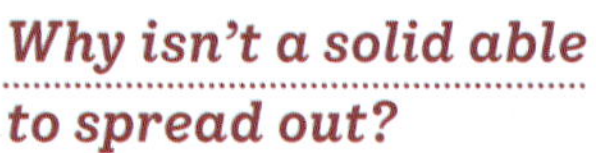
Why isn't a solid able to spread out?

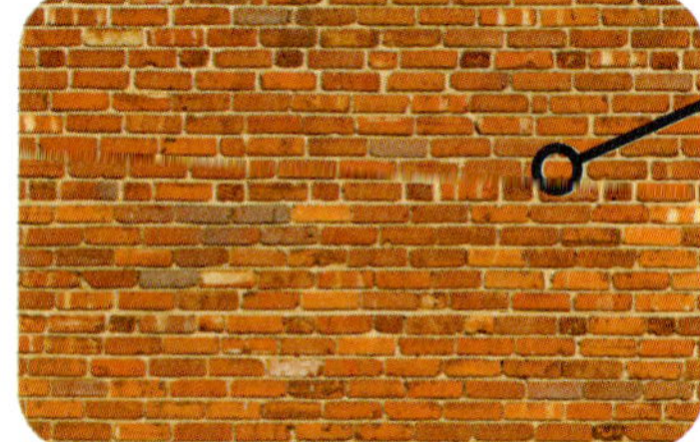

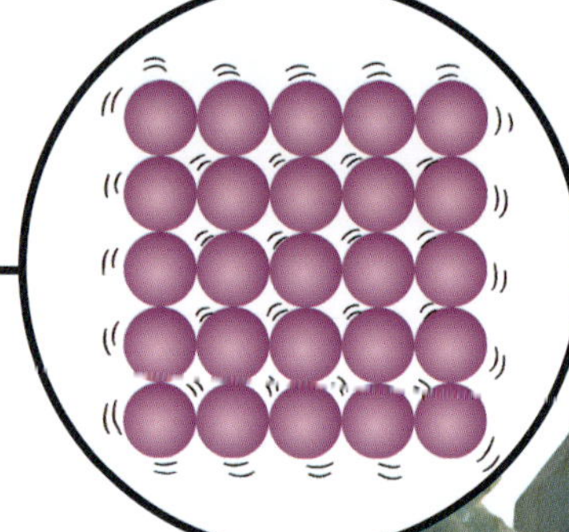

Figure 1.2 The particles in a solid are packed closely together, like bricks in a wall. They vibrate in their positions.

❷ The particles in a liquid can move past each other

The particles in liquids aren't as close together as they are in solids. The particles are still strongly attracted to each other, but there's room for them to move past each other.

A liquid has a fixed volume but not a fixed shape. A liquid will take on the shape of its container, because the attraction between the particles isn't strong enough to stop them from spreading out. A liquid can't be compressed very much, because there isn't much space between particles for them to squeeze closer together.

Why does a liquid take on the shape of its container?

Figure 1.3 The particles in a liquid can move past each other. They will take on the shape of their container.

❸ The particles in a gas have large gaps between them

The particles in a gas are very weakly attracted to each other, and so they can move around a lot. The particles have large gaps between them, and they are constantly moving in all directions. Some common gases on Earth are oxygen and carbon dioxide. Air is a mixture of gases including oxygen, nitrogen and carbon dioxide.

A gas has neither a fixed shape nor a fixed volume. Gases will spread out to fill up the container they are placed in. A gas can be compressed because there is space between the particles. In a smaller space the gas particles just have less room to move around.

Why can a gas be compressed?

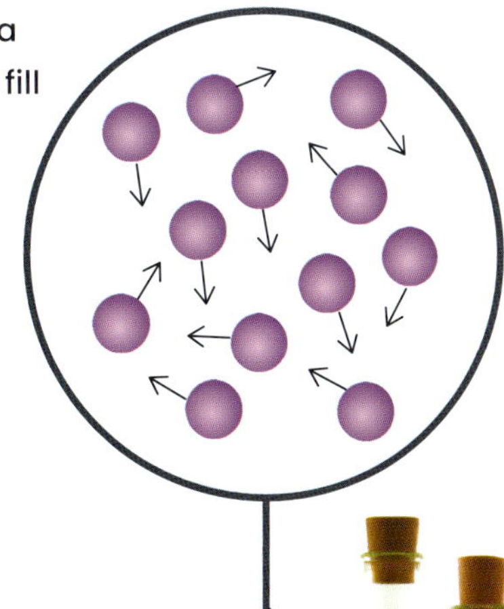

Figure 1.4 The particles in a gas are weakly attracted to each other. They can be squashed into a smaller space.

INVESTIGATION 1.1
Compressing liquids and gases

CHECKPOINT 1.1 ✓

1 List three solids, three liquids and three gases you have come into contact with today.

2 a My particles are constantly moving all over the place and have a weak attraction to each other. People always tell me that I will go far in life. I do not have a fixed shape or a fixed volume. Who am I?
  b My particles are constantly moving and have a very strong attraction to each other. I have a fixed shape and a fixed volume. As hard as you try, you can't give me the squeeze! Who am I?
  c My particles are constantly moving (usually over, under or past each other) and have a strong attraction to each other. I tend to go with the flow and do not have a fixed shape (although I do have a fixed volume). Who am I?

3 a How can a brick wall be used to explain the particles in a solid?
  b How can a jar of marbles be used to explain the particles in a liquid?
  c How can a pool table and balls be used to explain the particles in a gas?

CHALLENGE

4 Plasma is sometimes called the fourth state of matter. Use the internet to research plasma, and how it differs from the other three states of matter.

SKILLS CHECK

- I can state the three main states of matter.
- I can describe the behaviour of particles in each state.

1.2 HEAT ENERGY AND PARTICLES

At the end of this lesson I will be able to:

- **relate** an increase or decrease in the amount of heat energy possessed by particles to changes in particle movement.

KEY TERMS 

energy
the ability to do physical things such as move or change

heat
a type of energy

substance
matter that has a fixed chemical make-up

temperature
the measurement of how hot a substance is

transferred
moved from one thing to another

LITERACY LINK 

An analogy is a way of comparing one thing to another. Can you think of an analogy to describe the heating of gas particles?

NUMERACY LINK 

The temperature of a substance is 7°C.

If the temperature drops by 10°C, what is the new temperature?

The particle model states that all matter is made up of constantly moving particles. A solid object such as a statue doesn't appear to be moving, but its particles are.

The speed of the particles depends on the amount of **energy** they have. If you increase the energy of the particles, they'll move faster and further away from each other. How can you increase energy? Just add heat!

1 Heat is a type of energy

Energy is needed for matter to do things like move or change. **Heat** is one type of energy. The more heat energy a **substance** has, the faster and further apart the particles will move.

Heat can be **transferred** from one object to another. For example, in Figure 1.6 the heat energy in the electric kettle element is transferred to the water to boil it.

The same principle applies in solids. Have you ever rested a metal spoon in a hot drink? At first, the metal handle feels cool, but soon warms up and can even burn you. This shows that energy is transferred from the bottom of the spoon all the way to the top. The particles at the base of the spoon are heated and begin to move around more. They bounce up against the particles near them, which in turn gain energy, bouncing into the ones near them, and so it continues, all the way up the spoon to the handle.

How does the speed of particles change when heat is added to a substance?

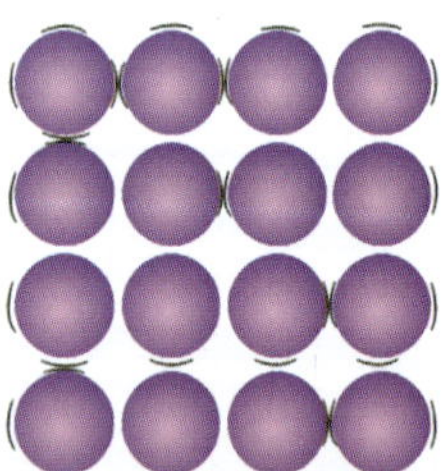

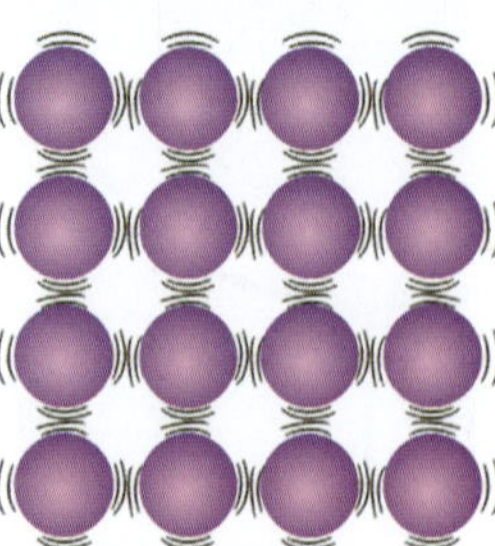

Figure 1.5 The more heat you add, the faster the particles in a substance move.

Figure 1.6 When you heat something, for example when you boil water, you're adding heat energy to it.

❷ Substances lose heat energy when they cool

There are many ways to cool a hot object. You could put it in the fridge, put it in cold water, or just leave it to cool in the air.

Heat energy is transferred from substances with lots of energy to substances with less energy until the particles in both substances have the same amount of heat energy.

When a substance loses energy in this way, the particles slow down and move closer together.

How does the speed of particles in a substance change when heat is removed?

❸ Temperature and heat are not the same thing

Temperature is a measurement of the average amount of heat energy the particles in a substance have. The higher the temperature of a substance, the more heat energy is held by the particles and the faster they move. If heat energy is added to a substance, the temperature will increase. If heat energy is removed from a substance, the temperature will decrease.

If a large iron block and a small iron block were heated in the laboratory to a temperature of 50°C, the larger block would contain more heat energy. This is because it has more particles, even though it is the same temperature as the smaller block.

What is temperature?

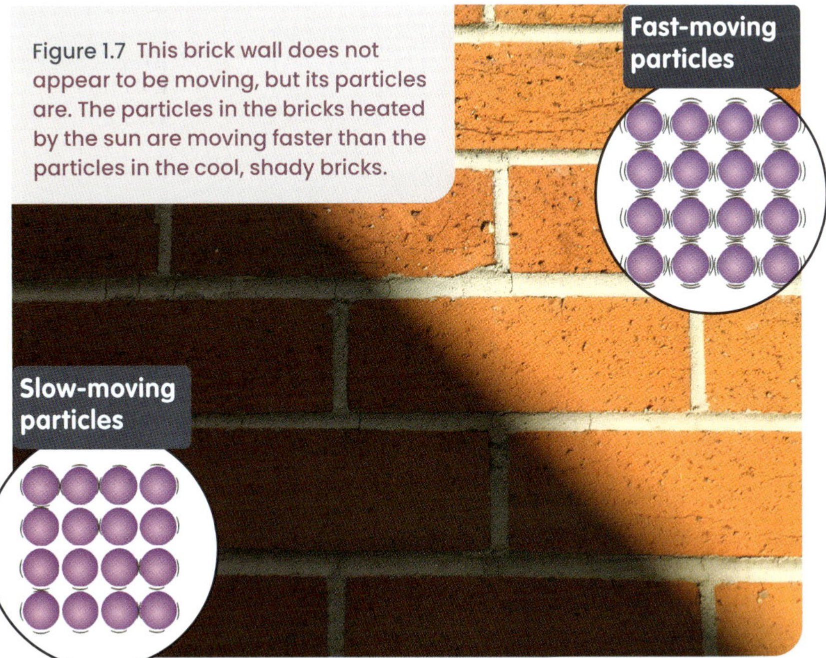

Figure 1.7 This brick wall does not appear to be moving, but its particles are. The particles in the bricks heated by the sun are moving faster than the particles in the cool, shady bricks.

INVESTIGATION 1.2
Heating materials

CHECKPOINT 1.2 ✓

1 Copy and complete these sentences.
The speed of particles is directly related to the amount of __________ they have. The more __________, the faster they move.

2 Would the particles in a cup of 30°C water move faster or more slowly than particles in a cup of 45°C water? Give evidence to support your answer.

3 In terms of the way the particles behave, how does heating a solid differ to heating a liquid?

4 Imagine you've decided to cook a steak for dinner. You remove it from the freezer and defrost it. Describe how the motion of the particles in the steak would change as it defrosts.

5 Choose an object. Draw a diagram that shows how the particles in that object behave when heat is added and when heat is taken away. Use the diagrams on this page to help you.

CHALLENGE

6 Particles move more slowly as heat energy is removed. Use the internet to find out if we can slow them down so much that they stop. If so, at what temperature does this happen?

SKILLS CHECK

- I can describe how particles act when heat energy is added and taken away.

1.3 ADDING OR REMOVING HEAT

At the end of this lesson I will be able to:

- **use** a simple particle model to predict the effect of adding or removing heat on different states of matter.

KEY TERMS 

contract
get smaller, shrink

expand
get bigger, increase in size

thermal
relating to heat

LITERACY LINK 

Write three questions for a classmate based on the information from this section. Swap with your classmate and answer their questions.

NUMERACY LINK 

A gas fills a volume of 110 m^3, but then contracts by 30% as the temperature changes. What volume does it now fill?

You now know that the speed of the particles in a substance depends on the substance's temperature. Particles in a hot substance will move quickly, while particles in a cold substance will move slowly.

Some other changes happen when an object is heated or cooled.

1 Solids expand when they are heated

If you add heat energy to a substance, the particles move faster. Because the particles have more energy and are moving faster, they will move further away from each other and so the substance will **expand**. This is called **thermal** expansion.

When a solid is heated, it will expand. Different solid substances expand at different speeds and by different amounts. Most solids do not noticeably change their size. Metals are one type of substance that expand noticeably when heated. When metals cool down again, the particles lose energy and so move closer together again. This is called thermal **contraction**.

What is thermal expansion?

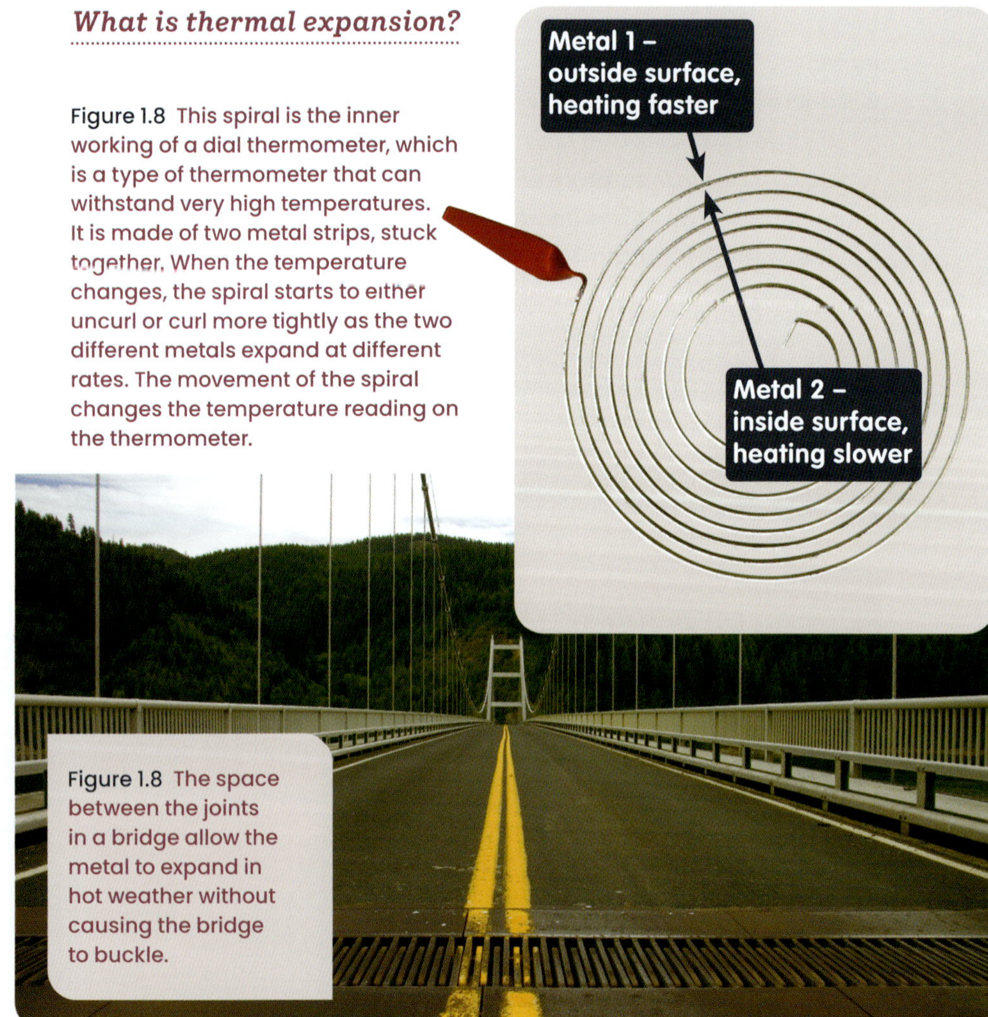

Figure 1.8 This spiral is the inner working of a dial thermometer, which is a type of thermometer that can withstand very high temperatures. It is made of two metal strips, stuck together. When the temperature changes, the spiral starts to either uncurl or curl more tightly as the two different metals expand at different rates. The movement of the spiral changes the temperature reading on the thermometer.

Figure 1.8 The space between the joints in a bridge allow the metal to expand in hot weather without causing the bridge to buckle.

2 Liquids expand when they are heated

Thermal expansion also happens in liquids. Heating a liquid causes the particles to move more quickly and further away from each other, so liquids will take up a larger volume as they become warm. They will contract as they cool back down again.

Water that is heated without any room to expand can be very damaging. For example, it can cause pipes to burst. This property of liquids can be used safely, in alcohol and mercury thermometers. The liquid inside them expands when the temperature increases.

Why can thermal expansion be a problem?

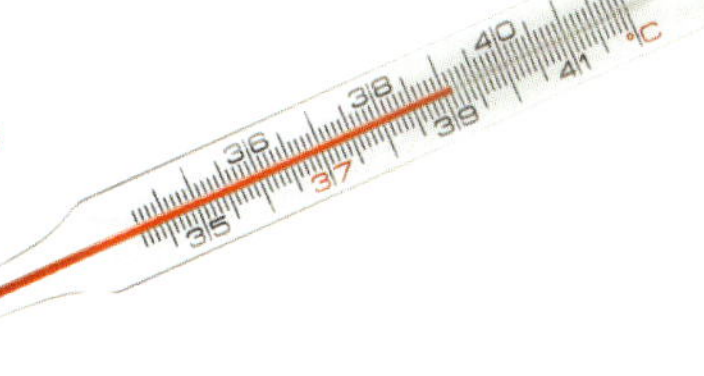

Figure 1.9 The liquid alcohol inside a thermometer takes up different amounts of space depending on how much heat energy it has, allowing the temperature to be measured.

3 Gases expand or contract with changes in heat

As with solids and liquids, thermal expansion and contraction also happens in gases. Because the particles in gases are already far apart, the expansion and contraction of gases can be very noticeable.

Although most gases are invisible, you can still see that a warm gas will take up more space than a cold gas. You can check this by blowing up a balloon and putting it in a refrigerator. The gases inside it will cool down and contract, making the balloon smaller.

How will the volume of a gas change if it is cooled?

Figure 1.10 The gas in a hot air balloon will expand when it is heated.

INVESTIGATION 1.3
Expanding gases

CHECKPOINT 1.3

1 What happens to the particles in a substance when heat energy is added?

2 What happens to the particles in a substance when they lose heat energy?

3 Identify which state of matter (solid, liquid or gas) would show the greatest change in volume if heat energy is added or lost? Use the particle model to help explain your answer.

4 An iron bar is heated from room temperature to 300°C. Would this make the bar expand or contract? Explain your answer using evidence from the text.

5 Describe and include a diagram showing what happens to the particles in the iron bar in question 4, above.

6 Design a safety feature in a building that would prevent the walls cracking when they expand after heating. You can choose to complete a written description or an annotated diagram.

CHALLENGE

7 Bridges have expansion joints to prevent buckling in hot weather, as shown in Figure 1.8. Use the internet to find out other ways that engineers protect against thermal expansion when building structures.

SKILLS CHECK

- I can predict the effect of adding or removing heat on:
  - solids
  - liquids
  - gases.

1.4 CHANGING STATES

At the end of this lesson I will be able to:

- **relate** changes in the physical properties of matter to changes in heat energy and particle movement that occur during observations of evaporation, condensation, boiling, melting and freezing.

KEY TERMS 

boiling point
the temperature when something changes from a liquid to a gas

condensation
changing from a gas to a liquid

evaporation
changing from a liquid to a gas

melting point
the temperature when something changes from a solid to a liquid

LITERACY LINK 

Write an aim for an investigation of your choice that investigates evaporation. (Hint: Start your aim with 'To investigate ...')

NUMERACY LINK 

A block of ice melts into 0.7 L of water.

Write 0.7 as a fraction.

Water is the only substance on Earth that exists as a solid, liquid and gas. Solid water is ice. If you heat it, the ice melts into liquid water. If you keep heating, it boils and becomes water vapour.

The particle model and your understanding of how heat affects particles can be used to explain these changes in state.

Figure 1.11 Water can change state from solid to liquid to gas as heat is added. If heat is taken away, the changes are reversed.

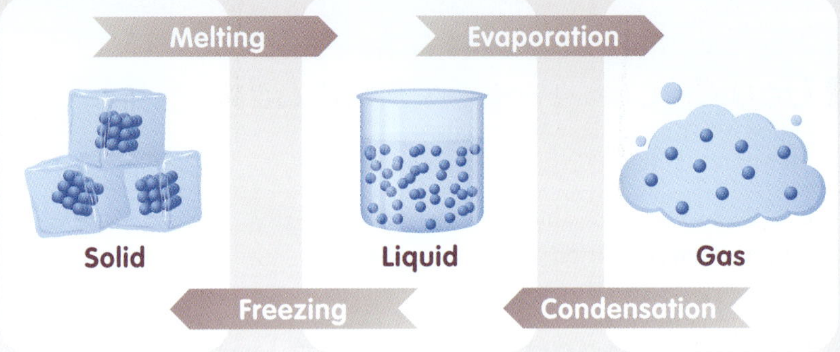

1 Solids can melt to liquids when heated

Melting is the change of a solid to a liquid. The **melting point** is the temperature when something changes from a solid to a liquid. This is different for every substance.

The particles of a substance in its solid state are strongly attracted and don't move very much – they just vibrate back and forth. As you add heat energy, the particles move faster and vibrate more. If you add enough heat, the particles have enough energy to break free of the solid structure and move around each other. The substance is now a liquid.

Which state of matter undergoes melting?

Figure 1.12 When particles in a solid such as wax speed up, the solid can melt to form a liquid.

❷ Liquids can freeze to solids when cooled

Freezing is the change of a liquid to a solid when cooled. It is the opposite of melting.

The particles in a liquid have enough energy to move around each other. If you cool a liquid, you remove heat energy and cause the particles to move more slowly. If particles lose enough energy, they no longer move around, but instead just vibrate in place. The substance is now a solid.

Freezing is when a liquid becomes which state of matter?

❸ Liquids can evaporate to gases when heated

Evaporation is the change of a liquid to a gas.

Evaporation can happen at any temperature. In any liquid, some of the particles will randomly have enough energy to break away from the others. These particles leave the surface of the liquid as a gas. Evaporation is the reason that a puddle will eventually shrink and disappear, even during cool weather.

Boiling is evaporation that happens when a liquid is heated to its **boiling point**. This point is different for every substance. When they have enough heat energy, the particles in the liquid move so quickly that they become a gas inside the liquid. This gas can be seen as bubbles that rise through the liquid and escape.

Figure 1.13 Particles near the surface of a liquid become a gas during evaporation.

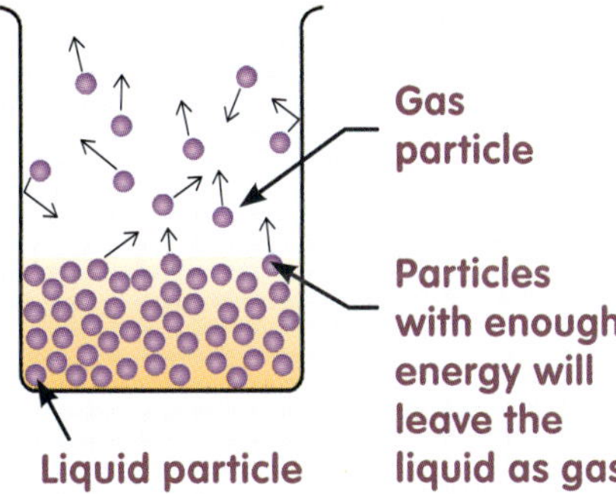

What is an example of evaporation?

❹ Gases can condense to liquids when cooled

Condensation happens when a gas changes to a liquid when cooled. It is the opposite of evaporation.

Gas particles move very quickly and so keep separate from the other particles. However, if you remove heat energy and slow the particles down, they no longer have enough speed to overcome the attraction of the other particles. They move closer together to become a liquid.

Have you ever seen water drops form on the outside of a cold glass of water? That's condensation.

What happens when gas particles slow down?

INVESTIGATION 1.4
Exploring melting points

CHECKPOINT 1.4 ✓

1 Copy and complete these sentences.
Changes in ________ happen when heat ________ is ________ or ________ from a substance.

2 From each pair, identify which state of matter has particles with more energy.
a liquid or gas
b gas or solid
c solid or liquid

3 Describe how heat energy affects what happens to the particles in:
a a solid that is melting
b a liquid that is evaporating
c a gas that is condensing
d a liquid that is freezing.

4 Explain the difference between evaporation and condensation.

5 Create a table to compare the amount of energy and distance between the particles in solids, liquids and gases. (Hint: Use words such as high, low, close, far).

CHALLENGE

6 Sublimation is when a substance changes from solid to gas, without becoming a liquid. Use evidence from the text to propose how this happens. Use your internet research skills to see if you were correct.

SKILLS CHECK

- I can describe what happens to particles during:
  - melting
  - freezing
  - boiling
  - condensation.

1.5 DENSITY

At the end of this lesson I will be able to:

- **explain** density in terms of a simple particle model.

KEY TERMS 

density
how heavy something is for its size; mass divided by volume

mass
the amount of matter in an object

volume
the amount of space an object takes up

LITERACY LINK 

Use these terms in a sentence: *dense, density*. Share your sentence with a classmate and then together come up with an explanation of the difference between the two words.

NUMERACY LINK 

The mass of a substance is 4900 grams, and it takes up a volume of 70 mL. What is its density?

Formula:
density = mass ÷ volume

What determines whether something floats in water? You might suggest that the **mass** of the object is important. But there must be more to it. Icebergs are extremely heavy, yet are able to float, while coins are much lighter but they sink.

The answer has to do with a property called **density**.

1 Density is how heavy an object is for its size

You can calculate the density of an object by dividing its mass by its **volume**. In other words, density is a measure of how heavy an object is compared to its size.

Solids usually have higher densities than gases and liquids as their particles can pack more closely together. A coin may be lighter than a log, but a coin-sized piece of the log would be lighter than the coin because metals are denser than wood. So, when comparing two objects of the same size, the heavier one has a greater density.

What two properties is density related to?

Figure 1.14 How is a heavy battleship able to float on water?

❷ Density can be explained using the particle model

In terms of the particle model, density is related to two things.

The first thing to consider is the amount of space between particles. The more tightly packed the particles are, the denser the object.

Second, it depends on the mass of the particles. The heavier the particles are, the denser the object. Imagine a bike with an aluminium frame and the same bike made of steel (iron metal). Although they are the same size, the steel bike is heavier because iron atoms are heavier than aluminium atoms.

What factors affect the density of an object?

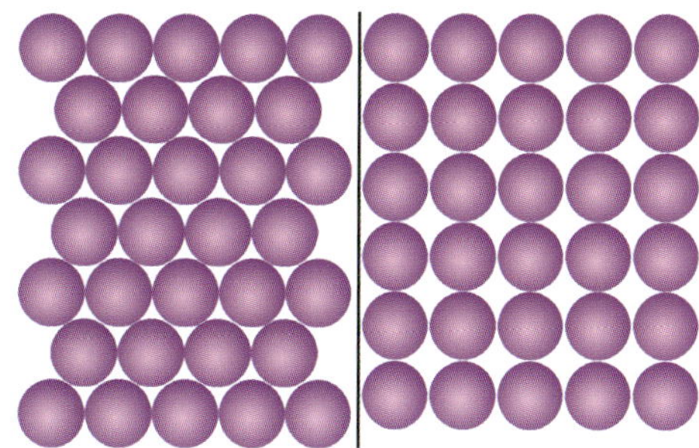

Figure 1.15 The particles on the left are more closely packed than the particles on the right. A substance with particles that are closely packed has a high density.

Higher density Lower density

❸ Liquids and gases with lower density float on those with higher density

Liquids and gases with a lower density float on top of substances with a higher density.

Have you ever heard that 'hot air rises'? This is because hot air is less dense than cool air. The space between the particles in the hot air is larger than the space between the particles in the cool air. This is what keeps a hot air balloon in the sky.

Consider the battleship in Figure 1.14. It floats on water, which must mean a battleship has a lower density than water. How can this be, if the ship is made of metal?

Even though a ship is made of metal, most of the inside of a battleship is air. The air takes up a lot of volume but has very little mass, and so overall the battleship has a density that is less than water.

What does a battleship contain that allows it to float on water?

Figure 1.16 Liquids of different densities in a container will arrange themselves so that the liquid with the highest density is at the bottom and the liquid with the lowest density is at the top.

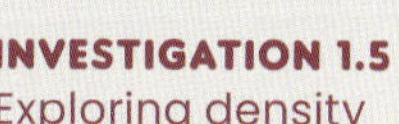

INVESTIGATION 1.5
Exploring density

CHECKPOINT 1.5 ✓

1 What is the formula for calculating density?

2 a List three objects that float in water and three objects that sink in water.
  b Using your knowledge of the particle model, explain why the objects in part **a** float or sink.
  c Draw a diagram showing high density and low density particles to illustrate the explanation you wrote in part **b**.

3 In terms of density and particles, explain why a soccer ball weighs much less than a bowling ball of the same size.

4 A fresh egg will sink in a glass of water, but a rotten egg will float. Which of the two eggs has a higher density? Give evidence from the text to support your answer.

5 Predict what would happen if you dropped a six-sided die, a Lego minifigure and a gold earring into the container shown in Figure 1.16. Explain how you came up with your prediction.

CHALLENGE

6 A type of star known as a neutron star contains some of the densest material in the universe. Use the internet to find out about the density of a neutron star.

SKILLS CHECK

- I can describe what density is.
- I can draw a diagram showing high density and low density particles.

1.6 USING MODELS

At the end of this lesson I will be able to:

- **identify** the benefits and limitations of using models to explain the properties of solids, liquids and gases.

KEY TERMS

accurate
very close to the true, exact value

gravity
a force that attracts objects towards each other

limitation
a factor that limits or restricts

phenomenon
an observed event

property
a characteristic or attribute of a substance

scientific model
an idea used to explain and predict occurrences that we can't directly interact with

LITERACY LINK

Write a letter to your teacher, trying to persuade them to agree with one of these statements.

- Models are the best way to demonstrate scientific knowledge.
- Models are too inaccurate and should not be used to demonstrate scientific knowledge.

NUMERACY LINK

Larissa examines a sample containing 12 iron particles. What are the numerical factors of 12?

Scientific models are important. They help scientists understand and communicate how things work. Some models that you might use this year in your science course include models of the solar system, the human body and the water cycle. Models have **limitations** as they cannot explain or communicate all variations. Different models have different limitations.

The particle model is a useful way to understand the **properties** of substances. The way particles actually behave is much more complex.

Figure 1.17 The particle model doesn't explain how a magnet can attract iron.

1 Many forces act on the particles in matter

The particle model explains that particles attract each other. This is true, but there are many more forces acting on these particles.

One of these is **gravity**, the force that pulls us towards the centre of Earth. It is a force that acts on any object with mass, so it must act on all particles. Gravity can affect the movement of particles and cause them to move towards each other.

What is one force that acts on particles?

2 Different particles have different properties

The particle model assumes that all particles in all substances are the same. In reality, particles in different substances can have very different properties. Over the centuries, scientists have mixed or merged substances to create new materials that combine different properties.

Iron is a metal that is a very useful material, as it is strong and can be shaped in a variety of ways. Carbon is a non metal. These two substances can be combined to create steel. Steel contains both iron particles and carbon particles, and because of the way the two interact is much stronger than pure iron.

What does the particle model assume about the particles in a substance?

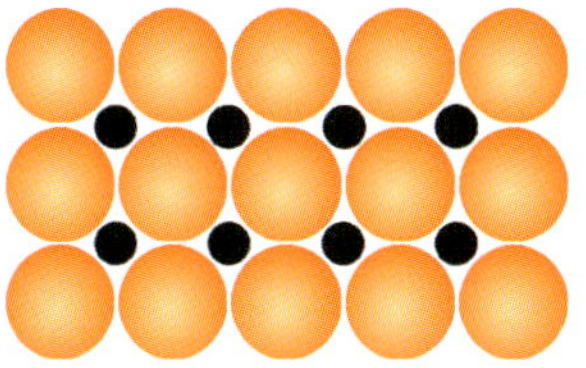

Iron atom

Carbon atom

Figure 1.18 Steel is made up of different types of particles

INVESTIGATION 1.6
Limitations of the particle model

3 There are large spaces between particles

When you draw particles using the particle model, you probably draw them quite close together. In reality, the particles have quite a lot of space between them, even in a solid.

If you removed all of the space between particles, Earth would be the size of a baseball, and every person in the world could fit into a teacup!

What is matter mostly made up of?

4 Models make concepts easier to understand

Why do people use the particle model if it contains all of these inaccuracies? The particle model can explain many of the ways substances behave, and it is relatively easy to understand.

A model can help to explain scientific **phenomena** and make predictions that are reasonably **accurate**.

Why do we use models if they have limitations?

Figure 1.19 We can use model kits to represent the particles in a solid.

CHECKPOINT 1.6

1 Copy and complete these sentences.
   a Models help scientists to understand and __________ how things __________.
   b Models have __________ as they cannot __________ or communicate everything.
2 Identify three advantages of using the particle model to explain the properties of solids, liquids and gases.
3 Identify three disadvantages of using the particle model.
4 What other models can you think of that are used to explain concepts in science?
5 As a class, divide into three groups. In your group, model the behaviour of solid, liquid or gas particles. You may elect to film your model behaviour and create a class video model.

CHALLENGE

6 Another model related to matter is the Bohr model of the atom. Use the internet to find out about the creator of this model, Niels Bohr.

SKILLS CHECK

- I can explain some of the benefits of using models to explain the properties of solids, liquids and gases.
- I can explain some of the limitations of using models to explain the properties of solids, liquids and gases.

CHAPTER SUMMARY

Matter
particles that make up all physical substances. It has mass and takes up space.

The three main states of matter

- Solid
- Liquid
- Gas

The particle model
a simple way of describing matter

Solid
particles in a solid are:
- arranged in a regular pattern like bricks in a wall
- constantly vibrating in a fixed position
- packed very close together with a very strong attraction to each other

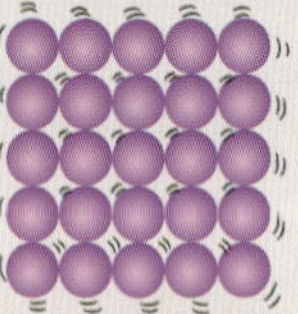

Liquid
particles in a liquid are:
- randomly arranged
- move over and around each other
- packed close together with a strong attraction to each other

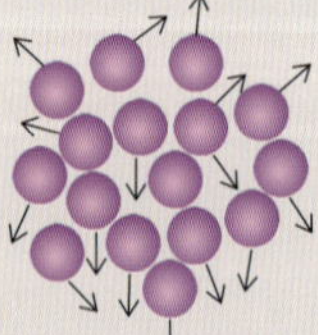

Gas
particles in a gas are:
- randomly arranged
- move freely in all directions
- far apart with a weak attraction to each other

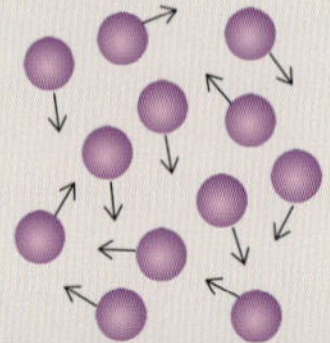

Particles vibrate more when heat is added.

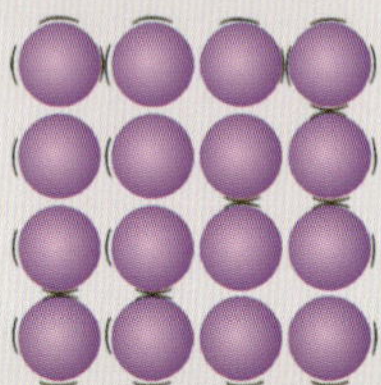
Less heat energy

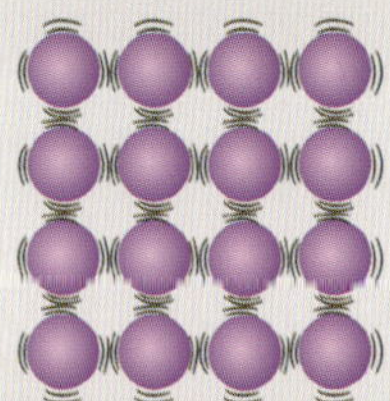
More heat energy

Thermal expansion can happen in solids, liquids and gases

Melting → Evaporation

Solid

Liquid

Gas

Freezing ← Condensation

Density is how heavy an object is for its size

Mass ÷ volume

Low density substances have more space between particles than high density substances.

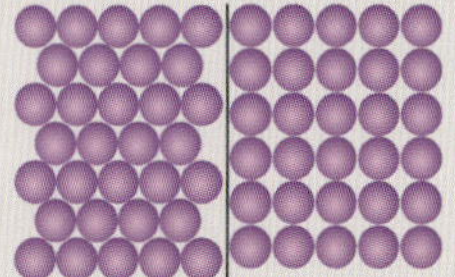

★ FINAL CHALLENGE ★

LEVEL 1

50xp

LEVEL UP!

1. Define *matter* in your own words.
2. Copy and complete this sentence: The ________ heat energy that particles possess, the faster they move. If you remove heat energy from the particles, they will ____________.
3. Liquids can become gases in two ways. What are they?

4. Match these terms to their definitions.

freeze	Heat a solid until it becomes liquid.
melt	Heat a liquid until it becomes a gas.
evaporate	Cool a liquid until it becomes a solid.
condense	How hot something has to get before it melts.
melting point	Cool a gas until it becomes a liquid.

5. Label each of these diagrams as 'solid', 'liquid' or 'gas'.

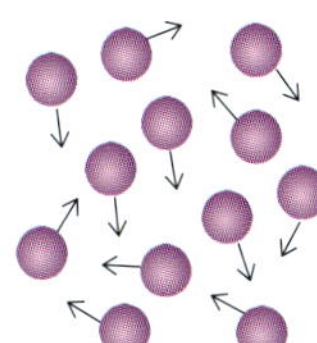

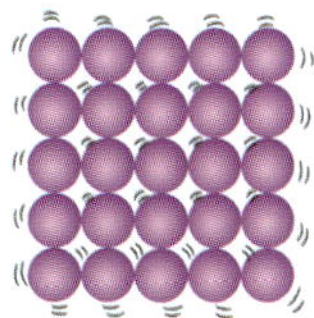

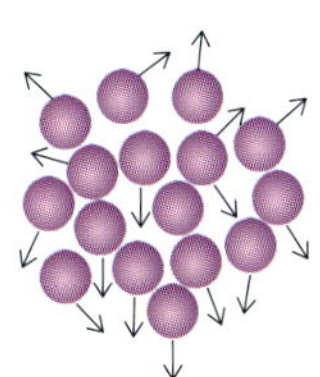

6. Explain the difference between the particles in a liquid and the particles in a gas.
7. Are particles in a solid always moving? Justify your answer.
8. Describe the effect of adding heat to a:
   a. solid
   b. liquid
   c. gas.

LEVEL 3

150xp

LEVEL UP!

9. Draw a diagram that shows the behaviour of particles that are heating, and another that shows the behaviour of particles that are cooling.
10. Explain how and why condensation happens.
11. Write a 'pros and cons' list for using the particle model to describe the properties of solids, liquids and gases.

12. You can calculate density by dividing an object's mass by its volume. Explain why mass and volume are used to calculate density.
13. Using an explanation involving the particle model, suggest why the bridge on page 8 is designed with this metal feature.

2 THE STRUCTURE AND PROPERTIES OF MATTER

Matter is something that takes up space and has mass. The tiny particles of matter are called atoms. Within these are even smaller particles, such as electrons. An element contains atoms of just one type – for example, the element carbon is made of carbon atoms. The periodic table displays all the known elements.

The properties of substances depend on the element they are made of. Some are good at conducting electricity, such as copper. Some are good at bonding with other elements, such as hydrogen and oxygen, which can bond to become water.

1 LEARNING LINKS

What do you already know about the structure and properties of matter?

What are some different properties of materials that make them useful?

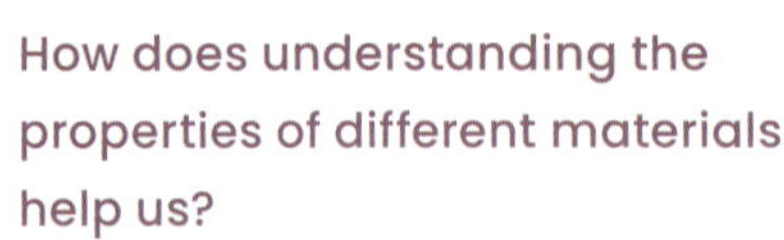

How does understanding the properties of different materials help us?

2 SEE-KNOW-WONDER

List three things you can **see**, three things you **know** and three things you **wonder** about this image.
(Hint: It is a pure metal but it's not mercury!)

3 CRITICAL + CREATIVE THINKING

Variations: What strategies can you use to tell if something is a metal or a non-metal? List as many as you can.

Commonality: Find some features that oxygen and gold have in common.

Prediction: Both coal and diamond are made of carbon. If people could change coal into diamonds in their own homes, predict how life on Earth would change.

4 THE LIGHTEST!

The lightest element in the periodic table is hydrogen. Hydrogen is even lighter than helium, which is what we fill balloons with so that they float in the air (because air is heavier than helium).

Why don't we use hydrogen instead of helium? Because … *boom*! Hydrogen is very explosive. The *Hindenburg* was an airship inflated with hydrogen. In 1937 it caught fire, tragically killing 36 people, in a disaster that was caught on film.

2.1 COMMON ELEMENTS

At the end of this lesson I will be able to:

- **describe** the properties and uses of some common elements, including metals and non-metals.

Elements are pure **substances**. Every object in the world is made up of either one type of element or a combination of elements.

Each type of element has different **properties**, and so you can use different substances for different purposes. You can think of elements in two groups: metals and non-metals.

Figure 2.1 Diamond and graphite are both forms of carbon. Carbon is a non-metal element.

KEY TERMS 

brittle
not able to be bent; will break if stressed

compound
a substance containing atoms of two or more elements bonded together in fixed proportions

element
a pure substance

malleable
able to be bent and shaped

property
characteristic or attribute of a substance

substance
matter that has a fixed chemical make-up

LITERACY LINK 

Explain the relationship between these words: *atom, element, compound.*

NUMERACY LINK

If you measured the melting point of five different metals, would your measurements be categorical or numerical data?

1 Non-metal elements share many properties

Non-metal elements share a lot of the same properties. Non-metals have a low melting point, they don't conduct electricity or heat very well, they aren't shiny, and they are **brittle**.

Carbon is a very common non-metal element. Pure carbon has several different forms. Diamond and graphite (the 'lead' in your writing pencil) are both made of pure carbon.

Carbon forms **compounds** with many other elements. All living things contain large amounts of carbon. So does crude oil, which is used for manufacturing petrol, engine oil, candle wax and plastics.

Oxygen is another common non-metal element. Nearly every living thing on Earth needs oxygen to make energy for living cells.

In its pure form, oxygen is a gas. Oxygen can form many compounds with other elements. For example, water is made up of oxygen combined with hydrogen, and oxygen is the most common element in rocks. Oxygen is required for burning – without it, we wouldn't be able to use stoves, engines, rockets or gas heaters.

What are two pure forms of carbon?

Figure 2.2 Different types of coal contain different amounts of carbon compounds. Some types can be burned to produce electricity.

2 Metal elements have the opposite properties to non-metal elements

Metal elements generally have the opposite properties to non-metal elements. Metal elements have a high melting point, they conduct heat and electricity well, they are shiny and they are **malleable**.

Iron is a common and useful metal element, but in its pure form it is soft and rusts easily. To avoid these problems, iron is usually mixed with other elements such as carbon to create steel. Steel is very strong and rustproof, so it is very useful for tools and construction.

Aluminium, another common metal element, is useful because it's very strong but also relatively lightweight. This makes it an excellent material for making things such as aircraft.

Aluminium is also used in food packaging and storage, such as in aluminium foil and soft-drink cans. Aluminium helps to protect products such as some medicines from air, light and moisture.

Why is aluminium useful for making aeroplanes?

INVESTIGATION 2.1
Comparing metals and non-metals

Figure 2.3 Steel is strong and can be bent into useful and interesting shapes, such as in the Sydney Harbour Bridge.

CHECKPOINT 2.1

1 What is an element?

2 Identify four elements that you have heard of.

3 Identify two compounds that you have heard of.

4 List the main properties of metals.

5 List the main properties of non-metals.

6 Match each element with its common use.

carbon	making tools
aluminium	food storage
iron	making plastic

7 You have been asked to build a rocket for Australia's space program. Would iron or aluminium be a better choice of construction material? Explain your decision.

8 You can't light a fire in space. Why do you think this is?

CHALLENGE

9 Use the internet to find uses of one of these elements: tungsten, argon, lithium, sodium.

SKILLS CHECK

- I can explain what an element is.
- I can name three properties of metals and non-metals.
- I can describe two uses of metals and two uses of non-metals.

2.2 IDEAS ABOUT ELEMENTS OVER TIME

At the end of this lesson I will be able to:

- **identify** how our understanding of the structure and properties of elements has changed as a result of some technological devices.

KEY TERMS 

atom
the smallest unit of an element

electron
a negatively charged particle that moves around the nucleus of an atom

mass
the amount of substance within matter

neutron
a particle found in the nucleus of an atom that has no charge

nucleus
the centre of an atom, which contains protons and neutrons

proton
a positively charged particle found in the nucleus of an atom

LITERACY LINK 

Create a storyboard about the life of one of these scientists: Dalton, Thomson or Rutherford.

NUMERACY LINK 

10 blocks of iron are ready to be used in an investigation. Each one weighs exactly 17.43 grams. What is the total weight of the blocks?

As technologies change and scientific understanding increases, so does our understanding of matter.

The ancient Greeks first proposed that matter was made of particles they called *'atomos'*, meaning 'unable to be divided'. Later, scientists were able to separate matter into pure substances called elements, before discovering that matter was indeed made of tiny particles called **atoms**. Atoms contain even smaller particles.

1 Our understanding of matter, elements and atoms has changed over time

Early scientists identified that matter could be broken down into pure substances called elements. They were able to determine that each element had its own set of unique properties.

In 1803, John Dalton was the first modern scientist to propose that all matter was made up of tiny particles called atoms. His experiments showed that each element was made of different atoms that had different mass. He proposed that atoms were like tiny solid spheres.

In 1897, Joseph John (JJ) Thomson discovered that atoms had areas with negative charges in them that sat in a positively charged sphere like plums in a pudding. These negatively charged particles were later called **electrons**.

Ernest Rutherford was a student of JJ Thomson and conducted his own experiments to learn more about atoms. Rather than being a solid sphere, he found that atoms were made up of a **nucleus** orbited by electrons. He found that the nucleus contained positively charged particles called **protons**.

Neils Bohr collaborated with Rutherford to find out more about how the electrons orbited the nucleus, and found that they orbited in different energy levels.

In the 1930s, James Chadwick discovered that the nucleus of an atom also contained neutrally charged particles called **neutrons**.

As scientists learn more about the nature of atoms, they are able to refine models used to explain their structure.

What were atoms first thought to be like?

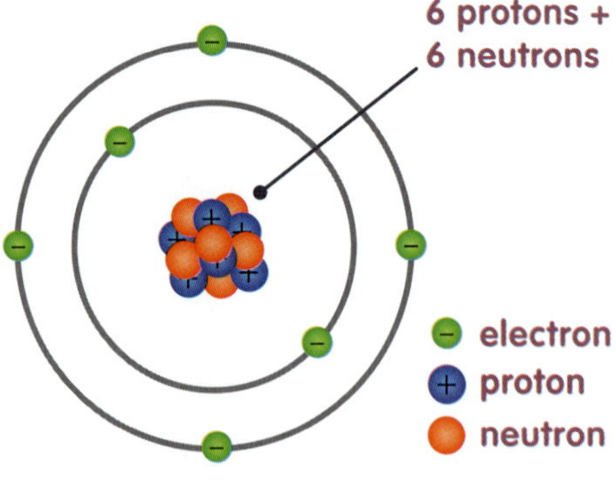

Figure 2.4 Today scientists know that atoms are mostly empty space. They have a nucleus that is made up of protons (positively charged) and neutrons (no charge), which is surrounded by electrons (negatively charged).

❷ The structure of atoms

Today, scientists know that atoms contain a nucleus that is made up of neutrally charged particles called neutrons and positively charged particles called protons. The nucleus is orbited by negatively charged electrons.

Atoms are mostly empty space! If the nucleus of an atom was the size of a pea and placed in the middle of a stadium, the electrons would be orbiting around the outside of the stadium.

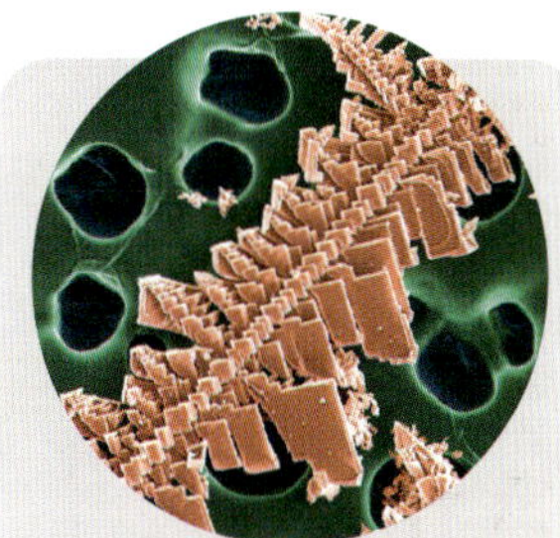

Figure 2.5 The electron microscope allows scientists to see how atoms are arranged.

What are the three main particles that make up an atom?

❸ New technologies can be used to learn more about matter

Scientists and engineers continue to study matter, elements and atoms. The electron microscope allows scientists to see how atoms are arranged in different substances and helps them learn more about their properties.

Devices called particle accelerators are used to accelerate tiny particles to enormous speeds and smash them into one another. By analysing these collisions and what remains afterwards, scientists are able to find out more about matter and the universe.

What does an electron microscope allow scientists to observe?

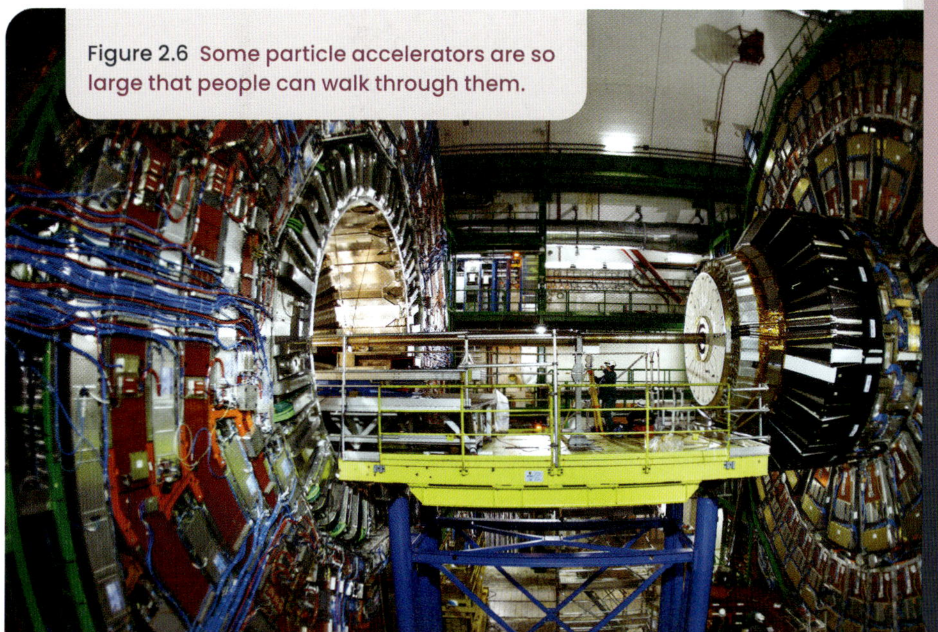

Figure 2.6 Some particle accelerators are so large that people can walk through them.

CHECKPOINT 2.2 

1 What was the name the ancient Greeks gave to the particles they thought made up matter?

2 What is an element?

3 Identify the modern scientist who first proposed that matter was made of atoms.

4 Compare Thomson and Rutherford's models of the atom.

5 Identify the particle/s of an atom that:
a is negatively charged
b is positively charged
c has no charge
d is located in the nucleus
e orbits the nucleus

6 Propose how the development of new technologies can allow an increase in scientific understanding of matter.

CHALLENGE

7 Use the internet to research the scientific work that led to our current understanding of matter. Use this information to make an annotated timeline. Include any breakthroughs or advancements you discover, such as the discovery electrons and the creation of the particle accelerator.

SKILLS CHECK

- I can describe some of the key breakthroughs that led to the current understanding about elements.
- I can describe at least two technologies that further scientific understanding of elements.

2.3 THE PERIODIC TABLE

At the end of this lesson I will be able to:

- **explain** why internationally recognised symbols are used for common elements.

KEY TERMS

chemical formula
uses chemical symbols to show the number of atoms of each element in the compound

chemical symbol
a symbol of one or two letters used to represent an element

chemist
a scientist who studies elements and compounds

element
a substance made up of only one type of atom

periodic table
a table of all known elements and their chemical symbols

LITERACY LINK

Choose one paragraph from this section and write it as a dialogue (a conversation between two or more people).

NUMERACY LINK

Magnesium (Mg) is the 12th element in the periodic table. Directly above it is beryllium (Be), the 4th element. What negative number would you add to 12 to get 4?

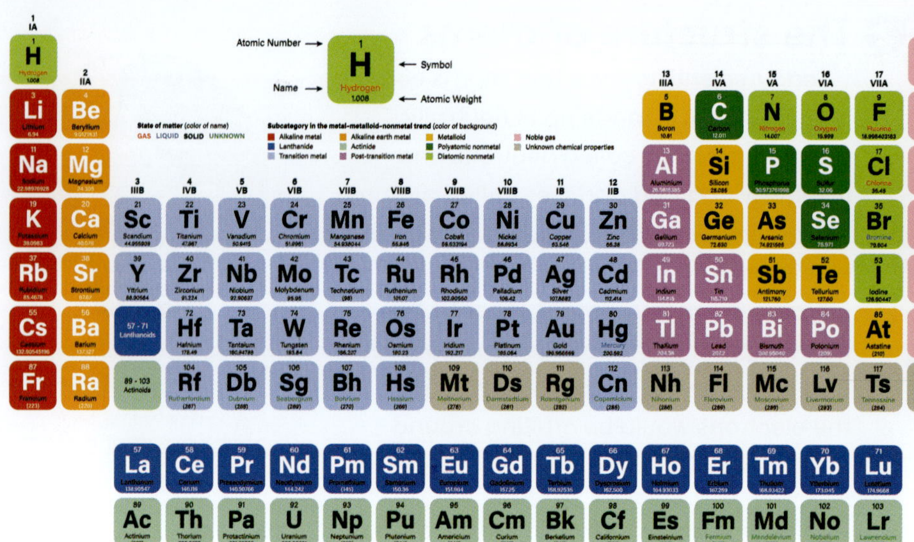

Figure 2.7 The periodic table shows all known elements and their chemical symbols. You can view a larger image of the periodic table on page 32.

The **periodic table** organises information for every known element in the universe. It has been designed to group elements with similar properties together. The table also provides information about the atoms of an element.

The symbols used in the periodic table are universal. This means that regardless of which language you speak or country you live in, you can understand the periodic table.

❶ Chemical symbols always stay the same

The English name for the 47th element on the periodic table is *silver*. In French it's *argent*, in Spanish it's *plata*, while in Russian it's *серебряный*. Although elements have different names in different languages, they always have the same chemical symbol. No matter where you are in the world, the **chemical symbol** for silver is always Ag.

In the case of silver, 'Ag' doesn't make much sense to English speakers, but it seems a perfect symbol if you are French and use the word *argent* for that element. Likewise, the K symbol for potassium makes sense if you know that the Latin word for potassium is *kalium*.

Table 2.1 Symbols and origins of some element names

Element	Symbol	Origin of symbol
Sodium	Na	Latin word *natrium*
Iron	Fe	Latin word *ferrum*
Strontium	Sr	Scottish Gaelic word *sron*

What is the chemical symbol for silver?

❷ The periodic table can tell you about the atoms of an element

Each element is represented by its own square on the periodic table. The information on this square can tell you a lot about the element.

The atomic number states how many protons are in the nucleus of an atom of the element. It also states how many electrons are in a neutrally charged atom of that element. This is because the positive charges of the protons will cancel out the negative charges of the electrons.

The atomic mass refers to the mass of an average atom of the element. This is measured in atomic mass units (amu).

Electrons do not weigh very much, so most of the mass of an atom is in the nucleus. The number of neutrons in an atom can be calculated by subtracting the atomic number from the atomic mass (to the nearest whole number).

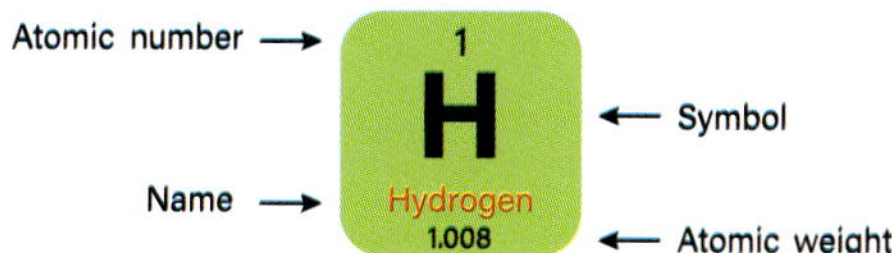

Figure 2.8 The information in the periodic table can tell you a lot about the element.

How can you determine the number of protons in an atom of an element?

❸ Chemical symbols are used to write chemical equations

In addition to helping scientists communicate around the world, chemical symbols can be used to make recording chemical reactions much easier.

Imagine you're taking a maths test and you have to write an equation. If you didn't use symbols for numbers and you had to write everything out in words it would be difficult and time consuming.

Chemists use equations to show how elements react with each other. Just like in maths, having symbols to work with makes writing equations much easier.

Chemical symbols can be used to write chemical formula of compounds. This tells us which elements are in the compound and what ratio the elements are combined in.

$$\text{carbon dioxide} + \text{water} \xrightarrow{\text{LIGHT}} \text{glucose} + \text{oxygen}$$

$$6CO_2 + 6H_2O \xrightarrow{\text{LIGHT}} C_6H_{12}O_6 + 6O_2$$

How can chemical symbols make recording chemical reactions easier?

CHECKPOINT 2.3 

1 What is a chemical symbol?

2 Copy and complete this table. (You may need to refer to the larger periodic table on page 32 to help you.)

Element	Symbol
Carbon	C
Oxygen	
	Na
Fluorine	
	Cu
Calcium	
Tungsten	

3 An atom was determined to contain 18 protons. Identify the element.

4 What does the atomic mass tell us about an element?

5 Use the periodic table to help you calculate the number of neutrons in an atom of nitrogen.

6 Identify two benefits of using chemical symbols instead of element names.

7 What can the chemical formula of a compound tell you about the substance that the name cannot?

CHALLENGE

8 Use the internet to find out where these elements got their symbols from.
a lead (Pb)
b tin (Sn)
c gold (Au)
d mercury (Hg)
e antimony (Sb)

SKILLS CHECK

- I can describe what a chemical symbol is.
- I can explain why we have universal chemical symbols.
- I can state the chemical symbols for at least five elements.

2.4 ELEMENTS, COMPOUNDS AND MIXTURES

At the end of this lesson I will be able to:

- **describe** at a particle level the difference between elements, compounds and mixtures, including the type and arrangement of particles.

KEY TERMS 

chemical bond
a force that holds atoms together

compound
a substance containing atoms of two or more different elements bonded together in a fixed ratio

element
a substance made up of only one type of atom

lattice
a three-dimensional shape made up of a repeating pattern of atoms

mixture
substances mixed together that can be physically separated

molecule
a group of atoms, chemically bonded together

LITERACY LINK 

Select one paragraph from this section and rewrite it in the past tense. Many scientists write their experiment methods in the past tense.

NUMERACY LINK 

Write 0.3 as a percentage.

Substances can be **elements**, **compounds** or **mixtures**. Elements contain one type of atom and compounds have more than one type of atom, bonded together. The atoms in compounds can be in molecules or arranged in a **lattice**.

A mixture can contain many types of particles, but the parts of the mixture are not bonded together.

1 An element contains only one type of atom

An element is a pure substance that is made up of only one type of atom. Gold is an element because it is made only of gold atoms. Likewise, aluminium is made of only aluminium atoms and hydrogen gas is made of only hydrogen atoms.

How many types of atom does an element contain?

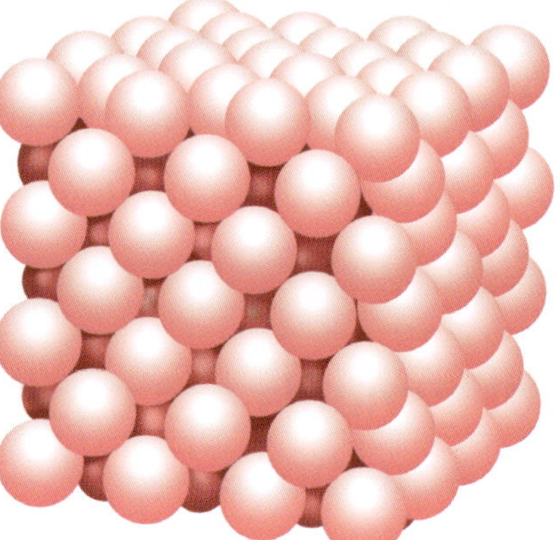

Figure 2.9 Pure gold is made of only gold atoms. Gold atoms are usually coloured pink in scientific illustrations.

2 A compound is made of different elements bonded together

A compound is a substance made up of more than one type of atom **chemically bonded** together in a fixed ratio. Water is a compound: every water particle contains one oxygen atom chemically bonded to two hydrogen atoms. Sodium chloride (table salt) is also a compound because it's made up of sodium and chlorine atoms bonded together.

Some compounds can be made up of molecules, which are structures where the atoms are chemically bonded together. Water is an example of a molecule.

Other compounds form a lattice, where the atoms are bonded together in a repeating pattern. Sodium chloride is an example of a lattice.

Why is water considered to be a compound?

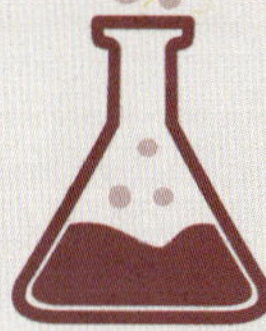

INVESTIGATION 2.4
Separating a mixture

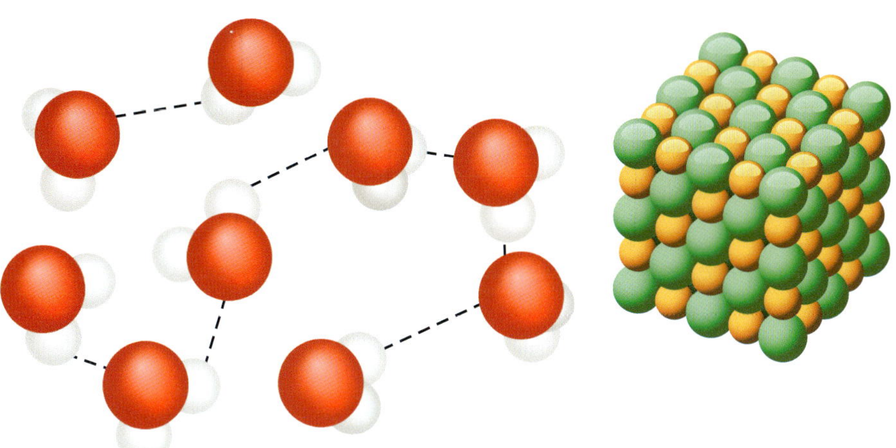

Figure 2.10 Compounds are usually molecules (left) or lattices (right).

❸ A mixture is made up of different substances that are not bonded together

A mixture contains two or more elements and/or compounds that are not bonded together and are able to be easily separated. Although a compound contains atoms of different elements, they are not able to be easily separated.

Sea water is a mixture of salt and water, and if you boil it the water will become a gas, leaving the salt behind. The hydrogen and oxygen atoms in water cannot be separated by boiling it, and the salt compound is not separated either.

What is the difference between a mixture and a compound?

Figure 2.11 This mixture has separated into layers. The particles are not bonded together.

CHECKPOINT 2.4 

1 Use diagrams to illustrate the difference between elements, compounds and mixtures.

2 Explain how you could tell if a substance is a compound or a mixture.

3 Describe the difference between a lattice and a molecule.

4 Categorise these substances as elements, compounds or mixtures.
a gold
b carbon dioxide
c milky tea
d soil
e oxygen
f water

5 You have been given a jar that contains salt water and sand. Propose a method that you could use to separate the sand, the water and the salt.

CHALLENGE

6 The air in Earth's atmosphere is a mixture of elements and compounds. Use the internet to find out the major components of the atmosphere and identify them as elements or compounds. Present this data in a suitable graph.

SKILLS CHECK

- I can describe the particles in:
  - elements
  - compounds
  - mixtures.
- I can summarise the difference between the particles in elements, compounds and mixtures.

2.5 COMMON COMPOUNDS

At the end of this lesson I will be able to:

- **identify** some examples of common compounds.

KEY TERMS 

compound
a substance containing atoms of two or more different elements bonded together in a fixed ratio

molecule
a group of atoms chemically bonded together

LITERACY LINK 

Consider the compound carbon dioxide. What clues are in the name that tell you what the compound may be made of? Explain your answer.

NUMERACY LINK 

Carbon dioxide is made up of $\frac{1}{3}$ carbon atoms and $\frac{2}{3}$ oxygen atoms. Write $\frac{1}{3}$ and $\frac{2}{3}$ as decimals.

Compounds are substances that contain atoms of two or more different elements bonded together in a fixed ratio.

Compounds are found throughout the universe because many atoms are unstable on their own and so bond with others. They are vital to life and survival – we could not exist without compounds such as water, carbon dioxide and sodium chloride.

❶ Water is a compound vital for life

A very well known example of a compound is water. Every living thing on Earth needs water to survive.

Each water **molecule** is made up of one oxygen atom bonded to two hydrogen atoms. This can be written as the capital letter H (for hydrogen), the number 2 (to show that there are two hydrogen atoms) and the capital letter O (for oxygen).

When you write compounds in this way, the number should be written as a subscript – so water is H_2O.

What types of living things need water to survive?

❷ Carbon dioxide is essential for plants

Carbon dioxide is another compound that's essential to life.

Humans and other animals breathe in oxygen and breathe out carbon dioxide as a waste product. Plants use carbon dioxide to make energy to survive. A product of this reaction is oxygen, which plants release into the atmosphere, and the cycle continues.

Carbon dioxide is a gas at room temperature. Each carbon dioxide molecule contains one carbon atom (written as C) bonded to two oxygen atoms – so its chemical formula is CO_2.

Why do plants need carbon dioxide?

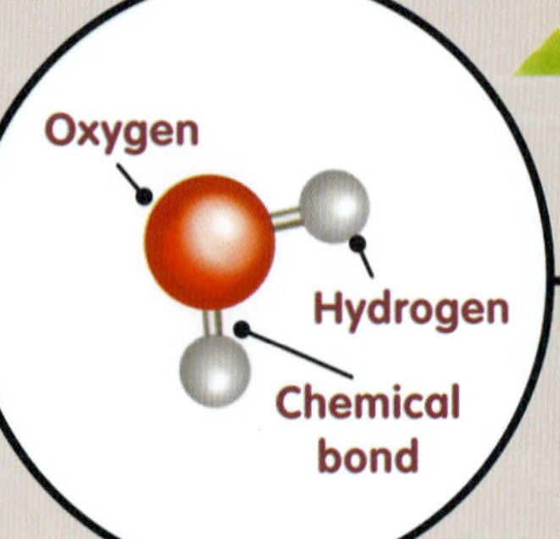

Figure 2.12 Plants need carbon dioxide to survive and grow.

3 Ammonia is used in many farm and household products

Ammonia is used to make products such as fertilisers to help plants grow, and to make clothes and cleaning products. At room temperature, ammonia is a gas that is poisonous to humans, which is why some cleaning products must be handled carefully.

Each ammonia molecule is made up of one nitrogen molecule (written as N) bonded to three hydrogen (H) molecules – so the chemical formula is NH_3.

What elements make up ammonia?

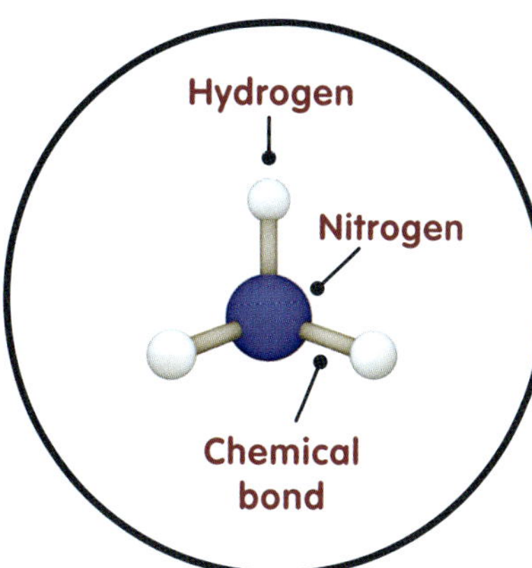

Figure 2.13 Ammonia is used to make fertilisers, which help farm crops grow.

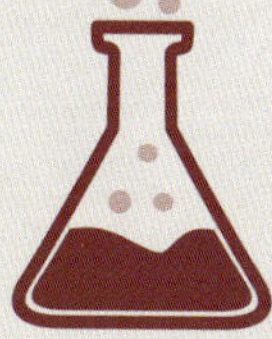

INVESTIGATION 2.5
Properties of compounds

4 Sodium chloride is used in many foods

Sodium chloride is better known as table salt – the same substance that is added to fish and chips.

The chemical formula for sodium chloride is NaCl – that's one atom of sodium (Na) and one atom of chlorine (Cl). Sodium chloride is a compound that does not form molecules but has a lattice structure, with alternating sodium and chlorine atoms.

What is sodium chloride better known as?

Figure 2.14 Table salt is a lattice containing sodium and chlorine, known as sodium chloride.

CHECKPOINT 2.5

1 Explain how a compound is different to an element.
2 Name three uses of ammonia.
3 What type of atoms make up table salt?
4 Classify the four compounds in this section as solids, liquids or gases at room temperature.
5 What is produced by plants that is so important for humans?
6 What do humans produce that is so important for plants?
7 Suggest how you could break compounds into individual atoms.
8 Water and carbon dioxide are essential to life on Earth. Explain why.

CHALLENGE

9 Use the internet to research other important compounds found in everyday life. Identify at least three and explain their role on Earth.

SKILLS CHECK

- I can explain what a compound is.
- I can identify some examples of common compounds.

2.6 ELEMENTS AND COMPOUNDS OVER TIME

At the end of this lesson I will be able to:

- **investigate** how people in different cultures in the past have applied their knowledge of the properties of elements and compounds to their use in everyday life.

KEY TERMS 

alloy
a mixture of a metal and one or more other elements

archaeologist
someone who studies history by examining sites and objects

ochre
a mixture of iron oxide, clay and dirt, used as a natural paint

LITERACY LINK 

Write a one-page research report on a historic culture of your choice, including how they used their knowledge of the properties of elements and compounds. This could relate to many things, including tools, weapons, art and technology.

NUMERACY LINK 

A 20c coin is 75% copper and 25% nickel.

If you flipped a 20c coin twice, what is the probability of getting heads both times?

Human technology is always developing. The first tools and weapons looked very different to those we use today. People in ancient times did not know about atoms, but they used their knowledge of the properties of substances. As metals were discovered and they replaced stone, tools became stronger, sharper and lighter.

1 Rocks contain compounds that make them useful tools

The Stone Age was named by **archaeologists** because that is the period of time where humans started making tools out of rock (stone). Prehistoric people found that different types of rocks were useful for different things.

Obsidian is a rock that breaks into sharp pieces that are useful for making arrow tips and scrapers. Very hard rocks with coarse textures were useful for grinding food. Very hard rocks with smooth textures were also useful for making axes.

Why was stone useful for making tools?

2 Metalworking began in the Bronze Age

The Stone Age ended in about 3700 BCE with the discovery of metalworking. In particular, copper was useful because it could be melted down and reshaped. After it cooled, it would solidify into its new shape. Unfortunately, copper isn't very strong, and not useful for tools that must be tough.

The ancient Sumerians discovered that adding tin to copper, making an **alloy**, would strengthen it. This alloy became known as bronze, and so this era of human history is called the Bronze Age.

What was the disadvantage of using pure copper to make tools?

Figure 2.15 Ancient peoples used bronze to make tools and weapons, including swords.

3 Steel was invented in the Iron Age

The Bronze Age lasted until about 1000 BCE, when iron was discovered. This metal was used to make more durable weapons and tools. Iron became the metal of choice, and so this period is called the Iron Age.

Around 400 BCE, early metalworkers discovered that, by adding carbon to iron, they created a stronger substance. We know it as an alloy called steel. Steel is much stronger than bronze and harder than iron. This meant that the metalworkers could use less metal for the same purpose. Their steel tools, weapons and utensils were lighter and stronger than their bronze and iron ones.

Why was iron better for tool-making than bronze?

4 Ochre is important in Indigenous Australian culture

People of many different cultures have developed and used dyes and paints from natural sources. A substance called **ochre** is very important in the cultures of Indigenous Australians, for example in paintings and body decoration.

Ochre is a mixture of clay, sand and a compound called iron oxide. Mixing different amounts of the ingredients results in different colours, such as reds, yellows and browns. Australia's rocks are rich in iron, which combines with moisture over time to form iron oxide (rust). This gives the outback soil its distinctive red colour.

What compound gives ochre its red colour?

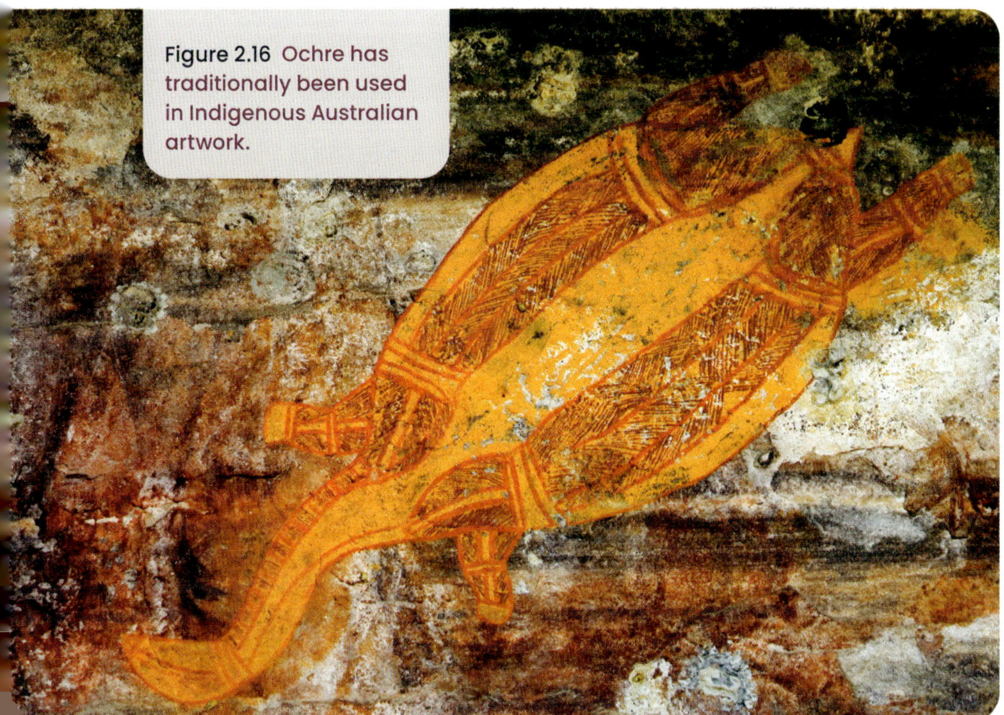

Figure 2.16 Ochre has traditionally been used in Indigenous Australian artwork.

CHECKPOINT 2.5

1 Copy and complete these sentences.
   a Human technology is always ___________ .
   b People in ancient times used their knowledge about ___________ of ___________ to make decisions about materials.
   c As ___________ were discovered, tools became more sophisticated.
   d People of many different ___________ have developed and used dyes and paints from natural sources.

2 What are the advantages and disadvantages of using stone to make tools and weapons?

3 How were the disadvantages of copper for making tools and weapons overcome?

4 Steel was found to be much stronger than bronze. What two main elements form steel?

5 Why is the earth in the centre of Australia red?

6 How do Indigenous Australians use the properties of ochre to create art?

CHALLENGE

7 Use the internet to find out what other cultures, such as Chinese, Roman and Sumerian, have used as paints.

SKILLS CHECK

- I can describe how people in the past from different cultures have used their knowledge of properties to utilise things in their everyday lives.

CHAPTER SUMMARY

An **element** is made of only one type of atom.

Pure gold is made of only gold atoms.

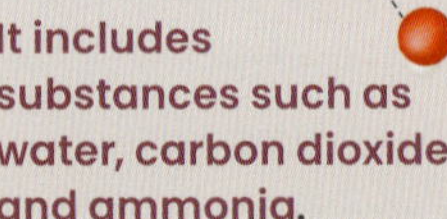

A compound has two or more types of atom bonded together.

It includes substances such as water, carbon dioxide and ammonia.

Non-metal elements (for example: carbon, oxygen)

- low melting point
- do not conduct electricity or heat well
- brittle
- not shiny

Metal elements (for example: iron, aluminium)

- high melting point
- conduct heat and electricity well
- malleable
- shiny

A mixture contains two or more types of atom without a chemical bond. It includes substances such as milk, dirt and air.

John Dalton was the first modern scientist to suggest the existence of **atoms**.

JJ Thomson discovered the charged particles called **electrons**.

Ernest Rutherford proved that atoms had a **nucleus**.

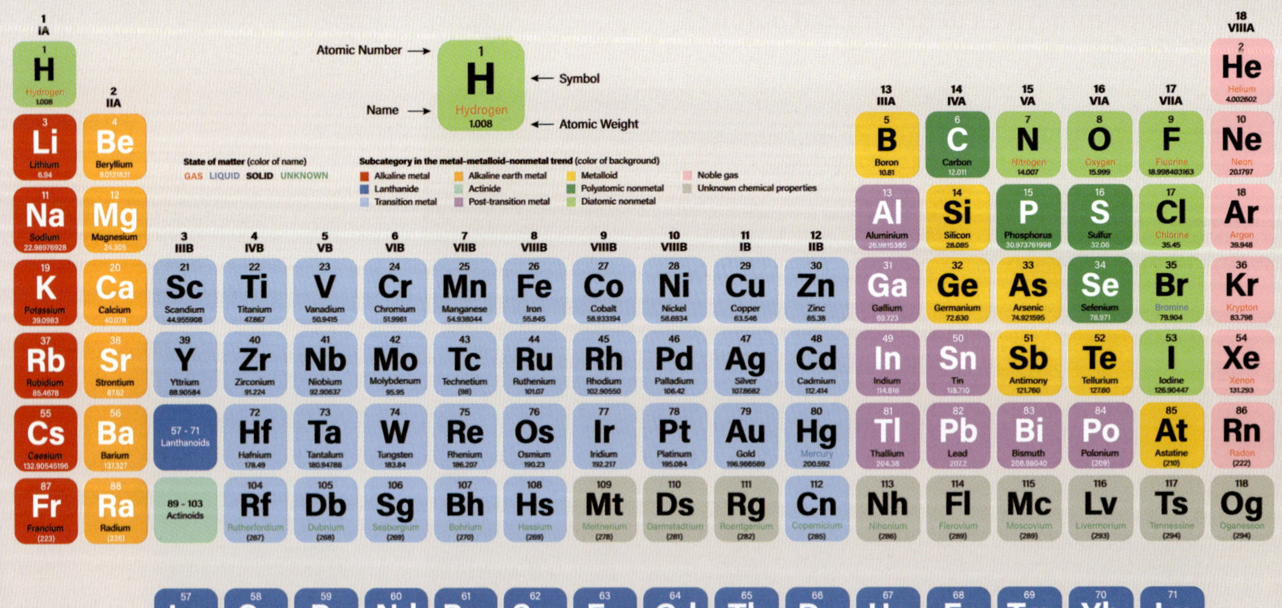

The periodic table shows all known elements and their chemical symbols.

★ FINAL CHALLENGE ★

1. Describe the difference between an element and a compound.
2. Name three compounds and three elements.
3. Copy and complete using the correct answers in the underlined parts of these sentences.

   Oxygen is a metal/non-metal that exists in gas/liquid form naturally. A common property of non-metals is that they have a high/low melting point, are brittle/strong and they do/don't conduct electricity. The opposite/same can be said of metals.

4. Copy and complete the table by stating whether the elements are examples of metals or non-metals.

Carbon	
Gold	
Oxygen	
Mercury	
Potassium	

5. Describe the contributions of Rutherford and Dalton to our knowledge of the properties and structure of elements.
6. List how many of each type of element is in each of the following compounds:
   a CO_2
   b H_2O
   c NH_3

7. Explain why the chemical symbol for potassium is not P.
8. Explain how the development of either the electron microscope or particle accelerator has contributed to our understanding of the structure and properties of matter.
9. Describe the two ways that particles are usually arranged in compounds.
10. A mixture is not the same as a compound. Explain why.

11. Justify the use of universal symbols in the periodic table.
12. Research and prepare a response to this research question.

    'How did the ancient Egyptians use their knowledge of the properties of elements and compounds in their lives?'

3 MIXTURES

If you eat pasta at home then you've probably seen mixtures in action. You may have filtered pasta from water or thickened a sauce by evaporation. If you had a soft drink with your pasta, you may have drank a mixture called a solution.

Mixtures can be separated. When we make foods, chemicals and medicines, we use separation techniques such as distillation and chromatography. We use separation to sort recycling from waste, clean up oil spills, and we separate the parts of blood and dyes to help solve crimes.

1 LEARNING LINKS

What do you already know about mixtures?

How many mixtures have you used today? List as many as you can.

Wet pour rubber (often used in playgrounds) is a mixture of rubber granules and polyurethane resin. How else can mixtures be used in innovative ways?

2 SEE-KNOW-WONDER

List three things you can **see**, three things you **know** and three things you **wonder** about this image.

3 CRITICAL + CREATIVE THINKING

Different uses: List some wildly different uses for filter paper. Think of as many as you can.

The alphabet: Think of a word about mixtures for every letter of the alphabet.

Ridiculous! Attempt to support this statement: 'From 2022, every home must process its own waste, including sewage'.

4 THE BIGGEST!

The biggest rubbish dump in the world probably isn't where you think it is – it's in the ocean. The Great Pacific Garbage Patch floats somewhere between Hawaii and California and is estimated to be at least three-quarters of the size of New South Wales. That's huge!

The ocean currents pull in more plastic every day, but there is some good news – a clean-up of the floating dump began in 2018 using floating 'ocean scrapers'. These scrapers are made out of recycled plastics from the garbage patch itself.

3.1 WATER AS A SOLVENT

At the end of this lesson I will be able to:

- **describe** the importance of water as a solvent in daily life, industries and the environment.

KEY TERMS

dissolve
a solid is mixed into a liquid to form a mixture called a solution

solute
a substance that is dissolved by a solvent

solvent
a substance that dissolves a solute

LITERACY LINK

Write a short story about living in a country that has very limited access to water.

NUMERACY LINK

A swimming pool full of water is 10 m long, 8 m wide and 1.5 m deep. What is its volume?

Formula: $V = lwh$

Water is vital for life on Earth – without water you couldn't survive. Our bodies need water, and so do the plants that we eat. We need water for daily activities such as cooking and washing.

Water is a **solvent** that can dissolve other substances. We use it to extract minerals from underground and to make medicines. Water is known as the *universal solvent* because it **dissolves** more substances than any other liquid.

1 Water is a solvent in daily life

A solvent is any substance, usually a liquid, that can dissolve other substances.

Water is a solvent, and it can dissolve many other substances – solids, liquids and gases. When these substances dissolve, they are called **solutes**. Water acts as a solvent in daily life in many ways:

- *human survival*: Water is an essential solvent for humans. Blood is mostly water, and it carries nutrients and oxygen throughout our bodies.
- *preparing meals*: Water dissolves spices, flavours and other ingredients as food cooks.
- *washing away dirt*: Water acts as a solvent for soaps and detergents when cleaning dishes, clothes and ourselves.

How does water act as a solvent?

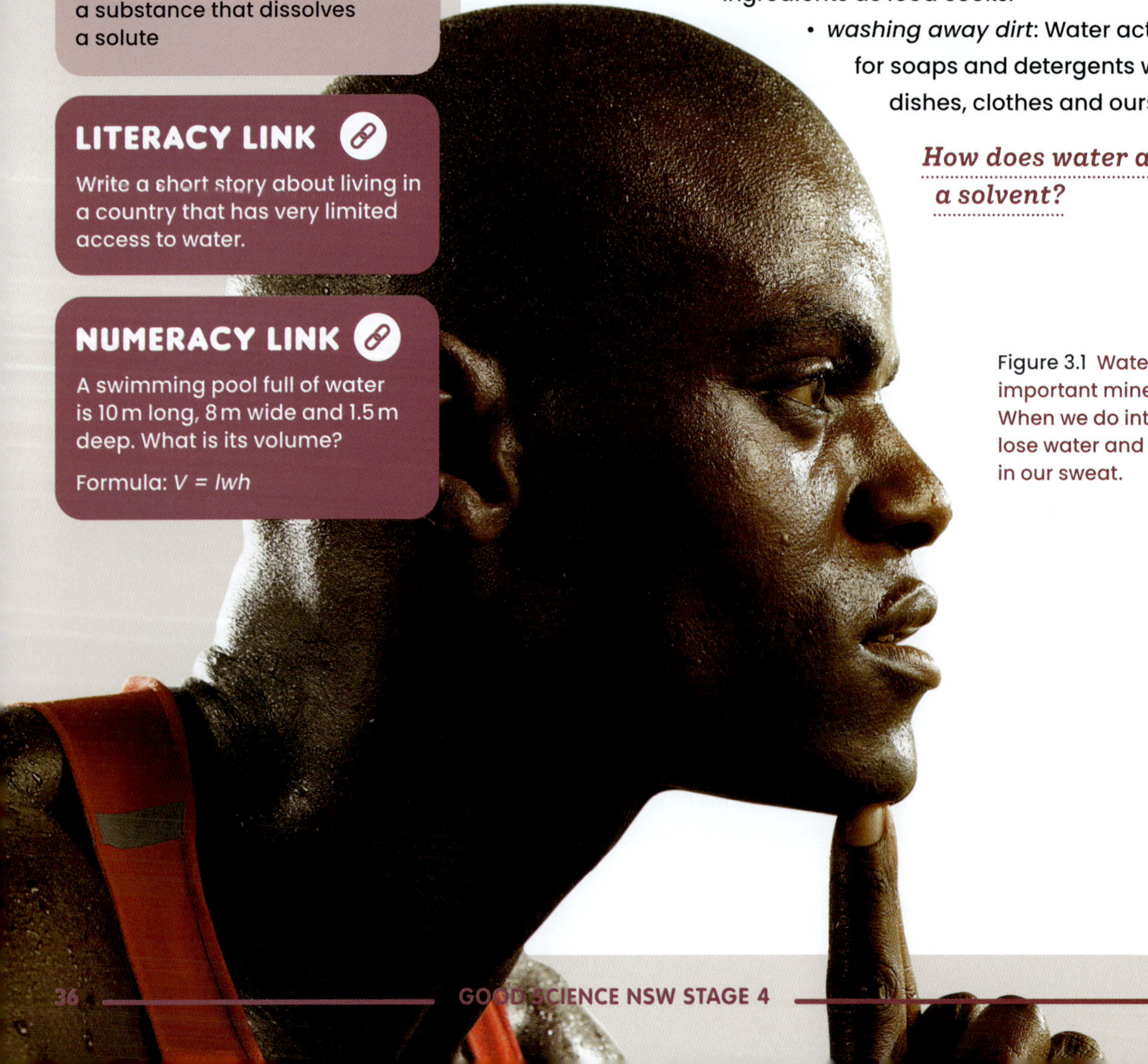

Figure 3.1 Water is a solvent for important minerals in our bodies. When we do intense exercise, we lose water and minerals dissolved in our sweat.

2 Water is a solvent in the environment

Many essential elements such as iron, zinc and calcium are needed by plants and animals. They exist as minerals in rocks and soil. Water can dissolve many of these minerals, so plants can absorb them from groundwater. Animals eat the plants, receive these essential nutrients and are able to survive.

The water in rivers, lakes and oceans contains dissolved oxygen that is vital to fish and other animals. Similarly, water contains the dissolved carbon dioxide that marine plants need.

What substances can water dissolve?

3 Water is a solvent in industry

Industries such as mining use water as a solvent to extract different minerals and elements.

Uranium is an element found in rocks located deep underground. It is used as a fuel in the reactors of nuclear power plants and in powering nuclear submarines. In Australia, the major uranium mines use water as a solvent to remove the uranium from the rocks. Instead of digging up the rocks to remove the uranium, acids are dissolved into water that is then pumped into the rocks that contain the uranium. This solution dissolves the uranium, and the acidic water is then pumped back to the surface where it is processed to remove the uranium.

How is water used as a solvent in industry?

Figure 3.2 Water is often the solvent used when medicines are manufactured.

CHECKPOINT 3.1

1 Explain the difference between a solvent and a solute.
2 List three uses of water as a solvent in daily life.
3 What is found in soil and rocks that water can dissolve?
4 Identify how water can be used in industries as a solvent.
5 Explain how water is important for washing away dirt.
6 Explain why dissolved oxygen and carbon dioxide in water is important.
7 Describe how water can be used as a solvent in mining.
8 Explain how water can be used as a solvent in medicine.
9 Water is critically important to the environment. Suggest why.

CHALLENGE

10 Use the internet to find out what properties of water make it a useful solvent. Use labelled diagrams to help you present your findings.

SKILLS CHECK

- I can describe what a solvent is.
- I can state how water is used as a solvent in daily life, industry and the environment.

3.2 SOLUTE, SOLVENT, SOLUTION

At the end of this lesson I will be able to:

- **describe** aqueous mixtures in terms of solute, solvent and solution.

KEY TERMS 

aqueous
containing water

insoluble
something that does not dissolve

mixture
substances mixed together that can be physically separated

soluble
something that dissolves

solute
a substance that is dissolved by a solvent

solution
a mixture made up of a solvent and a dissolved solute

solvent
a substance that dissolves a solute

LITERACY LINK 

Write a hypothesis for an investigation that explores what happens if a lot of sugar is dissolved in water compared to water with only a small quantity of sugar. Include the key terms *dilute, concentration, solvent, solute.*

NUMERACY LINK 

When mixing cordial, Habib uses one part cordial to five parts water. Express this as a ratio.

What's in a soft drink? What is dissolved and what are the ingredients that make these drinks so tasty?

It all starts with water, a solvent that can dissolve many substances. The ingredients of the soft drink – sugar, flavours and carbon dioxide for fizz – are solutes. They are dissolved in the **mixture**, which is a solution.

1 Solutes dissolve in other substances

A **solute** is any substance that can dissolve in another substance (the solvent). Solutes can begin as any of the three main states of matter:

- *solid*: For example, sugar, salt and instant coffee are all solutes that can dissolve in water.
- *liquid*: For example, cordial can dissolve in water to make a sweet drink.
- *gas*: For example, carbon dioxide can dissolve in soft drinks, which makes them fizzy.

Substances that dissolve are said to be soluble. Substances that do not dissolve are called insoluble. A substance that is insoluble in a solvent such as water might be soluble in a different solvent such as an acid.

What is a solute?

2 Solvents dissolve solutes

A **solvent** is any substance, usually a liquid, that can dissolve another substance (the solute).

Water is a liquid that can dissolve cordial (another liquid), sugar (a solid), and fizzy carbon dioxide (a gas) in soft drinks.

What does a solvent do?

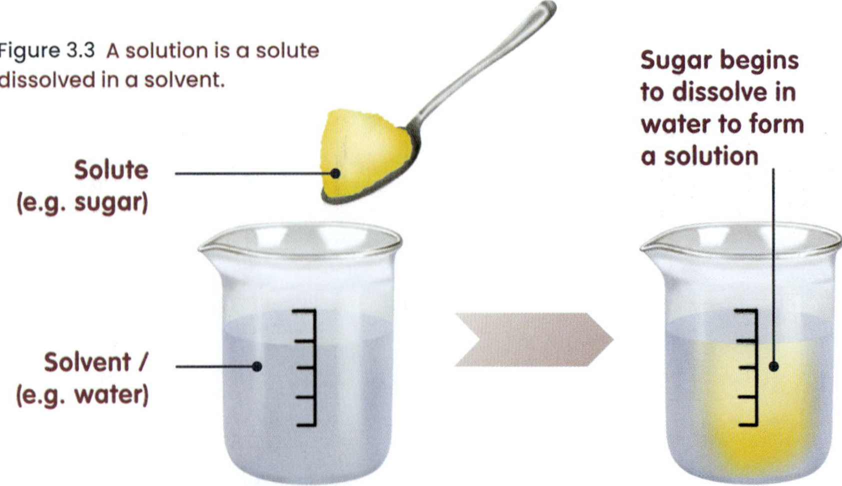

Figure 3.3 A solution is a solute dissolved in a solvent.

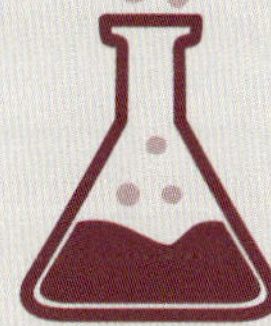

❸ Solutions are solutes in solvent

A **solution** is a mixture made up one or more solutes dissolved in a solvent. A cup of coffee, for example, is a solution made up of:

- solutes – powdered coffee and perhaps sugar
- a solvent – water.

Any solution that has water as a solvent is called an **aqueous** solution.

What is a solution made up of?

❹ Solutions, suspensions and colloids

Not all solutions are the same. Different solutes make different solutions. Also, the amount of solute affects the solution:

- *concentrated solution*: a solution with a lot of solute. Putting 20 tablespoons of cordial in one glass of water gives you a concentrated (and unpleasantly sweet) solution.
- *dilute solution*: a solution with a very small amount of solute. Putting a quarter of a teaspoon of cordial in a large glass of water makes a dilute (and not sweet enough) solution.

Suspensions and colloids contain particles that behave differently to those in solutions:

- *suspension*: a mixture with large, insoluble particles that are spread out evenly at first and eventually settle to the bottom. A snow globe, with a mixture of plastic 'snow' particles and water, is a suspension.
- *colloid*: a mixture with tiny particles spread out evenly that never settle to the bottom. Milk is a colloid because it contains tiny droplets of fat in water.

How do particles in a suspension behave differently to those in a colloid?

Figure 3.4 The liquid in a snow globe is a suspension of plastic particles in water.

INVESTIGATION 3.2
The Tyndall effect

CHECKPOINT 3.2 ✓

1 Name three solutes.

2 For each of the three solutes above, identify a possible solvent.

3 Describe the difference between a solution and a solvent.

4 Copy and complete this sentence. A solution is a ______________ dissolved in a ______________.

5 What is the difference between a concentrated solution and a dilute solution?

6 In each of these aqueous solutions, identify the solute and the solvent:
a a cup of sweet, black tea
b a glass of orange juice made from powdered concentrate
c sea water.

7 A mystery substance is added to a glass of water. Initially the substance makes the water blue and cloudy. The substance then settles at the bottom of the glass. What type of mixture is described here?

CHALLENGE

8 Is fog a colloid? Do research to justify your answer.

9 Investigation 3.2 explores the Tyndall effect. Before the investigation, type Tyndall effect into a search engine and click 'image search'.

SKILLS CHECK

- I can explain and give an example of a:
  - solute
  - solvent
  - solution.

3.3 FILTERING AND DECANTING

At the end of this lesson I will be able to:

- **relate** a range of techniques used to separate the components of some common mixtures to the physical principles involved in filtration and decantation.

KEY TERMS 

decanting
carefully pouring the liquid from a mixture, leaving the sediment behind

filtrate
the liquid that passes through a filter

mixture
substances mixed together that can be physically separated

residue
the solid that does not pass through a filter

sediment
the solid that settles to the bottom of a liquid

LITERACY LINK 

Draw and label a diagram showing how to separate the parts of a mixture using filtration.

NUMERACY LINK 

An identical mixture of chalk and water was separated by three students. Student A ended up with 5 grams of chalk, Student B ended up with 5.5 grams of chalk and Student C ended up with 4.2 grams of chalk. What is the average result?

Imagine you are stranded in the bush, and the only source of water is a muddy puddle. You scoop some water out with a container, and you notice how dirty it is. After a while, the mud sinks to the bottom of the **mixture**. You place a cloth over another container and, without disturbing the mud, carefully pour the liquid part of the mixture into the second container.

You've used decanting and filtering to separate the water and mud. You survive!

1 Mixtures can be physically separated

In science, a mixture is two or more substances mixed together that can be physically separated. Mixtures are all around us – concrete, air, mayonnaise, muddy water and sea water are all mixtures.

Some things are not mixtures. Oxygen, pure water and gold are not mixtures – they are pure substances.

Mixtures can contain:

- soluble substances, which will dissolve, such as salt
- insoluble substances, which won't dissolve, such as sand.

Some substances will partly dissolve, meaning that some but not all will dissolve into the solvent.

Scientists use information about solubility to help decide which technique to use to separate a mixture.

What is a the difference between a soluble and insoluble substance?

Figure 3.5 Salt dissolves in water but sand does not. They are both mixtures.

Salt water solution

Mixture of sand and water

2 Filtration separates liquids from solids

Filtration is used to separate insoluble substances from liquids. The filter acts like a sieve, using small holes to trap larger particles.

Figure 3.6 shows how filter paper can separate a chalk and water mixture. The solid chalk particles get trapped in the filter paper and the liquid water passes through the filter paper. The chalk left in the filter paper is called the **residue** and the liquid that passes through the filter paper is called the **filtrate**.

How does filtration separate solids and liquids?

Figure 3.6 A filter can be used to separate chalk particles from water.

3 Decanting separates sediments from liquids

Decanting can also be used to separate insoluble solids from liquids. After an insoluble solid in a mixture settles to the bottom of a container, the liquid can be poured out carefully, leaving the solid behind.

Think back to the 'lost in the bush' scenario at the start of this section. The solid mud settled to the bottom of the container as a **sediment**. When the water was poured out from the top, the mud was left behind.

How does decanting separate solids and liquids?

INVESTIGATION 3.3
Purifying muddy water

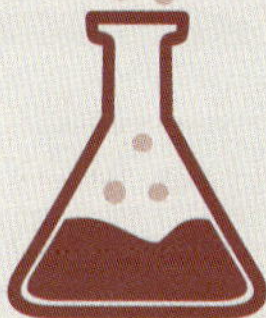

CHECKPOINT 3.3

1 Which of these is a mixture?
  a gold
  b concrete
  c air
  d oxygen
  e pure water
  f sea water
2 For your answers to question 1, describe how you can tell it is a mixture.
3 Name three soluble substances.
4 Sand is an insoluble substance – suggest how you can tell.
5 Using words and a diagram, explain filtration.
6 Describe the process of decanting in 15 words or less.
7 Compare the physical properties that allow filtration and decanting to work.
8 In a laboratory, you are given a mixture of salt, dust and pebbles. Design a method of separating these three substances. (Hint: Salt dissolves in water, pebbles sink to the bottom of the container and dust floats on the surface.)

CHALLENGE

9 How are filtering and decanting used to solve problems in everyday life? Write or illustrate your answer.

SKILLS CHECK

- I can describe what filtration and decanting are.
- I can suggest ways to separate mixtures using filtration and decanting.

3.4 WATER FILTRATION AND WASTE MANAGEMENT

At the end of this lesson I will be able to:

- **investigate** the application of a physical separation technique used in everyday situations or industrial processes, for example, water filtering and sorting waste materials.

KEY TERMS 

disinfection
destroying bacteria, often using special light or chlorine

sewage
semi-liquid human waste

sewerage
pipes that carry sewage

LITERACY LINK 

Some people oppose the addition of fluoride into the water system. Write a short letter that could be published in a newspaper, giving your opinion.

NUMERACY LINK 

A waste management plant combines $\frac{7}{12}$ litres of water with $\frac{3}{12}$ litres of chemicals. Add the fractions to find the total volume of liquid.

Many less economically developed countries do not have easy access to clean water, and the people there can get sick or die from diseases carried in water. These countries do not have good systems to manage their waste, creating serious health and environmental hazards in waterways and on land.

Australia has effective ways to purify water and manage wastes. This is better for our health and the environment.

1 Purification makes water safe to drink

Before water is safe to drink, it must be purified. This process kills bacteria and removes substances that are harmful.

In Australia, there are six main steps in water purification:

1. *screening*: Water passes through mesh screens to remove objects such as twigs and leaves
2. *flocculation*: A chemical called alum is added to the cloudy water to make the small floating particles clump together. These clumps are called floc.
3. *sedimentation*: The floc is heavy and settles to the bottom of the tank to form a sediment. This sediment is collected as sludge.
4. *filtration*: The water flows through tightly packed beds of different sized pebbles, sand and crushed coal to trap and remove the floc.
5. *chemical treatment*: Chemicals are added to the water. Fluoride is added to help prevent tooth decay. Chlorine is added to the water to kill harmful organisms such as bacteria. A chemical is added to reduce the acidity of the water.
6. *aeration*: Oxygen is added to the water to improve the smell and colour.

When the water is safe to drink, it's distributed through pipes to homes, schools, hospitals, businesses and countless other locations.

What steps are involved in water purification?

Figure 3.7 A water treatment plant has many different sections for the different stages, such as chemical treatment and aeration.

❷ Sewage is processed to reduce harm to the environment

Sewage is semi-liquid human waste. When you flush the toilet, the sewage goes into a **sewerage** system and is treated before it eventually goes into the ocean.

Some countries do not have proper sewerage systems. Untreated sewage is not good for human health and the natural environment, so in more economically developed countries like Australia it is processed before release.

Figure 3.8 In some less economically developed countries, human waste goes into open sewers and is not treated.

In Australia, there are six main steps in sewage treatment:

1 *sewerage*: A network of pipes moves sewage from homes and businesses to sewage treatment plants.
2 *screening*: Screens at the plant act as a sieve and catch large objects, which can be physically removed.
3 *aeration*: Air is pumped into tanks that hold the sewage. This feeds bacteria, which break down the sewage.
4 *settling*: Other chemicals are added that cause the bacteria and solids to settle to the bottom of the tank as thick sludge. This sludge is removed and is used in soil and fertiliser products.
5 *filtration*: The sewage passes through a filter made from pebbles. This traps more solids, which are removed.
6 ***disinfection***: Special ultraviolet light or chlorine is used to kill harmful bacteria in the sewage before it is released into the ocean.

What steps are involved in sewage processing?

CHECKPOINT 3.4 

1 Explain why sewage treatment is critical to a healthy environment.
2 Identify the purpose of flocculation in water purification.
3 Describe what happens during the sewage treatment phase called screening.
4 Identify two different ways of killing harmful bacteria.
5 Explain why the treatment and purification of water are such important processes.
6 What is the purpose of adding fluoride to water?
7 Explain why air is blown into tanks in a sewage treatment plant.
8 Create a table of similarities and differences between water purification and sewage treatment.

CHALLENGE

9 Investigate the processes used to separate items that go into yellow-topped recycling bins. How do these processes limit what can be put into recycling? What could happen if there is contamination?

SKILLS CHECK

- I can describe how water is purified.
- I can describe how sewage is treated.

3.5 EVAPORATION, DISTILLATION AND CRYSTALLISATION

At the end of this lesson I will be able to:

- **relate** a range of techniques used to separate the components of some common mixtures to the physical principles involved in each process, including evaporation, crystallisation and distillation.

KEY TERMS 

condensation
cooling of a gas to become a liquid

condenser
a glass tube cooled by water that cools a gas to become a liquid

crystallisation
separation of a solution by evaporating the solvent, leaving behind solute crystals

distillation
using heating to separate liquids with different boiling points

evaporation
heating a mixture to remove the liquid

LITERACY LINK 

Identify three words from this section that you are unfamiliar with. Write definitions for each.

NUMERACY LINK 

In a laboratory distillation process, 45 g salt is separated per 100 ml of water. If you begin the investigation with 500 ml of water, how much salt do you predict you will have at the end?

Australia is a country that experiences frequent droughts and water shortages. The country is surrounded by oceans; however, salt water cannot be used by living things. One way to use this resource is to separate the salt from the water.

Desalination is a process that uses separation techniques such as evaporation, distillation and crystallisation to purify salt water.

1 Solids can be evaporated and crystallised out of liquids

To separate a solid such as salt from water, without keeping the liquid, we can heat the mixture until the water boils and leaves the mixture as a gas. This method is called **evaporation**.

Crystallisation – the formation of crystals of the substance that was dissolved in the liquid – will start to happen when most of the liquid has evaporated. Smaller crystals form if the liquid evaporates quickly, and larger crystals form if it evaporates slowly.

What is the difference between evaporation and crystallisation?

Figure 3.9 In the laboratory, a solution of salt water can be heated until the water evaporates and the solid salt begins to crystallise.

Figure 3.10 Salt evaporation ponds are shallow artificial 'ponds' designed to extract salt from sea water. Natural salt ponds are geological formations that occur naturally when water evaporates and leaves behind a salt flat.

INVESTIGATION 3.5A
Evaporating a solution

INVESTIGATION 3.5B
Growing crystals

INVESTIGATION 3.5C
Distillation (Teacher demonstration)

❷ Distillation can be used to purify water

Distillation can be used to separate pure water from salt. First, the salt water is heated in a glass flask. The water evaporates and the gas is captured. It is then cooled in a tube called a **condenser**, which has cool water flowing around it. This changes the gas to a liquid. This change is called **condensation**. The condensed liquid is collected in a different beaker, while the salt crystals remain in the flask.

Distillation can also be used to separate two liquids, such as water and alcohol. The physical property that allows these substances to be separated is boiling point (which is different for every substance).

What is distillation?

Figure 3.11 In a laboratory distillation set-up, a Bunsen burner heats a mixture and evaporates the liquid. This gas changes to a liquid in the condenser.

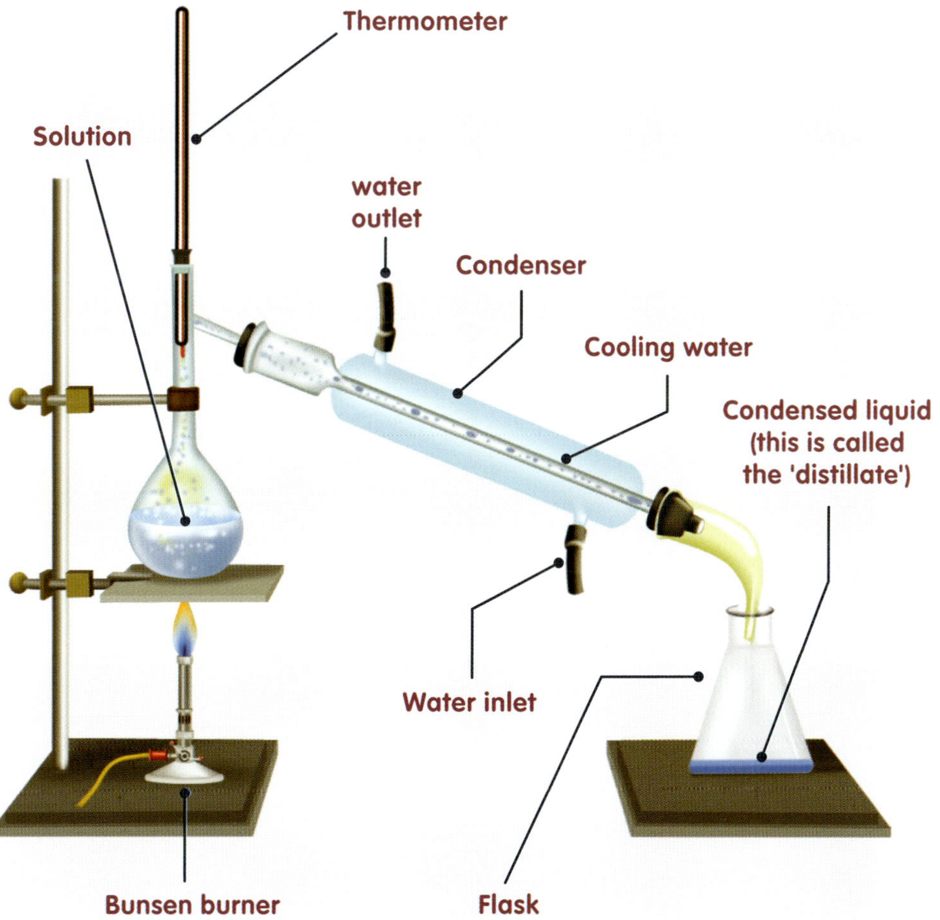

CHECKPOINT 3.5 

1 Evaporation and crystallisation are related – suggest how.
2 Give an example of a mixture that can be separated by evaporation and crystallisation.
3 Describe a situation in which you could separate something using distillation.
4 Identify the physical property of a substance that is used to separate mixtures in distillation.
5 Suggest how you would go about creating large crystals using crystallisation.
6 Scientific drawing is an essential laboratory skill. Redraw the apparatus shown in Figure 3.9 or Figure 3.11. Use a sharp pencil and label your diagram accurately.

CHALLENGE

7 Use the internet to research how a desalination plant uses the process of reverse osmosis to purify water. Use labelled diagrams to present your findings.

SKILLS CHECK

- I can describe the difference between evaporation, crystallisation and distillation.
- I can give examples of situations in which each technique could be used to separate substances.

3.6 SEPARATING EVERYDAY MIXTURES

At the end of this lesson I will be able to:

- **investigate** the application of a physical separation technique used in everyday situations or industrial processes, for example, extracting pigments or oils from plants, and separating blood products.

KEY TERMS 

centrifuge
a machine that spins fluid samples very quickly so that they separate into parts of different densities

condenser
a glass tube cooled by water that cools a gas to become a liquid

density
how heavy something is for its size; mass divided by volume

LITERACY LINK 

Explain the word *centrifuge* in a very simple way that a younger person would be able to understand.

NUMERACY LINK 

If a centrifuge extracts $\frac{2}{5}$ of the material from $\frac{2}{3}$ of a litre of liquid, calculate $\frac{2}{5} \times \frac{2}{3}$ = ____________

Many everyday mixtures need to be separated before we use them.

Oils need to be extracted from flowers, fruits and leaves before perfumes can be made. Blood can be separated into its different components of red blood cells, platelets, plasma and white blood cells. These can be used to treat people with certain medical conditions and to save lives in emergency situations.

1 Oils can be extracted from plants

The leaves or flowers of plants such as eucalypt and lavender contain sweet-smelling oils. These oils can be extracted, using distillation, to make perfumes.

When steam from boiling water is mixed with leaves or petals in a flask, the oils are released as a gas. The mixture of oil and water vapour is then channeled into a condenser and cooled, changing it into a liquid. It is then separated using a separating column, leaving the pure essential oil and a floral or herbal water.

How is oil extracted from plants?

Figure 3.12 Perfume is produced on a large scale, and the oils of different plants are separated using distillation.

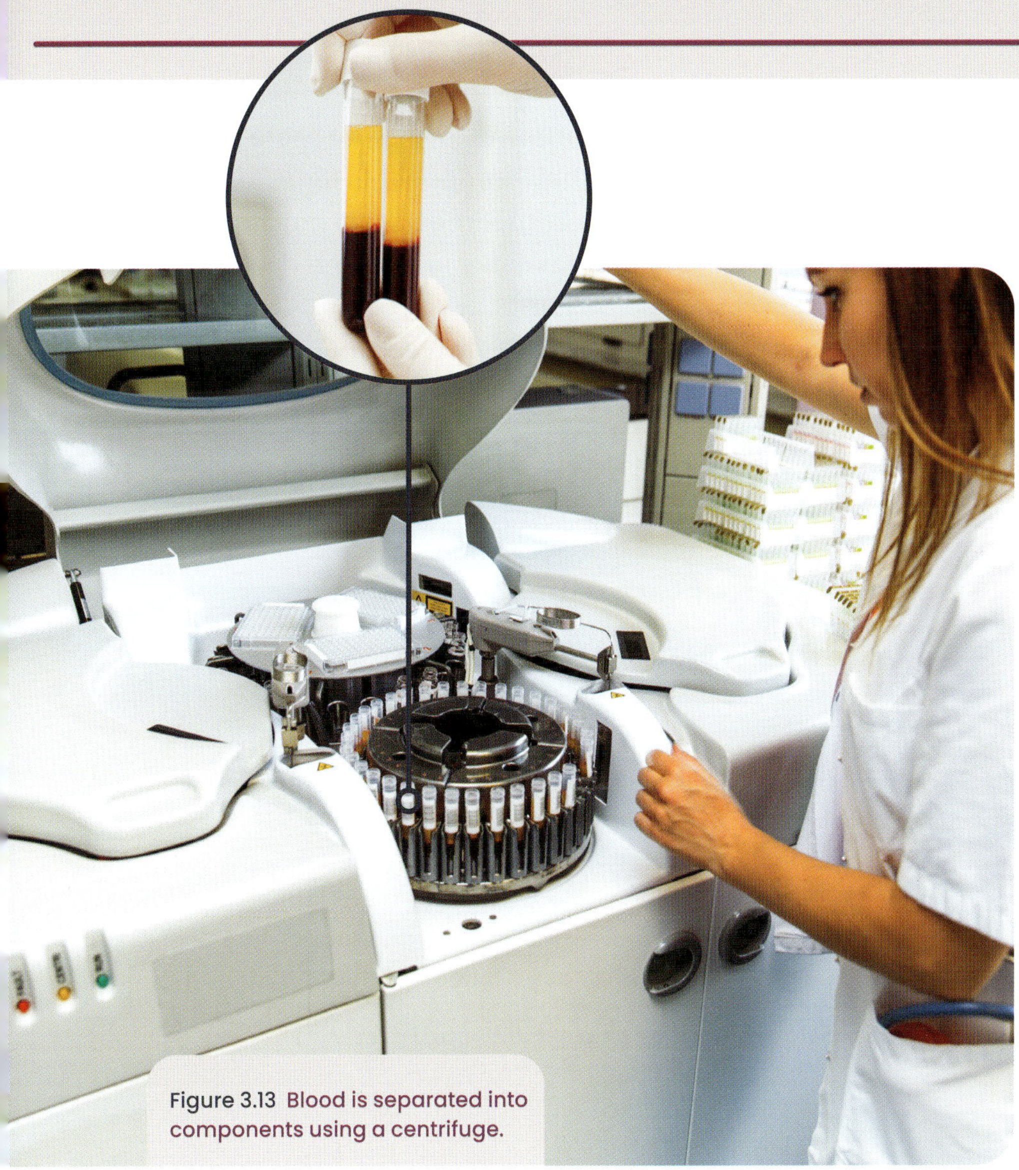

Figure 3.13 Blood is separated into components using a centrifuge.

❷ Many useful components can be separated from blood

Blood is a mixture that maintains life. It contains substances such as red blood cells (which carry oxygen), white blood cells (which fight infections), platelets (which help blood to clot) and plasma (which carries nutrients).

The components of blood can be separated using a **centrifuge**. When this machine spins quickly, the blood components separate into different layers based on **density**. The heavier particles such as red blood cells settle to the bottom, followed by platelets and white blood cells. The lightest particles, such as plasma, remain on the top. These individual components can be given to patients who have lost blood or who have certain blood diseases.

What is a centrifuge used for?

CHECKPOINT 3.6

1 Copy and complete the following sentences.
Blood is a ______________.
Blood is made up of __________, ____________, ____________ and ______________.
The process used by medical staff to separate blood is called ______________.
Pigments and ______________ from plants can be separated from the plant by using ______________.

2 Identify the state changes that happen when the oils from flowers are extracted for use in perfume.

3 What physical property allows oils to be separated from water when they are in liquid form?

4 Identify the physical property that allows the different components of blood to be separated.

5 The Australian Red Cross encourages people to be plasma donors. The procedure involves blood being taken into a machine that extracts the plasma, at the same time the non-plasma components of blood are returned to the patient. Suggest how the machine that separates the blood would work.

CHALLENGE

6 Use the internet to research how pigments can be extracted from plants. Then write a procedure that you could use to extract a pigment from a common plant.

SKILLS CHECK

- I can describe how oil can be separated from plants.
- I can describe how blood is separated into its components.

3.7 SEPARATING OTHER LIQUIDS

At the end of this lesson I will be able to:

- **relate** a range of techniques used to separate the components of some common mixtures, including chromatography, to the physical principles involved in each process.

KEY TERMS 

immiscible
not able to be mixed

paper chromatography
a technique used to separate coloured substances using a strip of paper and a solvent

LITERACY LINK 

Write a step-by-step method for an experiment that separates the colours in textas using chromatography.

NUMERACY LINK 

A separating funnel contains 150 mL of liquid. Convert 150 mL to litres.

Formula: 1 mL = 0.001 L

Different mixtures can be separated to find out more about their components or to use their components for different purposes. Forensic scientists sometimes use a process called chromatography to separate colours in inks and dyes to compare samples.

Sometimes mixtures do not mix well together and so a separating funnel can be used to separate the components.

1 Paper chromatography separates colours in mixtures

Many inks and dyes are made of a mixture of different colours. **Paper chromatography** is a separating technique that uses a solvent, such as water, to separate the different colours.

A spot of ink or dye is placed on a strip of paper that is touching a solvent, such as water or methylated spirits. As the solvent travels up the paper, it will dissolve the ink or dye. The more soluble the colour in the solvent, the further it is carried up the paper.

Inks or dyes that are not very soluble don't travel very far on the strip of paper. Dyes that are very soluble in water will travel further.

The result is that the ink or dye mixture separates into its pure components, each of which will have a different colour.

How can colours in inks or dyes be separated?

Figure 3.14 Paper chromatography uses a strip of paper to separate an ink or dye spot.

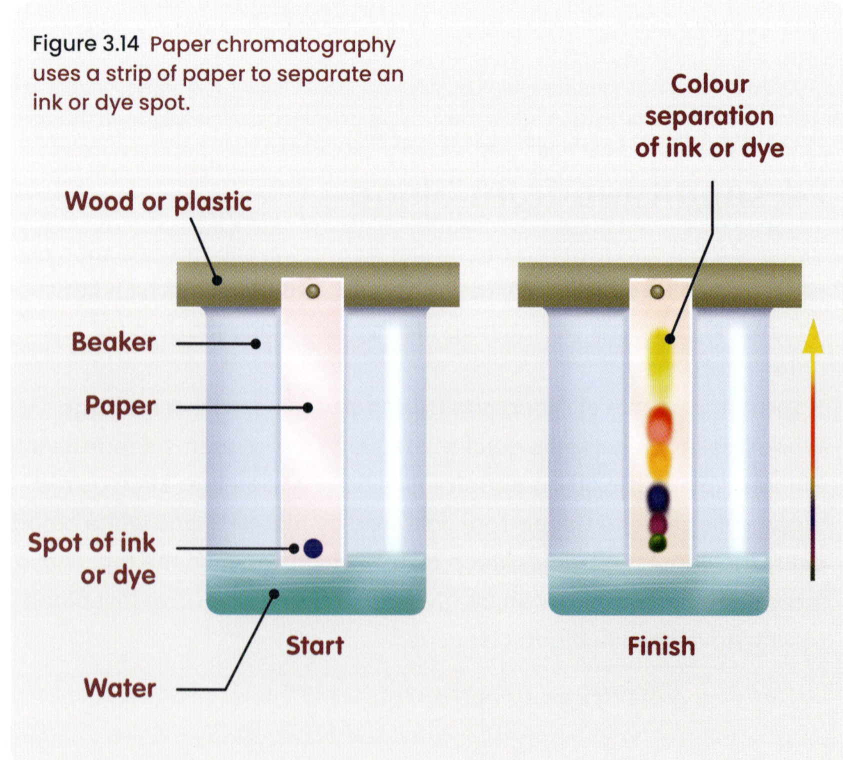

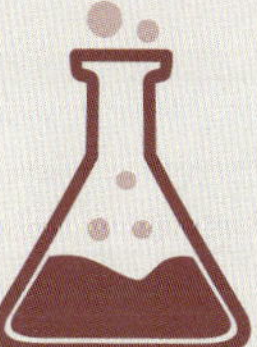

INVESTIGATION 3.7A Separating colours using paper chromatography

INVESTIGATION 3.7B Separating two immiscible liquids (Teacher demonstration)

❷ A separating funnel can be used to separate liquids that don't mix

Many liquids will mix together easily, but some will not. Oil and water do not easily mix – instead, the oil floats on top of the water. It's possible to shake them and get them to mix, but they soon separate.

Liquids that will not mix together are **immiscible**. When two or more immiscible liquids are different densities, the liquids can be separated using a separating funnel.

What can a separating funnel be used for?

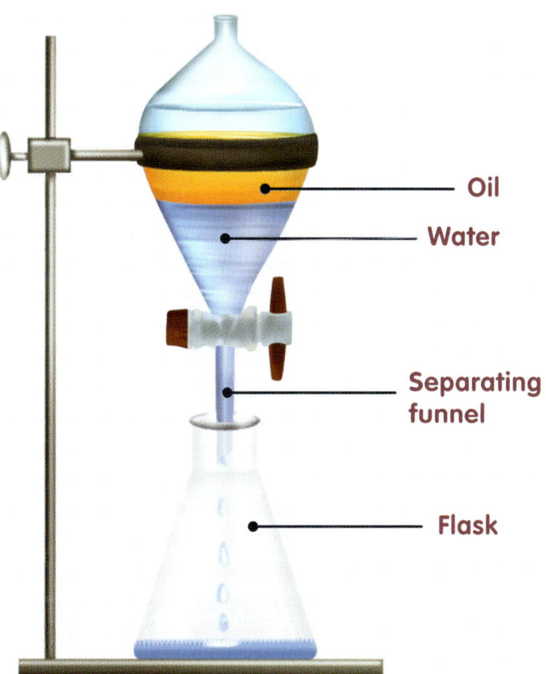

Figure 3.15 Two immiscible liquids can be put into a separating funnel and left to settle into layers. When the tap is opened, the liquid at the bottom runs into the flask.

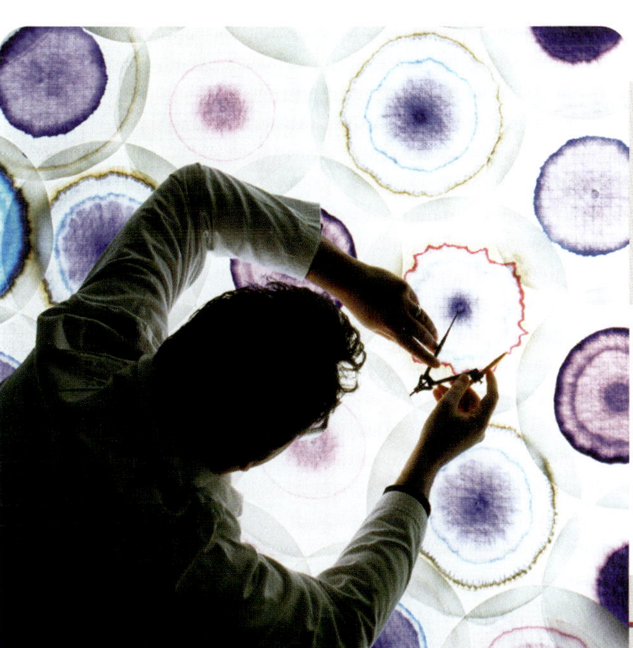

Figure 3.16 Forensic scientists can use chromatography to match pen and ink samples, or to compare dyes in fibre samples.

CHECKPOINT 3.7

1 Describe the process of paper chromatography.

2 Identify two solvents that can be used in paper chromatography.

3 You want to compare the components in some different inks.
   a What would you expect to observe if a dye was not a mixture?
   b What would you expect to observe if a dye was a mixture?

4 Explain why colours separate on a filter paper.

5 Identify a situation where a separating funnel could be useful.

6 Would a separating funnel be useful for separating a mixture that had a solute dissolved in a solvent? Explain why/why not.

CHALLENGE

7 Gas chromatography is a technique that is used by forensic chemists. Use the internet to find out how the process works and why it is used.

SKILLS CHECK

- I can explain how paper chromatography works.
- I can give some examples of things that can be separated using chromatography.
- I can describe how a separating funnel can be used.

3.8 CLEANING UP OIL SPILLS

At the end of this lesson I will be able to:

- **investigate** the application of a physical separation technique used in everyday situations or industrial processes, for example, cleaning up oil spills.

KEY TERMS 

crude oil
oil that has not yet been separated into useable petroleum products

immiscible
not able to be mixed

oil slick
a layer of oil on the surface of water

LITERACY LINK 

Write a 'how to' information sheet to help communities clean up after an oil spill. Include images and information that are easy to understand.

NUMERACY LINK 

An oil spill covers a rectangular region 20 m long and 12 m wide. What is the area of this region?

Formula: $A = lw$

Oil is one of the most important materials used today. When refined, it fuels the engines of vehicles, powers factories to produce electricity and is an ingredient in the manufacture of plastics. However, its extraction, transport and use have an impact on the environment.

The accidental release of crude oil into the environment is called an oil spill. Oil spills in the ocean can have a significant impact on ecosystems. Understanding the properties of oil and how it mixes with water and other chemicals has allowed scientists to work out methods to clean up oil spills.

1 Oil spills can be caused by different incidents

Oil spills into the marine environment can be caused by different events.

Crude oil is extracted by drilling it from reservoirs underground. Often this is done from oil rigs in the oceans. If an accident happens during the drilling or pumping process, crude oil can be released into the ocean.

Ships use oil as fuel. If a ship breaks down, runs aground, collides with another ship or their tanks are damaged, the oil can leak into the water.

Oil is transported by massive ships called oil tankers. If these ships are damaged, the oil that they are carrying can spill into the environment.

How do oil spills happen?

2 Oil spills can damage the environment

When oil is spilled into the ocean it forms a thin layer on the surface known as an **oil slick**. Over time, the oil slick will spread out to cover a large area. If it comes into contact with the shore line, the oil will stick to and mix in with the sediments.

The oil on the surface of the ocean stops oxygen and carbon dioxide gas from dissolving into the water from the atmosphere. If the levels decrease enough, this can kill marine plants and animals.

How do oil spills affect the environment?

Figure 3.17 Birds caught in oil spills can lose their ability to fly, swim and float, and they can be poisoned by the oil when they try to clean themselves.

3 Oil spills can be contained and filtered

Oil slicks float on top of water because oil is less dense, and **immiscible** with water. There are several different ways that oil slicks can be cleaned up by taking advantage of these properties.

Floating booms are used to contain oil slicks. Devices called skimmers are then used to scoop or suck oil from the surface of the water within the enclosed areas. This oil can then be processed in a factory to remove any sea water that was also collected.

Sorbent booms are a special type of boom that work in a similar way to a disposable nappy, but they only absorb oil. The booms can then be removed and disposed of.

If oil hits the shore line it can be difficult to clean up. One way to clean oil from the shore line is to wash the oil back into the water so that it can be easily skimmed off the top. Another way is to remove the contaminated sediment. This is either disposed of in landfill or processed in factories to separate the oil from the sediment.

How are oil spills contained and cleaned up?

Figure 3.18 Floating booms are used to contain oil spills on the surface of water. The oil can then be skimmed from the surface.

CHECKPOINT 3.8 

1 Describe one way that an oil spill can happen in the ocean.

2 Describe what happens to oil once it is spilled into the ocean.

3 Why does an oil spill reduce oxygen and carbon dioxide levels in water?

4 Identify two properties of oil that make it easy to clean up when a spill happens in the ocean.

5 Suggest why booms are important in cleaning up oil spills.

6 Explain why clean up crews wash oil on the shore back into the ocean, in reference to being able to collect oil.

CHALLENGE

7 The Montara oil spill in 2009 has been Australia's largest oil spill to date. Use the internet to research the causes of the spill, the area covered, the damage it caused and techniques used to clean it up.

SKILLS CHECK

- I can explain how separation techniques can be used to clean up oil spills.

3.9 SEPARATIONS IN INDUSTRY

At the end of this lesson I will be able to:

- **research** how people in different occupations use understanding and skills from across the disciplines of science in carrying out separation techniques.

KEY TERMS 

fractional distillation
a method that separates liquids by using their different boiling points

froth flotation
a method that uses special chemicals to separate minerals from their ores

mineral ore
a mineral that contains useful metals

LITERACY LINK 

Create a mind map that demonstrates your understanding of these terms: *froth flotation, magnetic flotation, ore, oil, fractional distillation.*

NUMERACY LINK 

A magnetic separator recovers 75% of the iron from ore. If 1000 kg of ore goes through the machine, how much iron does it recover?

We obtain many useful resources from Earth, such as metals and oil. Once raw materials have been extracted, they must be processed to separate the useful material from the waste.

1 Froth flotation uses water to separate metals

Copper is used for electrical wiring and plumbing pipes. It is rarely found in its pure form, instead it is found as a **mineral ore** called malachite. Before copper can be purified, it must be separated from the other rocks and material in the ore. This is done using **froth flotation**.

The mixture of unwanted material and copper is dug out of the ground in solid lumps. These are crushed into a fine powder, and mixed with water and some detergent-like chemicals. Air is blown into the bottom of the container, and the malachite is carried to the surface by the air bubbles as it sticks to the chemicals. The waste sinks to the bottom of the tank, unable to stick to the chemicals.

The malachite froth containing the copper is skimmed off the surface. It's then further treated to extract pure copper.

How does froth flotation separate metals from ore?

Figure 3.19 Froth flotation separates copper minerals from waste rock. Copper minerals are attracted to detergent-like chemicals and skimmed off for further processing.

❷ Magnets are used to process iron ore

Iron is another vital metal – it's one of the main components used to make steel. It is magnetic and this property can be used to separate minerals that contain iron from waste rock.

Rock containing iron minerals is crushed into a fine powder then dropped onto a conveyor belt. At the end of the conveyor belt, a magnetic roller attracts the pieces containing iron. As the waste rock reaches the roller, the non-magnetic pieces drop off the belt immediately. The pieces containing the iron are carried further around the belt and collected for processing.

How are magnets used to separate iron minerals from rock?

Figure 3.20 Magnetic separation can be used to process metals such as iron.

❸ Crude oil is processed using fractional distillation

Crude oil is a mixture of chemicals such as petrol, oil, kerosene and diesel. Each of these chemicals is useful, so a process called **fractional distillation** is used to separate them from the crude oil.

The crude oil is placed in a piece of equipment called a fractionating column and then heated. As each liquid evaporates, it rises up the column. As it rises, it cools, and when it cools to below its boiling point it becomes a liquid and leaves the column through a tube. Each liquid rises to a different height in the column, depending on its boiling point.

How is crude oil separated into different liquids?

INVESTIGATION 3.9
Froth flotation

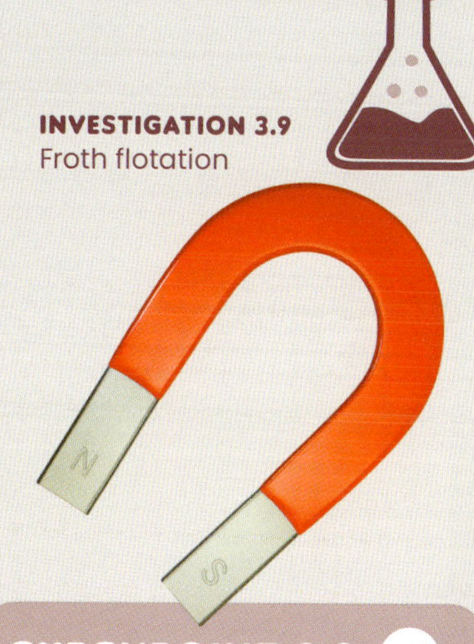

CHECKPOINT 3.9

1 Identify the difference between copper minerals and waste rock that allows them to be separated by froth flotation.

2 Identify the difference between iron minerals and waste rock that allows them to be separated by magnets.

3 Identify the difference between the components of crude oil that allows them to be separated by fractional distillation.

4 Suggest why it is important to crush rock into a fine powder before froth flotation or magnetic separation.

CHALLENGE

5 The agriculture industry also relies on various separation techniques, especially for the processing of food crops. Use the internet to research one example of separation being used in agriculture. Present your findings as a series of labelled diagrams.

SKILLS CHECK

- I can describe separation techniques are used in the mining industry.
- I can suggest other industries where knowledge of separation techniques is important.

CHAPTER SUMMARY

solute
a substance that is dissolved by a solvent

solvent
a substance that dissolves a solute

Solute (e.g. sugar)

Solvent (e.g. water)

Sugar dissolves in water

solution
a mixture made up of a solvent and a dissolved solute

Solution

Water is a **solvent** that can dissolve other substances. This makes it vital in everything from digesting food in your stomach to mining sulphur from underground rocks.

PROCESS	USED TO SEPARATE
Filtering	Solid from liquid
Decanting	Solid sediment from liquid
Distillation	Liquids with different boiling points
Evaporation and crystallisation	Solid from a liquid solution, without keeping the liquid
Paper chromatography	Colours

Mixture
two or more substances mixed together that can be physically separated

Colloid
a mixture with tiny particles spread out evenly that never settle to the bottom. Milk is a colloid because it contains tiny fat droplets in water.

Suspension
a mixture with large, insoluble particles that are initially spread out evenly and eventually settle to the bottom

Physical separation is useful in the 'real world'.

Cleaning up oil spills

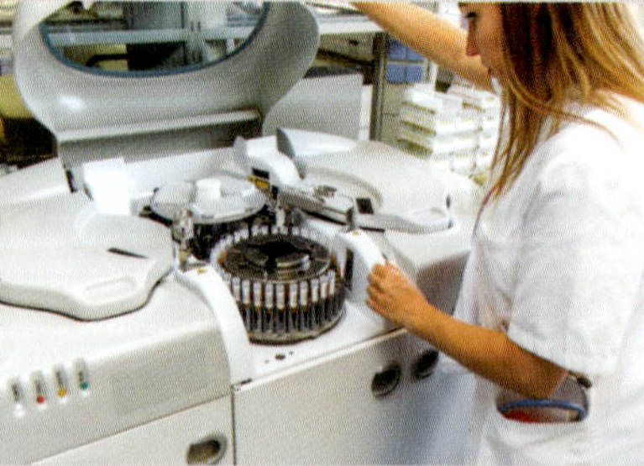

Separating blood products

Sorting waste material

Chromatography

★ FINAL CHALLENGE ★

1. Define the following terms: solute, solvent and solution.
2. Briefly summarise how water acts as a solvent in everyday life, in the environment and in industry.
3. Explain why decanting can assist filtration to ultimately get cleaner water.

LEVEL UP!
LEVEL 1
50xp

4. The following table lists separating techniques and different types of mixtures. Draw an arrow between the separating technique and the mixture that would allow its separation.

Filtration	Iron filings mixed with sand.
Decanting	Salty water from which you want to extract pure water.
Evaporation	A mixture of fruit pulps floating in water.
Distillation	Oil and water.
Centrifugation	A solution containing dissolved salt, from which salt needs to be kept.
Magnetic separation	Muddy water containing some sand.
Separating flask	A solution containing sand and water.

5. Select one of the separation techniques above and write a step-by-step method for its use.

6. Identify two ways in which water purification and waste treatment are similar, and two ways in which they are different.
7. Compare and contrast distillation and evaporation separating techniques.
8. A snow globe is a suspension but a drink of cordial is a solution. Explain why.

9. Write a numbered method detailing the steps to extract oil from plants.
10. The components of blood are separated using a centrifuge. Suggest what physical properties of blood make this possible.

11. Identify an occupation that would utilise the following processes:
    a froth flotation
    b magnetic separation
    c fractional distillation.

CHEMICAL CHANGE

Chemical changes can happen on a very large scale. Explosions are huge chemical reactions.

On a smaller scale, chemical changes take place in your body all the time. They happen in your mouth to break down your food and continue in the rest of your digestive system to change the food into energy your body can use.

From using fire to the latest scientific research, humans have learnt to use chemical changes and to create their own.

1 LEARNING LINKS

What do you already know about chemical change?

Some changes are reversible, such as when water changes state. Brainstorm some other reversible changes.

Some changes to materials are irreversible. You cannot recover burnt wood. List some other irreversible changes.

Glass is ideal to drink from but it's not suitable for many other uses. Why are materials chosen for specific tasks?

2 SEE-KNOW-WONDER

List three things you can **see**, three things you **know** and three things you **wonder** about this image.

3 CRITICAL + CREATIVE THINKING

Five questions: Write five questions that have 'chemical change' as the answer.

Variations: How many different ways can you change the way something looks without altering what it's made of?

Combinations: List the features of a rotten banana and of a plastic ball, and then combine some of these features to invent a new object.

4 THE MOST EXPENSIVE!

The most expensive thing ever made by humans is the International Space Station (ISS), which cost around 220 billion dollars! The ISS is the size of a five-bedroom house, but it's a giant laboratory that moves in low orbit around Earth.

The ISS is one of the best examples of scientific collaboration in human history. Many countries contributed towards the cost of making it, and over 15 countries have sent their scientists up to the ISS. The longest stay on the ISS was 665 days!

4.1 PHYSICAL AND CHEMICAL CHANGE

At the end of this lesson I will be able to:

- **identify** when a chemical change is taking place by observing a change in temperature, the appearance of new substances or the disappearance of an original substance
- **demonstrate** that a chemical change involves substances reacting to form new substances.

KEY TERMS 

chemical change
a change in properties with a new substance formed

physical change
a change in appearance with no new substance formed

properties
what a substance looks like and how it behaves

reversible
can be taken back to its previous state

LITERACY LINK 

Describe physical and chemical change in only three words each.

NUMERACY LINK 

Mixing salt and sand results in a mixture of 10 parts salt to 35 parts sand. Write the ratio of salt to sand in its simplest form.

Changes happen around us all the time: the ice caps of the north pole melting, explosions during New Year's Eve fireworks, toast burning, nails rusting and food rotting.

These changes are all either physical or chemical changes. When no new substance is formed, and the process is **reversible**, it's a physical change. If a new substance is produced and the process is not reversible, then this is a chemical change.

Figure 4.1 The sugar in the left pan is melting, which is a physical change – no new substances are formed. The sugar in the right pan has been heated more and it has chemically changed into a new substance: caramel.

1 Physical changes do not affect the properties of a substance

When a substance undergoes a **physical change**, the substance keeps its original **properties**. Physical changes can include changes of state, such as melting and evaporation, or other changes, such as crushing or stretching.

When water changes state from solid (ice) to liquid (water) to gas (steam), this is a physical change. All three states of water still consist of molecules of water.

Physical changes can be reversed. Think about hot, melted wax – when the melted wax cools and hardens, it still looks the same. The wax has undergone a physical change.

Some examples of physical changes are:

- salt and sand being mixed
- paper being cut
- evaporation
- a ruler breaking
- salt being dissolved in water.

What is a physical change?

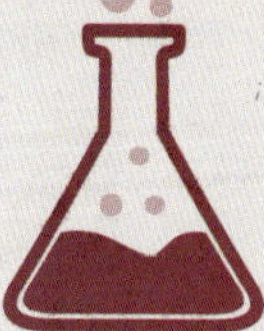

❷ Chemical changes can make new substances

A **chemical change** happens when a new substance is formed during a chemical reaction. Chemical reactions cause the atoms in the substance to rearrange and form new substances.

We can look for evidence that chemical change has happened. Some observations may include:

- there is a colour change
- a gas (bubbles) is released
- a new substance is formed
- heat is absorbed or released
- a precipitate (solid bits within a liquid) is formed.

Chemical changes are not reversible. When you cook a pancake, it cannot be 'uncooked' and become batter again. This is a chemical change.

Some other examples of chemical changes are:

- burning wood
- a battery working
- rusting iron
- a banana rotting
- fireworks
- cookies baking.

What is a chemical change?

Figure 4.2 When metals burn, they form new substances. Each burning metal gives off a different colour of light. This is a chemical change.

INVESTIGATION 4.1
Physical and chemical changes

CHECKPOINT 4.1 ✓

1 How are physical and chemical changes different?

2 Chemical change produces new substances. How are the new substances made?

3 List three pieces of evidence that could show that a chemical change has happened.

4 Put a P next to the physical changes and a C next to the chemical changes.
a ice melting
b a cake baking
c paper burning
d chocolate powder mixing into milk
e cardboard being cut
f wax melting

5 One sign that a chemical change has taken place is that gas is released and forms bubbles. Suggest why this is a sign of chemical change.

6 Give an example of a chemical change not already mentioned and justify why it is a chemical change and not a physical one.

CHALLENGE

7 Fireworks involve chemical reactions that produce coloured light. Use the internet to find out what chemical reactions take place and what produces the different colours.

SKILLS CHECK

- I can explain the difference between a physical and a chemical change.
- I can identify evidence for a chemical change.

4.2 CHEMICAL CHANGES IN EVERYDAY LIFE

At the end of this lesson I will be able to:

- **describe** the chemical changes that occur in everyday life.

KEY TERMS

chemical reaction
a process in which one or more substances are changed to new substances

corrosion
breaking down or destruction of a substance, especially a metal, through a chemical reaction

product
a substance formed during a chemical reaction

reactant
a substance that takes part in a chemical reaction

LITERACY LINK

The word *photosynthesis* can be broken into two smaller words – 'photo' and 'synthesis' – to understand its meaning. Explain how the meaning of each of these smaller words gives clues to understanding what *photosynthesis* means.

NUMERACY LINK

A chemical reaction affects 3 molecules, then 6, then 9 and 12. Write the next four numbers in the series 3, 6, 9, 12, ___________

Figure 4.3 Stain removers contain ingredients that break down stains into molecules that can be washed away.

A **chemical reaction** produces a chemical change. In a chemical reaction, one or more substances, the reactants, are changed to one or more new substances, the products. This happens as the reaction causes the atoms to rearrange to create new substances as products.

1 Photosynthesis is a chemical reaction within plants

Photosynthesis is a chemical reaction in which plants use energy from sunlight to transform water and carbon dioxide into oxygen and glucose.

We always use a word or chemical equation to represent a chemical reaction. In a chemical equation, the reactants are shown on the left hand side of the arrow and the products on the right hand side. The **reactants** are substances that change during a reaction, while the **products** are substances that form as a result of the reaction.

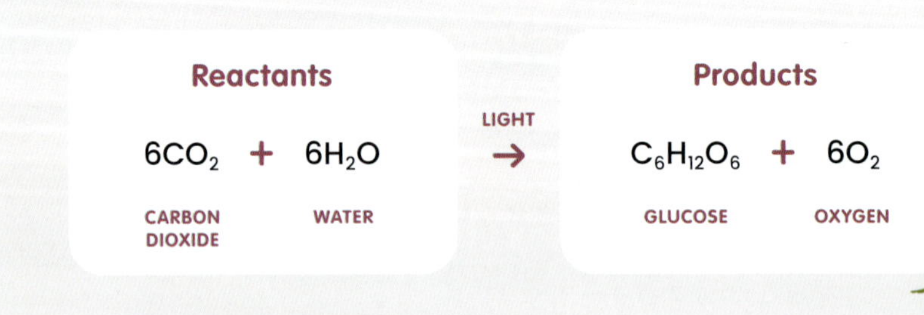

This equation shows that atoms in carbon dioxide and water have reacted and been rearranged to form glucose and oxygen. The products are very different from the reactants – they are completely new substances.

What are the reactants and products of photosynthesis?

INVESTIGATION 4.2A
Corrosion of iron

INVESTIGATION 4.2B
Preventing corrosion

❷ Cellular respiration combines oxygen and glucose

Cellular respiration is the way that cells of all living organisms obtain energy. During cellular respiration, oxygen (from the air) and glucose (from food) combine. This chemical reaction produces carbon dioxide, water and energy.

Reactants			Products		
$C_6H_{12}O_6$ +	$6O_2$	→	$6CO_2$ +	$6H_2O$ +	ENERGY
GLUCOSE	OXYGEN		CARBON DIOXIDE	WATER	ENERGY

Does this reaction look familiar? Remember that, during photosynthesis, carbon dioxide and water change into oxygen and glucose, in the presence of sunlight. This means that cellular respiration is the opposite process of photosynthesis. The atoms in the reactants are rearranged to form new substances (products).

What are the reactants and products of respiration?

❸ Rust is a chemical change

When some metals come into contact with oxygen in the air, they combine in a chemical reaction. This type of reaction is known as **corrosion**. Corrosion can cause metal objects to become weaker and break down. In a corrosion reaction, the reactants are the metal and oxygen. The product is a new substance called a metal oxide.

When objects made from iron react with oxygen they produce iron oxide. This is commonly known as rust.

Some metals are less likely to corrode than others. Galvanised iron is made by coating iron with a thin layer of zinc. The zinc is less likely to react with oxygen, so this coating protects the iron.

What is rust?

Figure 4.4 Rust is also known as iron oxide. It is produced when iron comes into contact with oxygen in the air and reacts. This is an example of a corrosion reaction.

CHECKPOINT 4.2 

1 What must happen for a chemical change to occur?
2 Explain the difference between a reactant and a product in a chemical reaction.
3 Identify the products formed during photosynthesis.
4 Identify the products formed during cellular respiration.
5 Identify the reactants in the reaction that forms rust.
6 A herbivore is an animal that eats plants. Suggest where the glucose in the herbivore's cells originally came from.
7 Suggest why using the chemical formula of substances when writing chemical equations is useful.

CHALLENGE

8 Research and explain the difference between breathing and cellular respiration.

SKILLS CHECK

- I can explain what a chemical change is.
- I can identify at least three examples of chemical changes that occur in everyday life.

4.3 COMPARING PHYSICAL AND CHEMICAL CHANGE

At the end of this lesson I will be able to:

- **compare** physical and chemical changes in terms of the arrangement of particles and reversibility of the process.

KEY TERMS 

electrolysis
passing electricity through a substance to break it up

model
using objects or a diagram to describe something

molecule
two or more atoms chemically bonded together

theory
an explanation that can be supported or disproved using evidence

LITERACY LINK 

Brainstorm three theories about chemistry. Make sure they are things that can be supported or disproved using evidence.

NUMERACY LINK 

If you had 8 molecules of H_2O (water) how many H atoms do you have in total? How many O atoms do you have?

Figure 4.5 Particles exist all around us as solids, liquids and gases.

Now that you know how physical and chemical changes differ, it's time to look at how these changes take place.

Like a scientist, you can use a **model** to understand and explain your observations, because the particles in chemical reactions are too small to see.

1 The particle theory models the nature of matter

There are five main ideas in the particle **theory**. This model is helpful to remember as you study chemical reactions:

1 All matter is made up of tiny particles.
2 All the particles are constantly moving.
3 Forces of attraction hold the particles together.
4 The further apart the particles are, the weaker the forces of attraction.
5 Particles at higher temperatures move faster and with more energy than those at lower temperatures.

What is one of the main ideas in the particle theory?

2 Particles do not change during a physical change

In a physical change, no new products form and the reaction is generally reversible. There is no change to the make up of atoms or molecules of the substance.

You can easily change the state of water by freezing, melting or boiling it. The water molecules stay the same each time, whether the water is a solid (ice), liquid (water) or gas (steam). The molecules will move faster or slower depending on their state, but they don't stop being water molecules.

What happens to water molecules as they change from one state to another?

Figure 4.6 Water molecules can exist as a solid, liquid or gas.

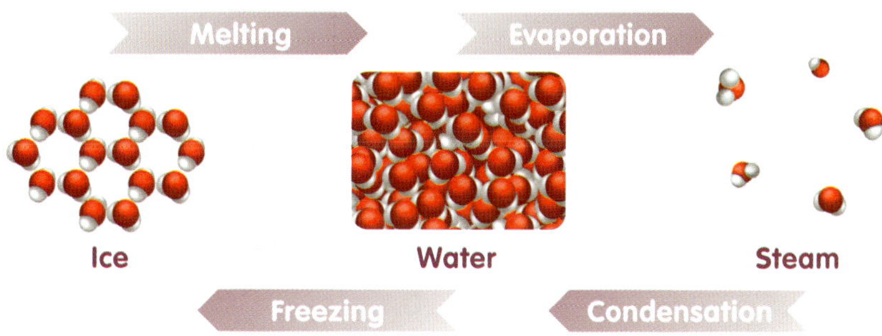

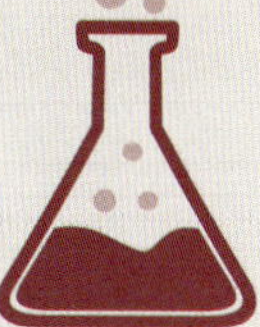

INVESTIGATION 4.3
Burning steel wool (Teacher demonstration)

3 Particles rearrange during a chemical change

In a chemical change, new substances are formed and the reaction is not reversible. The particles at the beginning of the reaction are *not* the same as those at the end. For example, as a reaction happens, the original molecules are broken down into atoms that are then rearranged to form molecules of new substances.

Water molecules are made up of two hydrogen atoms and one oxygen atom, so the molecular formula of water is H_2O. Breaking apart water molecules isn't easy, but it can be done using **electrolysis**. This involves passing electricity through the molecules. The electrical energy breaks the strong bonds holding the oxygen and hydrogen atoms together. The atoms of these reactants then rearrange and combine to form hydrogen gas and oxygen gas, which are the products of the reaction.

What products are formed when water goes through a chemical change?

Figure 4.7 The bonds inside water molecules break when electrical energy passes through them. Oxygen and hydrogen gases are formed.

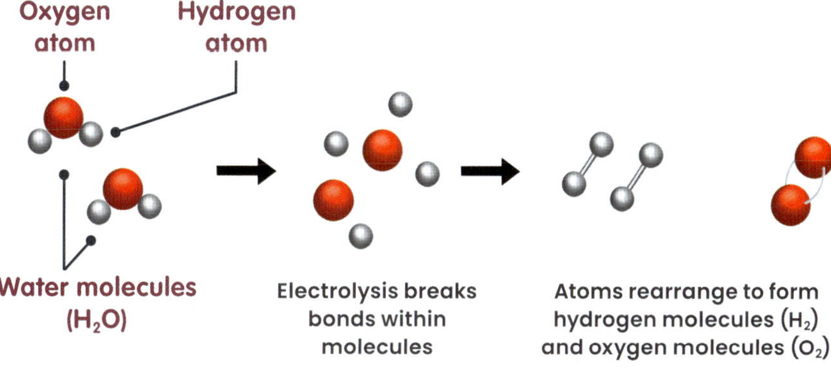

CHECKPOINT 4.3

1 What is the particle theory of matter?

2 Explain in terms of particles what happens when ice changes to water.

3 What is the molecular formula of water?

4 Copy and complete these sentences.
Physical changes are __________, and no new __________ are produced. Chemical changes are __________. In a chemical change __________ substances are __________.

5 Use diagrams of particles (such as molecules) to show the difference between a chemical and a physical change.

6 Write a word equation for the decomposition of water into hydrogen and oxygen gas.

CHALLENGE

7 Consider the physical change of water from liquid to gas. Now consider the chemical change of water when it breaks down to form hydrogen and oxygen. Compare the differences in terms of energy for the physical changes in water and the chemical changes in water.

SKILLS CHECK

- I can explain the difference between a physical and chemical change.
- I can use the particle model to compare and contrast physical and chemical changes.

4.4 MAKING NEW MATERIALS

At the end of this lesson I will be able to:

- **propose** reasons why society should support scientific research into making new materials.

KEY TERMS 

biodegradable
able to be broken down, for example by bacteria

fossil fuel
fuel formed underground from plant and animal remains

polymer
a substance consisting of large molecules made up of many small repeating units

LITERACY LINK 

Write a one-page persuasive piece that aims to convince people that society should support scientific research into making new materials.

NUMERACY LINK 

Nadia designs a model train using new polymers. It has six carriages, each of which has four windows. How many windows does Nadia need to build?

For many years, scientists have made new materials such as plastic from **fossil fuels**. But fossil fuels are a limited resource that take millions of years to form. One day, those fossil fuels will run out.

Scientists need to research ways of producing and testing materials that will be better, longer lasting and easier to use.

1 Plastics are useful but not environmentally friendly

Plastics are materials used in many ways, in millions of different products. They are a kind of **polymer**, which is made up of many tiny, repeating units attached to a 'backbone' of carbon atoms.

Plastics are made from petroleum, a non-renewable resource. The first scientists to make plastics wanted them to last for a long time. This is good when the products are being used, but not when they are no longer needed. Many plastics take thousands of years to fully break down in the environment. When not disposed of properly, plastics cause pollution by cluttering rivers, seas and beaches, killing fish and marine organisms. The production of plastics releases carbon dioxide into the atmosphere, which contributes to global warming.

Because of the huge problems with disposing of plastic, scientists have had to come up with ways of making it more environmentally friendly.

Why aren't plastics environmentally friendly?

Figure 4.8 Many toys that were once made of wood are now made of plastic, which is more colourful but is a non-renewable material.

❷ Bioplastics are more environmentally friendly

Bioplastics are a type of plastic invented about 100 years ago. They are made from plant products such as starch and cellulose, which come from corn and sugarcane. The starch is broken down into glucose, which is made into lactic acid. Many molecules of lactic acid are joined to make the polymer polylactic acid (PLA), which is used in bioplastics.

Bioplastics are used in products such as food containers, grocery bags, some cutlery and food packaging. They have many uses in electronics, farming, clothes making and health care.

Many bioplastics are **biodegradable** – they break down quickly in the environment. Bacteria and fungi are quickly able to break down these plastics if they are placed into an industrial composter. The bioplastics absorb water, swell up and break into small pieces that bacteria can easily digest – sometimes in just a few weeks.

Even if not properly composted, bioplastics have less environmental impact than plastics formed from petroleum. If PLA is burned, it doesn't give off toxic fumes. If it ends up in a landfill, it will break down a lot faster than traditional plastics.

How are bioplastics better for the environment?

Figure 4.9 Bioplastic wrapping breaks down much faster than older types of plastic.

❸ Scientists are working to improve bioplastics

There are many other benefits of bioplastics compared to petroleum plastics. They are cheaper and take less energy to make, and they create far fewer greenhouse gases. They are easier to recycle into other products. However, they still have disadvantages and can be improved.

One disadvantage is that bioplastics can't be broken down in a backyard compost heap. They need to be processed in special industrial composters. Australia does not have a recycling system that uses these composters, so bioplastics all end up in landfill.

Another issue is that bioplastics are made from plants such as corn. Using these plants for making plastics means that they can't be used as food. It also means that any chemicals used on those plants to kill pests or help them grow can make their way into the bioplastic. This can have unpredictable effects on both the bioplastic and the people who come into contact with it.

What are the positives and negatives of bioplastics?

CHECKPOINT 4.4 

1 Making plastics and polymers from fossil fuels is not sustainable. Explain why.

2 What are the benefits of supporting scientists in their research of new materials?

3 Explain how bioplastics are different from ordinary plastics.

4 PLA is a bioplastic that is biodegradable. Explain how it is created and why it is biodegradable.

5 Identify three advantages and three disadvantages of bioplastics.

CHALLENGE

6 Research what traditional plastics are, what they are made from and the problems they cause.

SKILLS CHECK

- I can describe at least two reasons why society should fund more research into new materials.

4.5 NANOTECHNOLOGY

At the end of this lesson I will be able to:

- **propose** reasons why society should support scientific research.

KEY TERMS

biopsy
tissue removed from a living body for testing

chemotherapy
a treatment of diseases such as cancer using chemical substances

malignant
cells dividing uncontrollably and perhaps spreading to other parts of the body

nanotechnology
the field of science that deals with individual atoms and molecules

tumour
a swelling caused by an abnormal growth of tissue

LITERACY LINK

Summarise the information in this section as a series of dot points to help you organise your thinking about nanotechnology.

NUMERACY LINK

A nanometre is one-billionth of a metre, while a centimetre is one-hundredth of a metre.

a Convert 30 cm to m
b Convert 1 nm to m

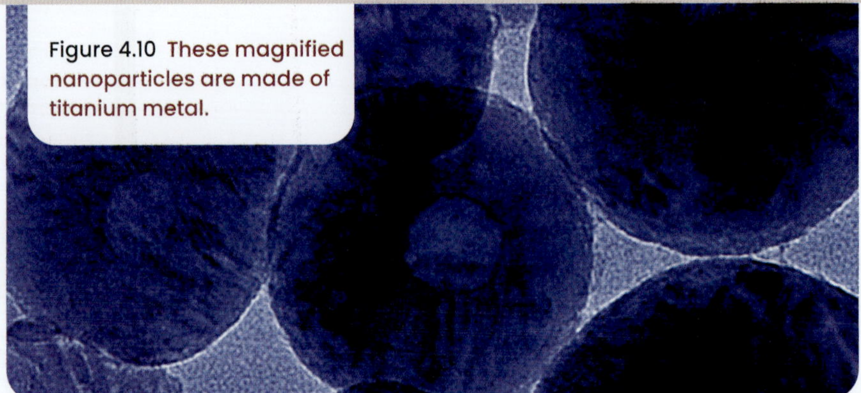

Figure 4.10 These magnified nanoparticles are made of titanium metal.

As recently as a hundred years ago, many people died of what are now considered minor illnesses because there were no cures. Today we have treatments available for many different illnesses, and some diseases have been eliminated entirely. This is only possible because society has supported scientific research in these areas.

One of the newer areas of research in health and medicine is **nanotechnology**.

1 Nanotechnology is the creation of extremely small particles

You know that in the metric system *centi*- means 'one-hundredth'. The prefix *nano*- means 'one-billionth'. If a centimetre is one-hundredth of a metre, a nanometre is one-*billionth* of a metre – an incredibly short distance.

Nano is also used more generally to mean something that is *very* small. The field of nanotechnology means 'small technology' – creating extremely tiny particles (or even tiny machines) that have a special purpose. These 'nanoparticles' are incredibly small, up to about 100 nanometres across. A human hair is about 800 times wider! The abbreviation commonly used for nanometre is nm.

Nanoparticles are created from existing materials such as gold and silicon. Engineers use chemicals, lasers and other technologies to design the physical and molecular structure of the particles, creating materials that aren't found in nature.

What is nanotechnology?

2 Breast cancer is a very common disease

Breast cancer is the most common **malignant** disease in women worldwide. It also affects a small number of men. Breast cancer is linked to certain genes, which can pass from generation to generation. Scientists are trying to identify and locate these genes so that they can cure or slow down the cancer process.

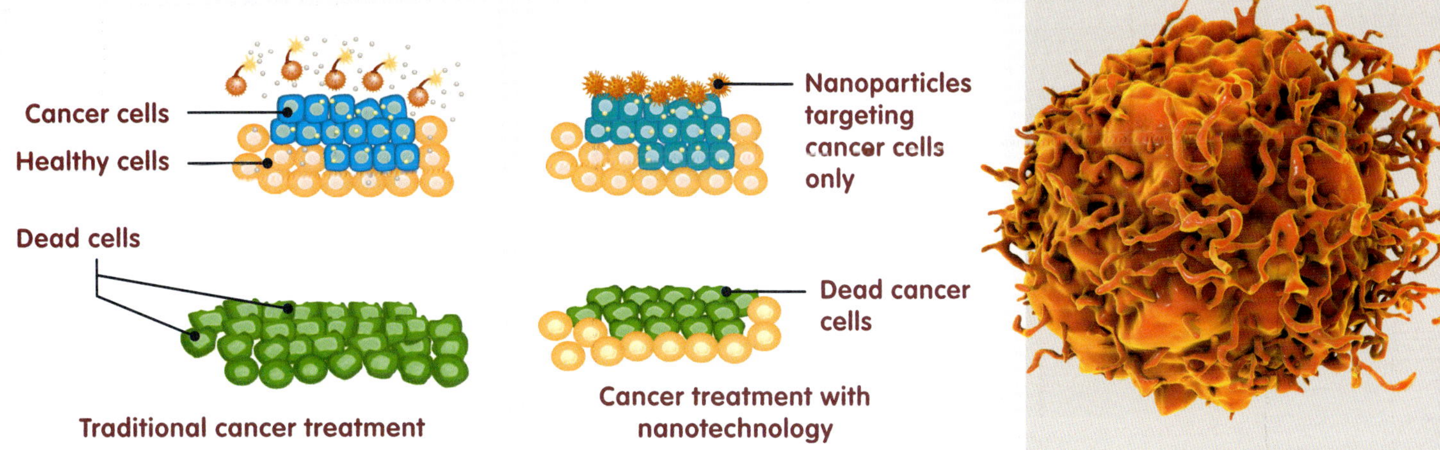

Figure 4.11 Nanotechnology treatments can target cancer cells without attacking healthy cells.

Treatment for early breast cancer aims to remove the cancer and reduce the risk of it spreading or coming back. At the moment, the most successful options are surgery, **chemotherapy** and radiotherapy (targeting cancer cells with X-rays or other radiation). Often a patient has more than one type of treatment.

Unfortunately, these treatments don't always work, and they can cause other health problems, such as a patient having to heal from surgery, or damage that harsh drugs and radiation cause to healthy cells.

Why is breast cancer treatment difficult?

3 Nanotechnology could lead to better breast cancer treatment

Using nanotechnology, doctors have new and better ways to diagnose and treat breast cancer.

An important part of treating breast cancer is detecting the cancerous cells. Usually a **biopsy** must be done to identify the **tumour**. In some new techniques, patients are injected with nanoparticles that bind to certain parts of breast cancer cells. These nanoparticles can be detected using a scan. This means cancerous areas can be seen earlier, and without harming patients.

Nanotechnology cancer drugs are used as part of chemotherapy. The chemotherapy drugs are coated with nanoparticles that prevent the immune system from attacking. This allows the medicine to reach the cancer cells without damaging healthy cells. In future, the nanoparticles may be able to detect cancer cells. After they attach to the cancer cells, the nanoparticles could be activated with a special light to destroy the cancer.

How can scientists use nanoparticles to treat breast cancer?

CHECKPOINT 4.5

1 Explain what nanotechnology is in your own words.

2 How large are nanoparticles?

3 Describe the difference between traditional treatment of cancer and nanotechnology treatments.

4 Identify one application of nanotechnology.

5 Using the example of nanotechnology in breast cancer treatment, explain why it is important for society to support scientific research.

CHALLENGE

6 Research and find three other examples of new technology and scientific research that improve human life. Summarise these and present your findings to the class.

SKILLS CHECK

- I can explain what nanotechnology is and give an example of how it is used in medicine.
- I can provide at least two reasons why society should support scientific research.

4.6 SCIENTIFIC COLLABORATION IN AUSTRALIA

At the end of this lesson I will be able to:

- **describe** how science knowledge can develop through collaboration and connecting ideas across the disciplines of science.

KEY TERMS 

collaboration
working cooperatively together

hydrophilic
water-attracting

hydrophobic
water-repelling

nuclear medicine
the branch of medicine to do with the use of radioactive substances in research, diagnosis and treatment

LITERACY LINK 

Create a one-page flyer or brochure that illustrates the work of the CSIRO in Australia. Include examples of their projects and the kinds of scientists that work there.

NUMERACY LINK 

At a conference of 1000 scientists, 200 work at ANTSO, 500 work at CSIRO and the rest work for universities.

Draw a pie chart to show who attended the conference.

In movies and television shows, inventors and scientists seem to create and discover great things on their own. In real life, almost all scientific advances are the result of **collaboration** – groups of scientists and researchers working together.

Working with scientists from other countries, or with specialists from different scientific fields, is the best way for Australia's scientists to create new things and improve their understanding of the natural world and the universe.

1 ANSTO is Australia's national nuclear research organisation

The Australian Nuclear Science and Technology Organisation (ANSTO) is Australia's national nuclear research and development organisation. It produces and uses nuclear radiation for medicine, science, industry, business and agriculture. ANSTO collaborates with international scientists.

The Open Pool Australian Lightwater (OPAL) reactor is a research reactor at Lucas Heights in Sydney. OPAL is used to make **nuclear medicines** as well as for research. It can create neutron beams – streams of high-energy particles – that are used to solve problems in many different fields of science.

Figure 4.12 OPAL is one of the world's most effective research reactors.

ANSTO's other major Sydney facility is the National Research Cyclotron. This machine shoots particles along a spiral path, accelerating them to very high speeds. Cyclotrons and particle accelerators are used to analyse different materials, helping to advance knowledge in areas such as water management and climate science.

Another ANSTO facility is the Australian Synchrotron in Melbourne. This huge machine creates light many times brighter than the Sun. This light can be used to see the otherwise invisible structure and composition of materials, with a level of detail not possible in other laboratories.

What does ANSTO stand for?

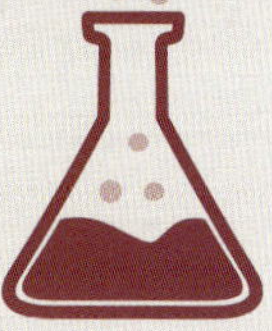

❷ CSIRO is Australia's national science organisation

CSIRO is the Commonwealth Scientific and Industrial Research Organisation. This government group partners with thousands of companies, universities and other organisations to develop new technologies. CSIRO's scientific breakthroughs and inventions benefit billions of people around the world each day, in fields such as health, manufacturing, mining, agriculture and sport.

Wearing the right type of clothing helps elite athletes to improve their performance. To help Australia's sportspeople compete, CSIRO scientists produced a fabric that enables athletes to stay dry during exercise. They developed Sportwool, a two-layer wool fabric.

Sportwool consists of a layer of superfine wool on the inside, next to the skin, and a layer of tough polyester on the outside. Sportwool is worn by Australia's Olympic cycling and cricket teams.

Because of this design, Sportwool is **hydrophobic** on the inside and **hydrophilic** on the outside. Sweat moves quickly from the inside of the garment to the outside, keeping the wearer dry. On the outside, the sweat spreads, which increases evaporation and helps keep the wearer cool. The fabric is very light, provides protection from ultraviolet light and doesn't retain any smells.

Sportwool is just one of thousands of inventions that have come out of CSIRO's collaborations. Because of CSIRO's work with other nations and organisations, Australian scientists play a major role in research and invention all around the world.

How does Sportwool keep athletes dry?

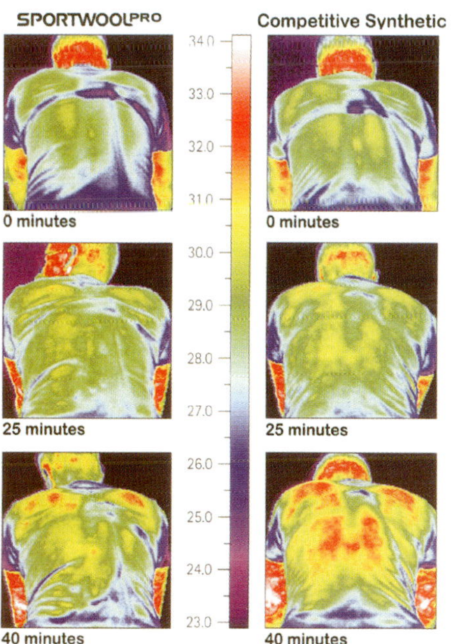

Figure 4.13 Sportwool fibre has wool on the inside and polyester on the outside.

INVESTIGATION 4.6
Suitability of sports fabrics

CHECKPOINT 4.6 ✓

1 Describe ways that ANSTO scientists work collaboratively.

2 Briefly describe the work of CSIRO.

3 Give evidence from the text that CSIRO collaborates with other organisations.

4 Explain how ANSTO helps us.

5 Why is the development of Sportwool an example of scientific collaboration?

6 Justify why Sportwool is the best fabric to use for the Australian cricket uniform.

7 Your school probably produces a very large amount of waste (rubbish) every day.
   a Imagine you are tasked with solving the problem of waste at your school. Think of three ways you could improve the amount of waste.
   b Join with two other students to compare your individual notes. What other ideas did they come up with that you did not? In what ways is collaboration more powerful than working on your own?

CHALLENGE

8 Another example of scientists working together collaboratively is the Human Genome Project. Research the *what*, *where*, *who*, *when*, *why* and *how* of the Human Genome Project.

SKILLS CHECK

- I can explain what collaboration is and why it's so important in science.
- I can give examples of when scientists have worked collaboratively.

CHAPTER SUMMARY

Chemical change
a change in properties with a new substance formed

Chemical changes
- burning wood
- an egg cooking
- rusting iron
- a banana rotting
- fireworks
- cookies baking.

Physical change
a change in appearance with no new substance formed

Physical changes
- salt and sand being mixed
- paper being cut
- evaporation
- a ruler breaking
- salt being dissolved in water.

Chemical changes can occur in nature. Photosynthesis is a chemical reaction in plants.

Reactants			Products	
$6CO_2$ +	$6H_2O$	$\xrightarrow{\text{LIGHT}}$	$C_6H_{12}O_6$ +	$6O_2$
CARBON DIOXIDE	WATER		GLUCOSE	OXYGEN

▼ **Rust** (iron oxide) is produced when iron reacts with oxygen in the air.

◀ Substances do not change during a physical change.

Particles split or join during a chemical change.

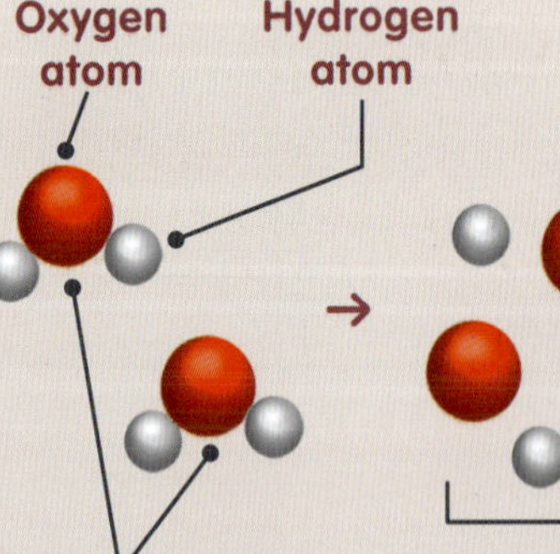

Electrolysis breaks bonds within molecules

Atoms rearrange to form hydrogen molecules (H_2) and oxygen molecules (O_2)

▲ Scientific research has led to the creation of new materials, such as **bioplastics**, that have much less impact on the environment

FINAL CHALLENGE

1. Explain the difference between a physical and a chemical change.
2. Describe two examples of a physical change and two examples of a chemical change happening in our daily lives. In each case explain how it is a physical change or chemical change.
3. What are four signs or observations that a chemical change has occurred?

LEVEL 1
50xp
LEVEL UP!

4. Define what reactants and products in a reaction are.
5. Outline the differences between photosynthesis and cellular respiration.
6. What is corrosion? How does it occur and what is the result of corrosion?

LEVEL 2
100xp
LEVEL UP!

7. Explain the particle theory of matter, in your own words.
8. Model an equation showing what happens to particles in a chemical change, then describe in words what happens.

LEVEL 3
150xp
LEVEL UP!

9. Plastics are useful but not environmentally friendly. Support or reject this statement by providing evidence from the text as well as your own opinions.
10. Explain how bioplastics are made and list the advantages and disadvantages of using bioplastics.
11. What are nanoparticles and how have they helped improve areas such as health and medicine?

12. How has technology improved the health of human beings so that they are able to live longer?
13. Collaboration in science is extremely important. Suggest why.
14. Name two Australian organisations who collaborate with other scientists and explain how their work has helped Australians and other people around the world.

5 EARTH AND THE ROCK CYCLE

Have you ever dug a hole and wondered what would happen if you just kept digging? Perhaps you have imagined what the area that you live in was like millions of years ago. Geologists are scientists who study Earth. By making careful observations of the rocks on Earth's surface and gathering a range of other evidence, geologists have been able to work out the structure of the inside of Earth and how that and the landscape around us have changed over time.

1 LEARNING LINKS

What do you already know about Earth and the rock cycle?

What is in our solar system? (Hint: don't just list the planets!)

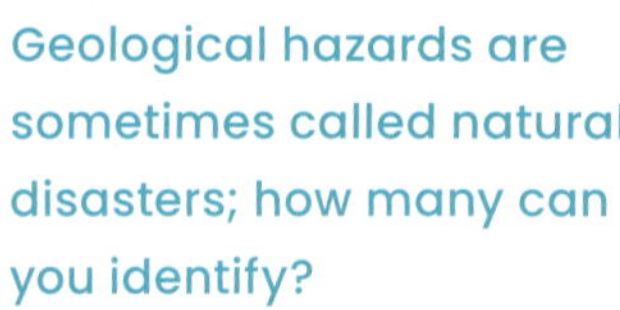

Geological hazards are sometimes called natural disasters; how many can you identify?

How do landscapes change over time?

2 SEE-KNOW-WONDER

List three things you can **see**, three things you **know** and three things you **wonder** about this image.

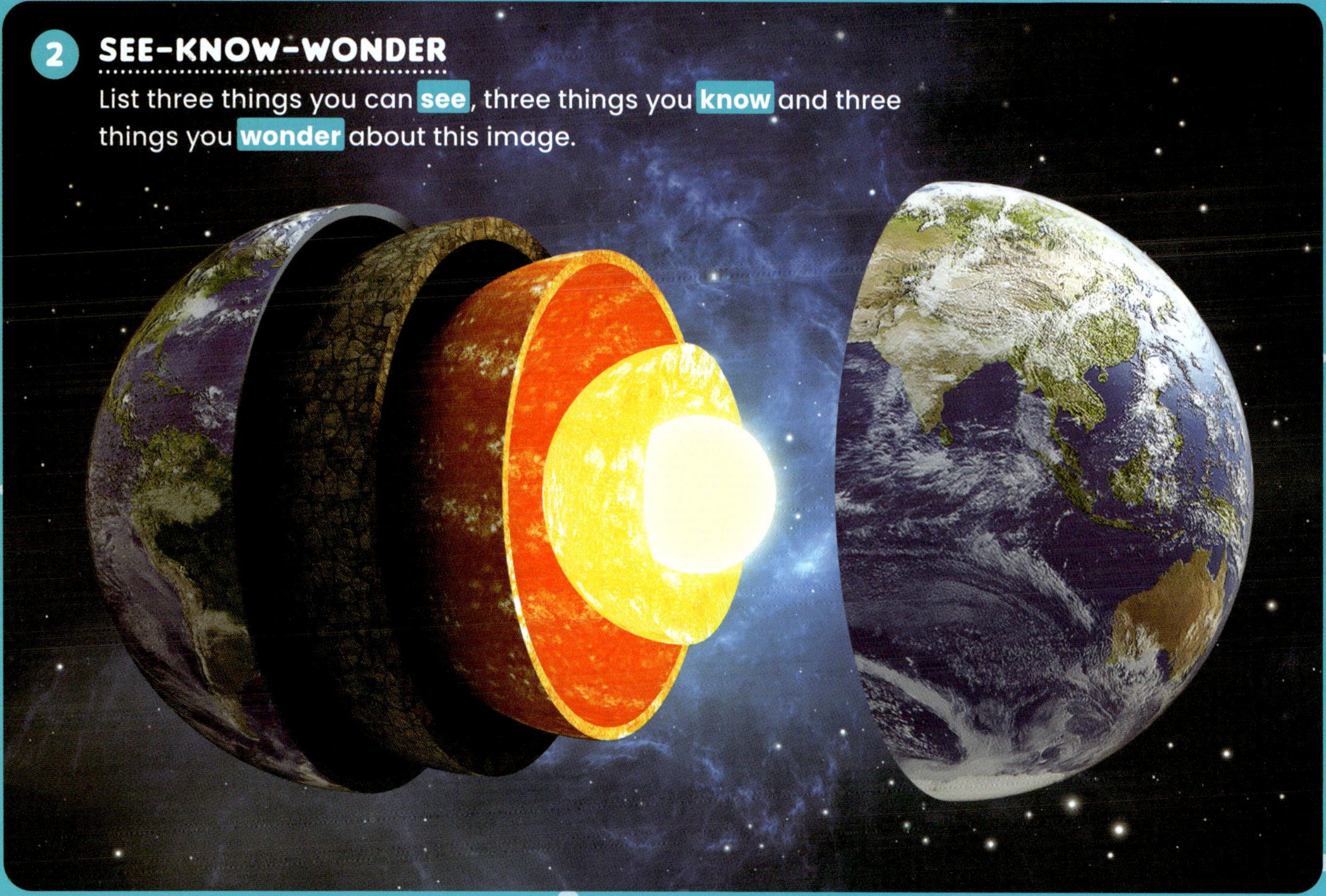

3 CRITICAL + CREATIVE THINKING

Alphabet: Think of a word about geology (Hint: rocks, fossils, Earth) for each letter of the alphabet.

What if ... the outer core of Earth was solid instead of liquid?

Five questions: Write five questions that have the answer 'fossils'.

4 THE OLDEST

The oldest rock in Earth's crust is a metamorphic rock known as the Acasta Gneiss. It is approximately 3.96 billion years old. The rock was originally granite, formed when molten rock solidified about 4.2 billion years ago. About 3.9 billion years ago, this rock was pushed down several kilometres underneath the surface. The heat and pressure changed (metamorphosed) the original rock into the gneiss. Movement of Earth's crust later brought the rock back up to the surface.

5.1 EARTH'S STRUCTURE

At the end of this lesson I will be able to:

- **describe** Earth's structure in terms of the core, mantle, crust and lithosphere.

KEY TERMS 

core
Earth's central layer, which has a liquid outer core and a solid inner core

crust
Earth's thin outer layer

density
the amount of matter in a certain volume

lithosphere
Earth's rigid outer zone (crust and upper mantle), made up of tectonic plates

mantle
Earth's middle layer, made up of an upper mantle and a lower mantle

seismic wave
a wave of energy caused by an earthquake or explosion

LITERACY LINK 

Create a mind map using the key terms and at least two additional terms of your choice.

NUMERACY LINK 

A section of the continental crust is 34 km thick.

Convert 34 km to metres.

When Earth formed approximately 4.5 billion years ago it was a ball of molten (melted) rock. As it gradually cooled, this molten rock separated into three main layers: the core, the mantle and the crust.

The most dense metallic elements moved to the centre to form Earth's core. Elements of medium **density** formed the mantle around the core, and the least dense elements moved to the surface to form its crust.

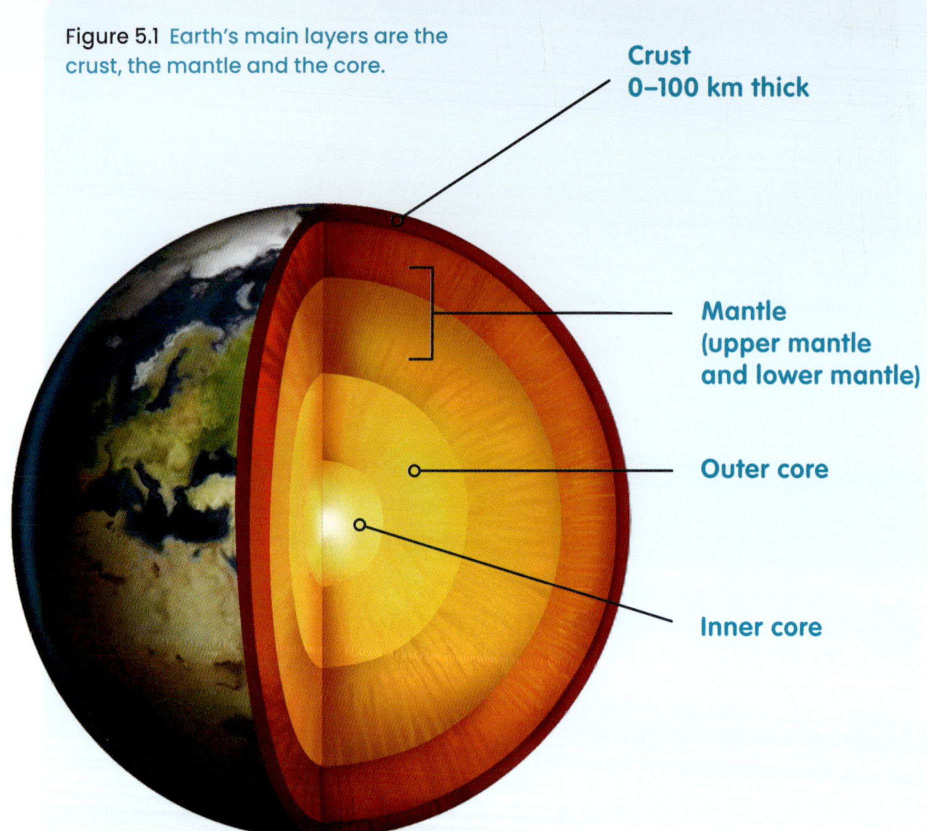

Figure 5.1 Earth's main layers are the crust, the mantle and the core.

1 Earth's core is solid on the inside, liquid on the outside

Earth's centre is the **core**, made up of a liquid outer core and a solid inner core. Scientists used information from **seismic waves** generated by earthquakes to determine this. The core is made up of a mixture of metals, mostly nickel and iron. Because it is so hot, the outer core is made of liquid metal. The pressure in the inner core is so immense that the atoms are forced together to form a solid.

What is Earth's core made up of?

❷ The mantle sits between Earth's core and the crust

Earth's thickest layer is the **mantle**. It lies between the core and the crust. Even though the mantle is made of solid rock, the very high temperatures and pressure enable the rock to flow very slowly over time. This process is similar to what you might see when you put a blob of silly putty on the edge of a desk. The processes in the mantle cause a lot of the change and movement on Earth's surface. The mantle can be thought of as two parts: the upper mantle and the lower mantle.

Are most rocks in the mantle solid or liquid?

INVESTIGATION 5.1
Modelling Earth's structure

❸ The crust is Earth's thin outer layer

The **crust** is Earth's hard outer layer, and its thinnest layer. If Earth were an apple, the crust would be thinner than the apple's skin. The crust is made of a variety of rocks. There are two different types of crust: continental crust and oceanic crust.

The continental crust forms the continents and the shallow seas around the continents. It covers about 40% of Earth's surface. It is 10–100 km thick.

The oceanic crust is formed in Earth's ocean basins. It covers about 60% of Earth's surface, and is 5–7 km thick.

What is Earth's crust made of?

❹ Earth's lithosphere is the rocky outer zone

The **lithosphere** is Earth's rigid, rocky outer zone. It includes the crust and the upper mantle. The lithosphere is made up of tectonic plates that 'float' and move around on a zone called the asthenosphere. There are 15 major tectonic plates and some smaller ones.

The asthenosphere is a thin zone of the mantle that sits just beneath the lithosphere. Rocks here are almost at their melting point (about 1300°C), so they flow more than they do in other parts of the mantle. This allows the tectonic plates of the lithosphere to move and act on each other, which can cause earthquakes and volcanoes at Earth's surface.

Which two layers of Earth make up the lithosphere?

Figure 5.2 Volcanoes are openings in Earth's crust.

CHECKPOINT 5.1

1 a Identify Earth's three main layers.
  b Describe each layer in five words or less.
2 Identify Earth's thickest and thinnest layers.
3 Identify the two types of crust.
4 Explain why the mantle is able to move.
5 If you started digging directly through Earth in Sydney, you would emerge in the Atlantic Ocean near the Azores Islands. List, in order, the layers of Earth that you would travel through.
6 Distinguish between the lithosphere and the asthenosphere.

CHALLENGE

7 Turn back to the See-Know-Wonder image on page 73. Redraw it and label Earth's layers. (Hint: The image shows the upper and lower mantle.)

SKILLS CHECK

- I can name the three main layers of Earth.
- I can describe each layer.
- I can draw and label a cross-section of Earth.

5.2 MINERALS

At the end of this lesson I will be able to:

- **identify** that rocks contain different types of minerals
- **classify** minerals into groups using their observable properties.

KEY TERMS 

compound
a substance made up of two or more different elements bonded together

crystal
a solid substance made up of very ordered microscopic parts

element
a pure substance made of one type of atom

mineral
a naturally occurring inorganic (non-living) substance

physical property
a characteristic that can be seen or measured (e.g. colour, shape, hardness)

LITERACY LINK 

Question 4 states that minerals in rocks are 'like Lego bricks in a tower'. This comparison is called a simile. Create another simile for the same concept.

NUMERACY LINK 

Antonia collects 64 quartz crystals. Calculate both the cube root and the square root of 64.

All rocks are made up of minerals. Minerals are the building blocks of rocks. Identifying the minerals that make up a particular rock can be very useful for geologists. The types and amounts of minerals in a rock can suggest how the rock was formed and help geologists to classify it.

Identifying rocks that contain useful mineral resources means that the minerals can be mined and used.

1 Minerals are made up of elements

Minerals are inorganic (non-living) substances that are found in nature. Each mineral contains one or more of the 98 naturally occurring elements within Earth. **Elements** are pure substances, made of only one type of atom. Most minerals are **compounds**, made up of two or more elements. Quartz is made up of silicon and oxygen (SiO_2). Some minerals are pure substances, made of only one element. These are called native elements.

What is the difference between a rock and a mineral?

2 Minerals have specific physical properties

Geologists identify different minerals by observing and measuring their different **physical properties**. Because each mineral has a different chemical make-up, it will have a unique set of properties, including hardness, lustre, streak and crystal shape.

Lustre

Lustre refers to how light reflects off the surface of a mineral. Some minerals have a metallic lustre – they look shiny, like polished metal. Others have a non-metallic lustre – they look dull and earthy.

Figure 5.3 Some minerals, such as pyrite, have a metallic lustre. Others, such as jade, have a non-metallic lustre.

INVESTIGATION 5.2A
Observing minerals

INVESTIGATION 5.2B
Extracting copper

Hardness

Talc is a mineral so soft that you can scratch it with your fingernail. Diamond is also a mineral but is one of the hardest substances known. German geologist Friedrich Mohs developed a scale in 1812 to compare the hardness of minerals. His scale ranks minerals from 1 (very soft) to 10 (very hard). Hard minerals, with a larger number on the scale, will scratch softer minerals, with a smaller number on the scale. Scratching a crystal of an unknown mineral with a known mineral from a Mohs testing kit can determine the unknown mineral's hardness.

Table 5.1 Mohs scale of relative hardness

Index mineral	Hardness	Everyday material
Talc	1	Fingernail (2.5) Copper coin (3.5) Wire nail (4.5) Glass and knife blade (5.5) Streak plate (6.5)
Gypsum	2	
Calcite	3	
Fluorite	4	
Apatite	5	
Orthoclase	6	
Quartz	7	
Topaz	8	
Corundum	9	
Diamond	10	

Colour and streak

Colour is not usually a reliable way to identify a mineral because the mineral may be many different colours, or the same colour as another mineral.

Geologists will often use the streak of the mineral to help identify it. The streak is the colour of the powdered form of a mineral. The powder is made by scratching the mineral on an unglazed white tile.

Crystal shape

Minerals usually have a crystal structure. **Crystals** are solid substances that have a regular shape because their atoms are bonded together in a regular, repeating pattern. Different minerals have different chemical make-ups, which can be seen in the shape of their crystals.

What are five physical properties used to identify minerals?

Figure 5.4 Crystal shape can be used to identify minerals.

CHECKPOINT 5.2

1 Copy and complete these sentences.
Minerals are the ____________ ____________ of rocks. They are ____________ occurring substances. A native ____________ is a mineral that contains only one type of ____________.

2 Identify the main properties that geologists use to identify minerals.

3 What type of mineral is a gold nugget? Justify your response.

4 Explain why minerals in rocks can be thought of as Lego bricks in a tower.

5 Explain how a native element is different to other minerals.

6 Use the Mohs scale in Table 5.1 to identify the mineral(s) that can:
a be scratched by apatite
b be scratched by topaz, but scratch orthoclase
c be scratched by a human fingernail
d scratch corundum.

CHALLENGE

7 Can you imagine a drill bit or saw blade made with diamond? They exist! Research what a diamond saw is used for and describe in your own words why it would be useful in mining. Support your answer with two images.

SKILLS CHECK

- I can explain what minerals are and how they are different to rocks.
- I can describe some common physical properties used to classify minerals.

5.3 THE ROCK CYCLE

At the end of this lesson I will be able to:

- **describe** how sedimentary, igneous and metamorphic rocks are related through a variety of naturally occurring processes
- **classify** rocks into groups according to their observable properties.

KEY TERMS 

igneous rock
rock made by the cooling of molten rock

metamorphic rock
rock formed from another rock that has been changed by heat and pressure

sediment
small particles of rocks such as clay, sand and pebbles

sedimentary rock
rock formed by sediments that have been pressed together

LITERACY LINK 

Explain how the meanings of the words *igneous*, *sedimentary* and *metamorphic* give clues to how each of these rocks is formed.

NUMERACY LINK 

Gareth estimates that a rock sample is 13 170 000 years old. Write this number in words.

Rocks are constantly changing and being recycled. As the tectonic plates of the lithosphere move and act on each other, rocks can be pulled under Earth's surface or forced upwards.

Rocks beneath the surface can change due to extreme heat and pressure. Rocks at the surface can change due to natural processes such as weathering (rocks breaking down into smaller particles) and erosion (rocks moved to new locations).

1 There are three main types of rock

Geologists classify rocks into three main types based on how they were formed.

Igneous rocks are made by the cooling of molten (melted) rock, either on the surface or within Earth's crust. 'Igneous' comes from the Latin word *ignis* which means 'of fire'. The minerals in the rock can be seen as individual crystals.

Sedimentary rocks are made up of **sediments** such as clay, sand, pebbles, shells and other pieces of material. The sediment is usually deposited in layers by wind, water or gravity. Over time the sediment is buried and squashed, and the particles in it are stuck together to form rock. The minerals in the rock are within the sediments and may not be easily seen.

Metamorphic rocks form when other rocks change (metamorphose) because of high temperatures and pressure within Earth's crust. The minerals in the rock can be seen as individual crystals.

What are the three main types of rock?

Figure 5.5 There are three types of rock: igneous, sedimentary and metamorphic.

❷ Rocks can change over time

The three main rock types – igneous, sedimentary and metamorphic – are all related, because they can form from one another. Forces within Earth bring rocks to the surface or sink them back down deep within the crust. The rock cycle is a model used to show how different processes can form the three different types of rock. Because some of these processes take millions, sometimes billions, of years and often occur only in certain environments, it is very likely that a rock will not undergo all possible processes in the life of the planet.

These are some common processes that act on rocks:

- *weathering*: Rocks break down into smaller pieces called sediment.
- *erosion*: Sediment moves from one place to another.
- *deposition*: Sediment settles in one place.
- *burial and compaction*: As more sediment is deposited, the sediments below it are buried and squashed together.
- *cementation*: Sediments are chemically glued together into rock.
- *metamorphism*: Rocks change due to heat and pressure.
- *melting*: Rock melts to magma due to high temperatures.
- *solidification*: Molten rock cools and hardens.

What does the rock cycle describe?

Figure 5.6 The rock cycle shows how different, naturally occurring processes in Earth form different types of rock.

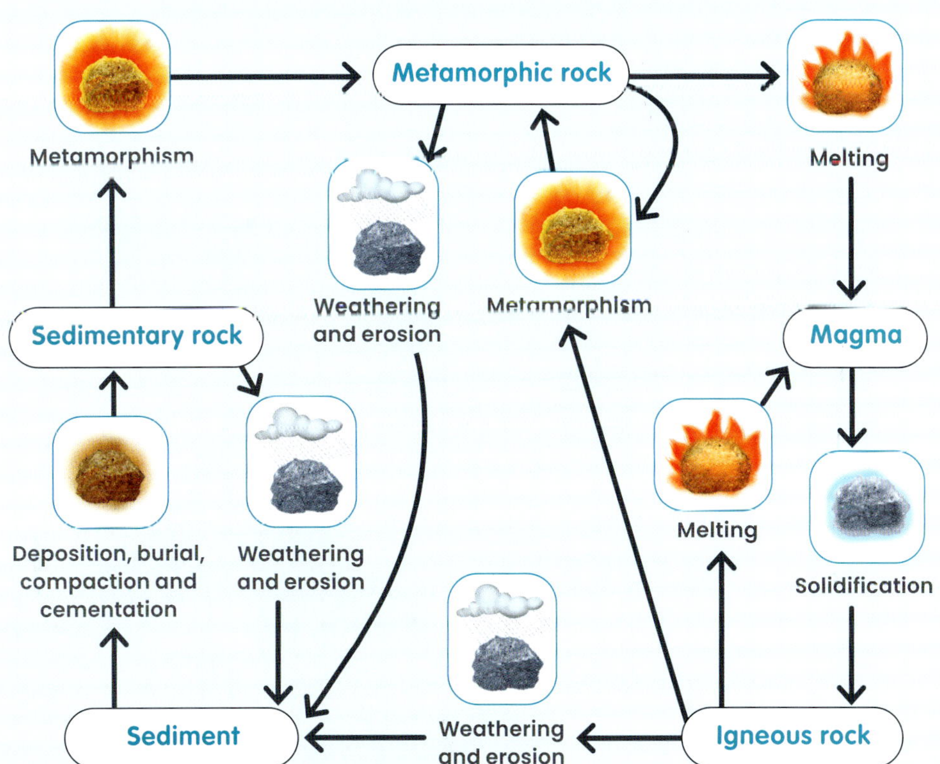

INVESTIGATION 5.3
Modelling the rock cycle

CHECKPOINT 5.3 ✓

1. Describe the rock cycle in your own words.
2. Identify the rock type(s) that are made up of crystals.
3. How is an igneous rock similar and how is it different to a sedimentary rock?
4. Are sediments formed above or below Earth's surface? Give evidence from the text in your answer.
5. What must happen to a sedimentary rock before it can become an igneous rock?
6. What must happen to a metamorphic rock before it can become a sedimentary rock?
7. Identify the processes that could happen to a rock if it is buried deep under Earth's surface.
8. Identify the processes that can only happen to a rock if it is on Earth's surface.

CHALLENGE

9. James Hutton (1726–1797), the Scottish founder of modern geology, first used the concept of the rock cycle to explain how some geological processes occur. Use the internet to find out more about the contributions that James Hutton made to our understanding of geology.

SKILLS CHECK

- I can identify the three types of rock.
- I can describe how each type of rock is created.
- I can explain what the rock cycle is.

5.4 IGNEOUS ROCKS

At the end of this lesson I will be able to:

- **describe** what an igneous rock is and explain how they are formed
- **classify** igneous rocks into groups according to their observable properties.

KEY TERMS 

extrusive igneous rock
igneous rock formed at Earth's surface

intrusive igneous rock
igneous rock formed under Earth's surface

lava
molten (melted) rock at Earth's surface

magma
molten (melted) rock under Earth's surface

solidify
become a solid

LITERACY LINK 

Think of alternative terms for these words: *molten, solidify, cool.*

NUMERACY LINK 

The materials for a model volcano are 1 tablespoon of liquid dishwashing soap, 1 cup vinegar, $1\frac{1}{2}$ cups water and 2 tablespoons baking soda per person. What quantity of each material will your teacher need to prepare for a class of 20 students?

Igneous rocks form when molten rock cools and becomes hard or solid. Most igneous rocks contain interlocking crystals (connected like a jigsaw) of the minerals that were in the molten rock.

The size of the crystals in igneous rocks depends on the time that the molten rock took to cool. Small crystals form in rocks that have cooled quickly. Large crystals form in rocks that have cooled slowly.

Figure 5.7 The mineral crystals in this sample of igneous rock interlock like jigsaw pieces.

❶ Rocks can melt to form magma

What happens to chocolate when it is heated? It melts! The same thing happens to rocks. As rocks are pulled deep below Earth's surface, towards the base of the lithosphere, the temperature increases. Rocks melt between 700°C and 1300°C, creating **magma**.

Magma is a liquid, so it is less dense than the surrounding rock and will rise to Earth's surface through faults, cracks and other weak spots, also melting the rock that it passes through. If it reaches the surface and erupts through a volcano it is called **lava**.

How is magma different to lava?

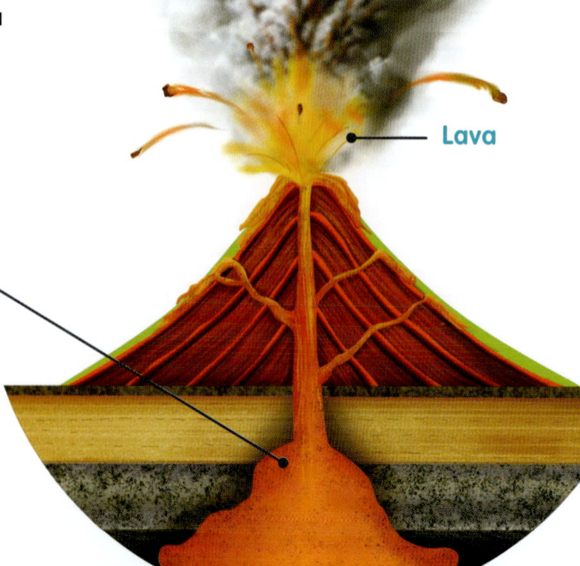

Figure 5.8 Magma collects in a chamber of a volcano before making its way to the surface to erupt as lava.

❷ Molten rock cools to form igneous rock

What happens to melted chocolate when it cools down? It **solidifies**! When molten rock cools and solidifies, the minerals form crystals that interlock like a jigsaw to form igneous rocks.

Rocks formed when lava cools on or near Earth's surface are **extrusive igneous rocks**. The magma has been extruded from (pushed out of) the crust. When magma reaches Earth's surface as lava, it will cool down very quickly, even faster if it is under water. Extrusive igneous rocks have very small crystals, so small that they are often not visible with the naked eye. This is because they did not have very long to form before the lava solidified.

Figure 5.9 Sawn Rocks in Mount Kaputar National Park near Coonabarabran in New South Wales are made of columns of basalt, an extrusive igneous rock. As the lava flow cooled it shrank, creating long hexagonal 'pipes'.

When magma cools under the surface, it forms **intrusive igneous rocks**. The magma has intruded on the rocks that were originally there. Magma under the surface cools very slowly, which means the crystals have more time to grow. The crystals in intrusive igneous rocks, such as quartz and granite, are much larger than those in extrusive igneous rocks and can be seen with the naked eye.

What is the difference between extrusive and intrusive igneous rocks?

❸ Crystal sizes tell us where an igneous rock was formed

The size of crystals (known as rock texture) in igneous rocks can tell geologists whether those crystals formed on the surface due to an erupting volcano, or deep underground. The bigger the crystals, the deeper the region where they formed. Geologists also look at the type of minerals that make up a rock to work out how and where the magma was formed.

Would a rock that formed on the surface contain large or small crystals?

INVESTIGATION 5.4A Cooling rate and crystal size

INVESTIGATION 5.4B Observing igneous rocks

CHECKPOINT 5.4 ✓

1 Copy and complete these sentences.
Molten rock below Earth's surface is known as __________.
Molten rock at the surface is known as __________.

2 Copy and complete these sentences.
Igneous rocks that have cooled quickly have __________ crystals.
Igneous rocks that have cooled slowly have __________ crystals.

3 Identify two key characteristics that could be used to classify a rock as an igneous rock.

4 Explain how the cooling rate affects the size of the crystals in an igneous rock.

5 Explain what is meant when magma is referred to as 'less dense than solid rock'.

CHALLENGE

6 When magma loses gas as it moves to Earth's surface, empty gas bubbles may be preserved as the lava solidifies. This creates igneous rocks that contain many air bubbles.
   a Use your research skills to find the name of a specific igneous rock with many air bubbles.
   b Create a diagram to illustrate this process.

SKILLS CHECK

- I can describe what an igneous rock is and how it forms.
- I can describe the difference between extrusive and intrusive igneous rocks.
- I can state the relationship between cooling and crystal size.

5.5 METAMORPHIC ROCKS

At the end of this lesson I will be able to:

- **describe** what a metamorphic rock is and explain how they are formed
- **classify** metamorphic rocks into groups according to their observable properties.

KEY TERMS 

contact metamorphism
the process of change that happens to a rock over small areas, often near volcanoes

metamorphism
the process of change that happens to a rock because of heat, pressure or both

regional metamorphism
the process of change that happens to rock over large areas

tectonic plate
a plate made up of the mantle and crust of the lithosphere

LITERACY LINK 

A butterfly is an example of something other than rock that metamorphoses. Think of three other examples of things that can go through a metamorphosis.

NUMERACY LINK 

One metamorphic formed 100 km underground, while another formed 80 km underground. Which is smaller: -100 or -80?

Metamorphosis means 'change'. Like a caterpillar changing into a butterfly, a metamorphic rock has changed its form.

Metamorphic rocks form from other rocks that have been changed by heat, pressure or both.

1 Rocks are changed by heat and pressure

Metamorphic rocks are formed from rocks that have been put under heat or pressure, or both heat and pressure. This process is called **metamorphism**. It happens when Earth's **tectonic plates** push together, move apart or slide past each other, burying rocks and making them very hot and putting them under pressure. As rocks are heated and put under pressure, the minerals inside them will rearrange. Some may even form new minerals by chemically reacting with each other or with fluids passing through the rocks.

Metamorphosis results in rocks that have crystals. Sometimes these crystals have rearranged into layers.

How are metamorphic rocks formed?

2 Rocks can be squashed, folded or 'cooked'

Rocks can be metamorphosed in two major processes.

Regional metamorphism happens over large areas, often when two tectonic plates push together. This puts rocks under much higher temperatures and pressures, and they may squash and fold. As they are pushed deeper and deeper they will keep changing. The amount of metamorphism depends where the rocks are.

Contact metamorphism happens when a body of rising magma meets rock, increasing the temperature of the rock and 'cooking' it, forming new crystals. It happens in a small, local area, often around volcanoes. The crystals in rocks formed by contact metamorphism will not be in layers. Rocks that have been formed by regional metamorphism will have their crystals arranged in layers.

How could you tell if a rock was formed by regional metamorphism or contact metamorphism?

Figure 5.10 These heavily folded and metamorphosed rocks at Narooma in New South Wales formed due to intense pressure when the Pacific tectonic plate first started colliding with the east coast of Australia.

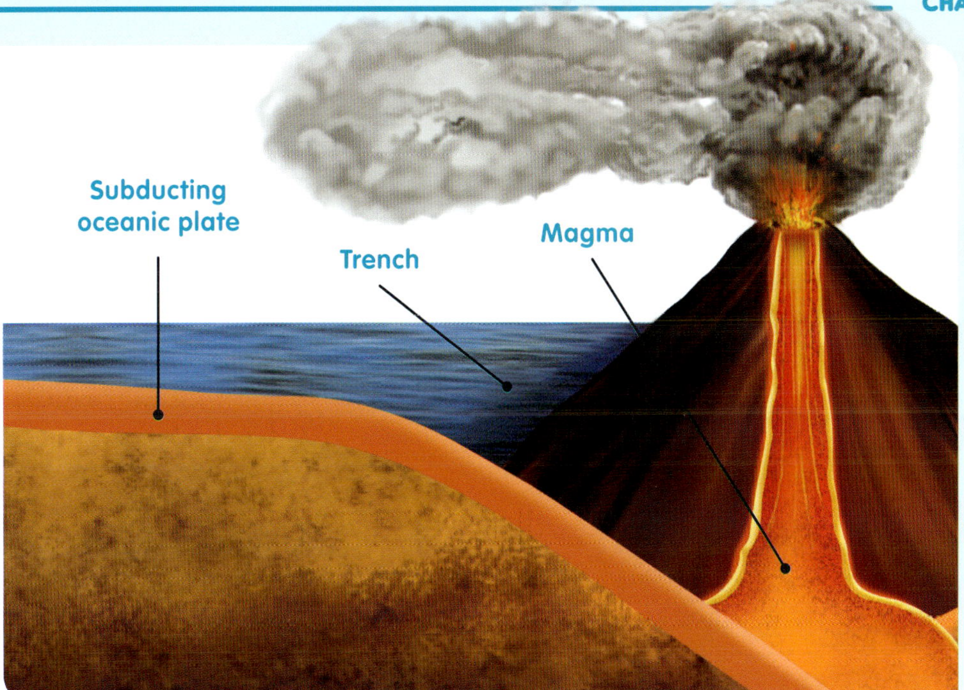

Figure 5.11 Regional metamorphism happens when two tectonic plates push together. As the plates collide, the increase in pressure and temperature will cause the rocks to change over a very large area.

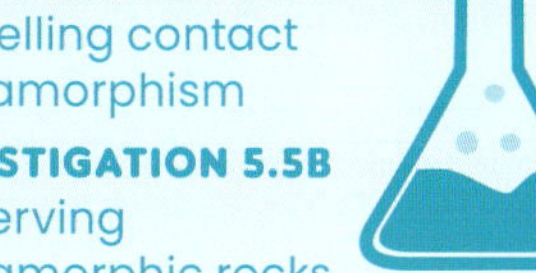

INVESTIGATION 5.5A
Modelling contact metamorphism

INVESTIGATION 5.5B
Observing metamorphic rocks

❸ Metamorphism can be an ongoing process

A rock can undergo different amounts of change, depending on what happens to it. Even metamorphic rocks can undergo more metamorphism! As rocks are buried deeper and deeper they will keep changing as the temperature and pressure increase. Think about what happens when you bake chocolate chip cookies. A cool oven and a short period of cooking time will result in soft, doughy cookies, and the chocolate chips keep their shape. A hot oven and a longer period of cooking time will result in hard cookies with chocolate that melts then resets.

Geologists compare the types of minerals in metamorphic rocks to work out where and how the rocks formed. This can help them to understand more about how Earth's tectonic plates move and act on each other.

What increases as rocks are buried deeper in Earth?

Figure 5.12 Shale is a sedimentary rock formed by layers of very small sediments. As it is buried deeper and deeper it will change into different types of rock. Slate is a metamorphic rock formed from shale at quite low temperatures and pressures. If it gets buried more and more it will change into phyllite, schist and, finally, gneiss.

Shale → Slate → Phyllite → Schist → Gneiss

CHECKPOINT 5.5

1 Copy and complete these sentences.
Metamorphic rocks are formed when __________ have been __________ due to __________, __________ or both.

2 Identify two key characteristics that could be used to classify a rock as a metamorphic rock.

3 Identify two locations on Earth where metamorphism could occur, and use evidence from the text to justify your answer.

4 Describe how heat and pressure cause the changes in metamorphic rocks.

5 Sandstone is a sedimentary rock that has no crystals. It is metamorphosed into quartzite, a metamorphic rock with crystals. Propose where the crystals come from.

6 Is it possible for metamorphic rocks to form on Earth's surface? Justify your response.

CHALLENGE

7 Use the internet to find out about how metamorphic processes create diamonds. You may like to use search terms such as 'diamonds + metamorphism'.

SKILLS CHECK

- I can describe what a metamorphic rock is and how these rocks form.
- I can describe the difference between regional and contact metamorphism.

5.6 THE FORMATION OF LANDFORMS

At the end of this lesson I will be able to:

- **relate** the formation of a range of landforms to physical and chemical weathering, erosion and deposition.

KEY TERMS 

deposition
a process in which sediment is left in a new place

erosion
a process in which sediments are moved from one place to another

landform
a natural feature of Earth's surface

sediment
small particles of rocks such as clay, sand and pebbles

weathering
a process in which rocks are worn down into smaller particles

LITERACY LINK 

Identify five adjectives (describing words) from this section and replace them with a word that has the same or a similar meaning.

NUMERACY LINK 

A layer of sediment on a river bed is 2^4 cm thick.

Write 2^4 in expanded form.

Igneous and metamorphic rocks are formed by changes in Earth's crust. Rocks on Earth's surface also change. Wind, water and changes in temperature can act on these rocks, wearing them down into smaller parts called sediments and depositing them in new locations.

Over a long time, these processes will make and change **landforms** such as mountains, riverbeds, deserts and coastlines, changing the face of Earth.

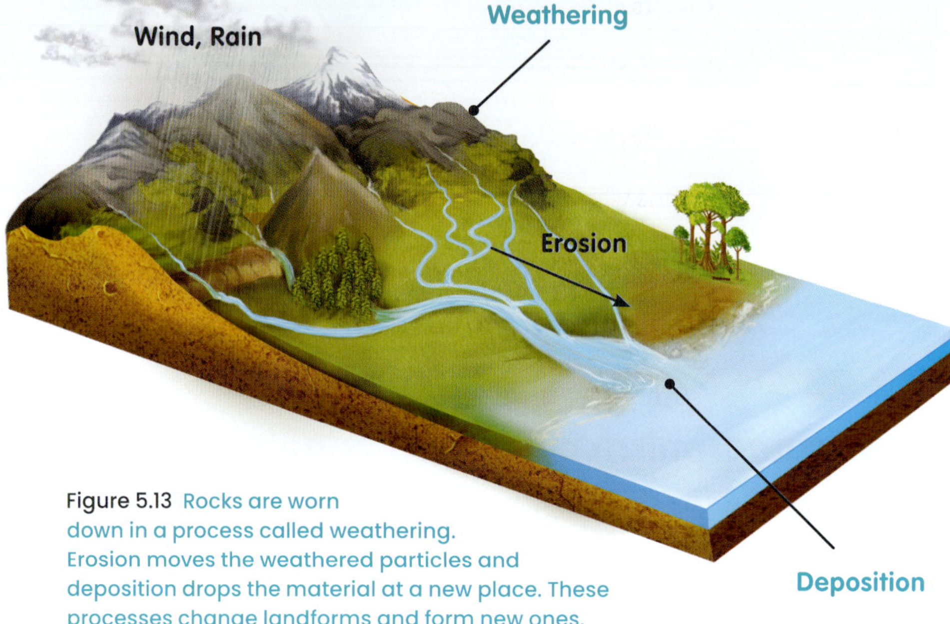

Figure 5.13 Rocks are worn down in a process called weathering. Erosion moves the weathered particles and deposition drops the material at a new place. These processes change landforms and form new ones.

1 Rocks are worn away by weathering

Weathering is the process in which rocks are worn down into **sediments** such as pebbles, sand and clay. There are two main ways this can happen.

Physical weathering breaks rocks into smaller particles through processes that do not change the chemical make-up of the minerals. This can happen when rocks expand, shrink and crack due to temperature changes, or if they are worn down by the actions of wind or water.

Figure 5.14 Frost wedging (left) and wind abrasion (right) are both types of physical weathering.

Chemical weathering breaks rocks into smaller particles through chemical reactions that change the minerals in the rocks. This can happen when rocks contact chemicals in the air or water. The new minerals will often separate from the original rock to form sediments.

How can weathering produce sediment?

2 Erosion moves sediment

Erosion is the movement of sediment from one place to another. The sediments can be transported by wind, water, ice or gravity – these are called the agents of erosion.

What are the four agents of erosion?

3 Deposition drops sediment in new areas

Deposition happens when erosion deposits (drops) sediments in a new place. Often this happens when there is no longer enough energy in the wind or water to carry them any further. If these sediments are not eroded they may eventually be compacted and cemented together to form new sedimentary rocks.

What is deposition?

Figure 5.15 As the water in a river slows down (for example where it meets the sea), the larger sediments it is carrying will be deposited first. Very small sediments such as clay will settle furthest from shore, in calm water.

4 Uluru was formed by sedimentary processes

Uluru is made of layers of very hard sandstone. This sandstone was formed when a nearby ancient mountain range similar to the modern Himalayas was weathered, eroded and deposited in a valley more than 500 million years ago. Over time, this sediment was buried under more sediment, compacted and cemented together, forming very hard rock. As tectonic plates pushed together, these rocks were folded and pushed up to the surface again. The softer rock weathered and was eroded, leaving only the hard rock of Uluru at the surface. It is just the tip of a large section of rock that may reach as far as 5 km below the surface.

How did sedimentary processes form Uluru?

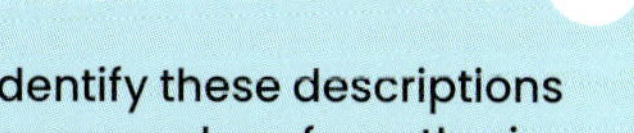

INVESTIGATION 5.6
Modelling weathering of rocks

CHECKPOINT 5.6

1 Identify these descriptions as examples of weathering, erosion or deposition.
  - **a** the process of transporting rock or sediment to a different location
  - **b** the breakdown of large rocks into smaller rocks
  - **c** the process of sediments, rocks and soil being added to a landform
  - **d** a stream carrying rock and sediment
  - **e** a landform being broken apart
  - **f** mud flowing from one place to another.

2 Explain the difference between physical and chemical weathering.

3 In November 2018 there was a significant dust storm over New South Wales that lasted for several days. Use your knowledge of the processes of weathering, erosion and deposition to explain what happened.

4 Very small sediments that are carried by a river will be deposited where the river meets the sea or other large body of water – suggest why.

CHALLENGE

5 Use the internet to research how weathering, erosion and deposition were important in the formation of Uluru.

SKILLS CHECK

- I can describe erosion, deposition and chemical and physical weathering.
- I can explain how erosion, deposition and weathering create landforms over time.

5.7 SEDIMENTARY ROCKS

At the end of this lesson I will be able to:

- **describe** what a sedimentary rock is and explain how they are formed
- **classify** sedimentary rocks into groups according to their observable properties.

KEY TERMS 

chemical sedimentary rock
sedimentary rock formed from layers of mineral crystals that have crystallised from water

clastic sedimentary rock
sedimentary rock made of sediments cemented together

organic sedimentary rock
sedimentary rock formed from the remains of plants or animals

sediment
small particles of rock, such as clay, sand and pebbles

sedimentary rock
rock formed by sediments that have been pressed together

LITERACY LINK 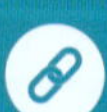

Underline or copy out as many nouns (names of a person, place or thing) as you can find in this section.

NUMERACY LINK 

The pressure in an underground lake is calculated using the formula $P = \frac{3m^2}{2}$, where m = the depth of the water in metres.

If the lake is 8 m deep, what is the pressure (P) at the bottom?

Sedimentary rocks are made of **sediments** such as clay, sand, pebbles, shells and other fragments of material. This is often deposited in layers by wind, water or gravity. These are gradually buried, compacted and cemented together into rock.

Sedimentary rocks can be classified into three main types based on how they were formed.

Figure 5.16 Sedimentary rocks are formed when sediments are eroded, deposited, compacted and cemented together.

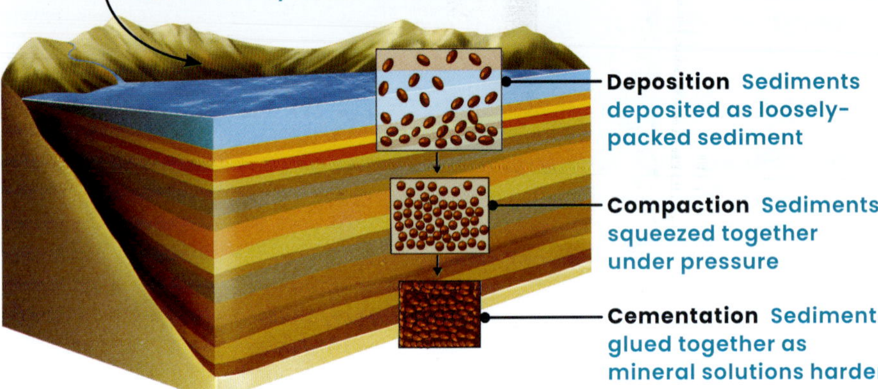

1 Clastic sedimentary rocks are made up of layers 'glued' together

Clastic sedimentary rocks are made up of sediments (clasts) formed by the weathering of other rocks. Wind, water, ice and gravity deposit these clasts into layers. As the layers build up, pressure on the sediments increases and they are compacted together. Chemicals in the groundwater that moves between the sediments will then cement the layers together, forming a hard rock. Clastic sedimentary rocks are classified by the size of their sediments.

What are clastic sedimentary rocks made of?

2 Pressure can create rocks from organic remains

Organic sedimentary rocks are formed from the remains of plants and animals. These remains have been deposited together, buried and compacted. Coal, limestone and chalk are examples of organic sedimentary rocks.

Coal is formed from the remains of ancient swamps. The plant and animal remains do not decompose. Instead, when they are buried the increasing pressure compacts and cements the remains and changes them into coal.

Limestone is made of the remains of ancient coral reefs. The shells of the corals, shellfish and other marine animals made of a mineral called calcium carbonate were buried, compacted and cemented together.

Figure 5.17 Limestone often contains lots of fossils.

Chalk is a type of limestone formed from the compacted remains of billions of microscopic marine organisms that had shells made of calcium carbonate. As the organisms died, their shells were deposited on the bottom of the ocean. Over time these built up, were buried, compacted and cemented together.

What are organic sedimentary rocks made of?

3 Crystals can form rocks after evaporation

Chemical sedimentary rocks are formed when water that contains dissolved minerals evaporates, allowing the mineral crystals to grow. Halite (rock salt) is formed in this way. Some limestones can also be formed in this way: water evaporates and leaves behind crystals of calcium carbonate.

How are chemical sedimentary rocks created?

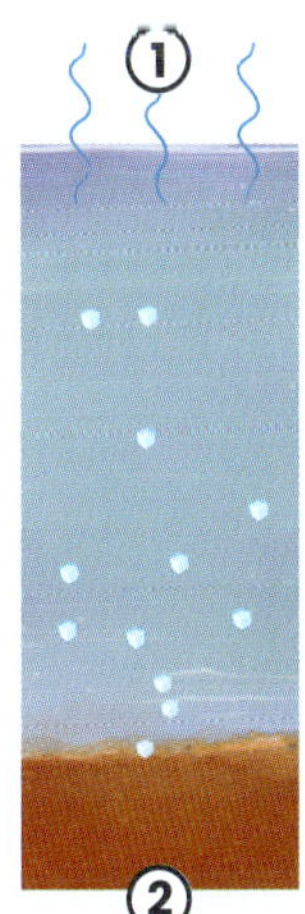

Figure 5.18 Halite is a chemical sedimentary rock that is formed when water containing salt evaporates, allowing salt crystals to form.

1 Water evaporates.
2 Crystals sink.
3 Evaporation continues.
4 Salt crystals form and grow.
5 Halite remains.

INVESTIGATION 5.7A
Modelling the formation of sandstone

INVESTIGATION 5.7B
Observing sedimentary rocks

CHECKPOINT 5.7

1 Identify at least three examples of types of sediment that can make up sedimentary rocks.
2 Identify the three main types of sedimentary rock and provide an example for each.
3 Describe two key features that could be used to identify a rock as being sedimentary.
4 Outline the difference between organic and chemical sedimentary rocks.
5 Explain why compaction and cementation are important steps in the formation of sedimentary rocks. What could happen if these steps did not occur?
6 How might you distinguish between a chemical sedimentary rock with crystals and an igneous rock?
7 Which kind of sedimentary rock do you think would be best at preserving fossils? Give evidence to support your answer.

CHALLENGE

8 Conglomerate, mudstone and sandstone are all types of sedimentary rock. Use the internet to research what sediments they are made from and the environments they are formed in.

SKILLS CHECK

- I can describe what a sedimentary rock is and how these rocks form.
- I can explain the difference between clastic, chemical and organic sedimentary rocks.

5.8 FOSSILS

At the end of this lesson I will be able to:

- **describe** how fossils form.

KEY TERMS 

cast
an object created when sediment or minerals fill a mould

fossil
preserved remains or traces of once-living things

fossilisation
the process of the formation of a fossil

mould
a hollow impression formed by an imprint of an organism, or when the original bone or shell has dissolved

palaeontologist
a scientist who studies fossils

LITERACY LINK 

Complete this sentence: 'The term fossil is different to fossilisation because ...' (Hint: Use the key terms if you need to.)

NUMERACY LINK 

Seo-yun starts her hunt for fossils by marking out a triangle 10 m long on each side. What is the angle of each corner of the triangle?

Fossils are the preserved remains or traces of once-living things. Fossils can be complete or part skeletons, shells and tree trunks, or traces that an animal existed such as footprints, burrows, and even poo.

Fossils and the rocks that they are in can help **palaeontologists**, who study fossils, to reconstruct information about past living things and the environments that they lived in.

❶ Fossils form in sediment

The process that results in a fossil is called **fossilisation**, and it is rare. Usually when an organism dies it decomposes or is eaten. Soft parts decay or are eaten quickly, and even hard parts such as bones and shells will eventually be eaten or weather away if they are left exposed.

For a fossil to form, a series of events that preserves an organism needs to happen.

How do fossils form?

Figure 5.19 The fossilisation process.

① The organism dies and its remains are in an environment, such as under water, where they will be quickly buried by sediment.

② The remains continue to be buried under more layers of sediment. These layers compact, and minerals in the groundwater cement them together into sedimentary rock.

③ Groundwater minerals can also dissolve and replace the minerals in the bones of the organism, often making them harder than the original bone.

④ Movement of tectonic plates bring the fossil layer closer to the surface, and weathering and erosion or digging can expose them.

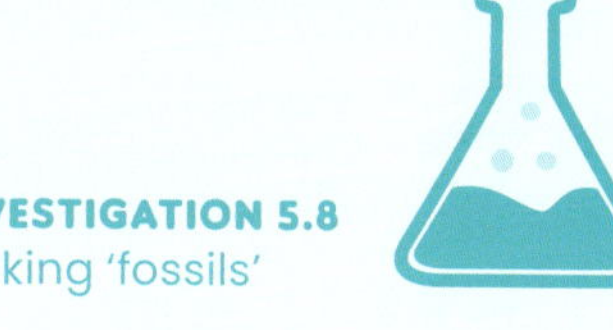

❷ Impressions of dead organisms can be moulds or casts

A **mould** is a hollow impression formed by an imprint of an organism, or when the original bone or shell has dissolved. A **cast** is created when sediment or minerals fill the mould.

What is the difference between a mould and a cast?

INVESTIGATION 5.8
Making 'fossils'

Figure 5.20 The hollow imprint, called a mould, left by a dead organism, can be filled with sediment or minerals to become a cast.

❸ Fossils can be found in other places

Most fossils are usually found in sedimentary rocks, although sometimes plant and animal remains can be found preserved in amber (fossilised tree sap), tar and permafrost (ground that is always frozen). These things often preserve the soft parts of organisms that would not usually be preserved in rock.

Footprints and burrows can also be preserved by fossilisation. These are called trace fossils. They can show palaeontologists how animals moved and lived, in a way that skeletons cannot.

What is a trace fossil?

Figure 5.21 This 100-million-year-old mosquito has been preserved in amber.

CHECKPOINT 5.8

1 Define the term *fossil*.

2 Provide an example of a trace fossil.

3 Identify which of these environments a fossil would most likely form in and justify your choice.
a sandy desert
b mountain range
c swamp

4 Describe the difference between a cast and a mould.

5 Fossils that include feathers and other soft body parts are rare – suggest why.

6 Palaeontologists are often just as interested in the rock that a fossil is found in. Why do you think this is?

7 Why are the bones of a fossilised organism sometimes harder than the original bone?

CHALLENGE

8 Use the internet to research the significance of one of these Australian fossil sites. Include in your response how old the fossils are, what organisms they came from, how they were formed and why they are significant.
- Naracoorte, South Australia
- Riversleigh, Queensland
- Ediacara, South Australia

SKILLS CHECK

- I can describe how fossils form.
- I can identify what kinds of things can become fossils.

5.9 HISTORY IN THE ROCKS

At the end of this lesson I will be able to:

- **explain** how geological history can be interpreted from a sequence of sedimentary layers.

KEY TERMS 

conglomerate rock
sedimentary rock made of large, rounded pebbles and fragments cemented together

geological history
how Earth has changed over time

relative age
which rocks are younger and which rocks are older when compared with one another

sequence
the order of something

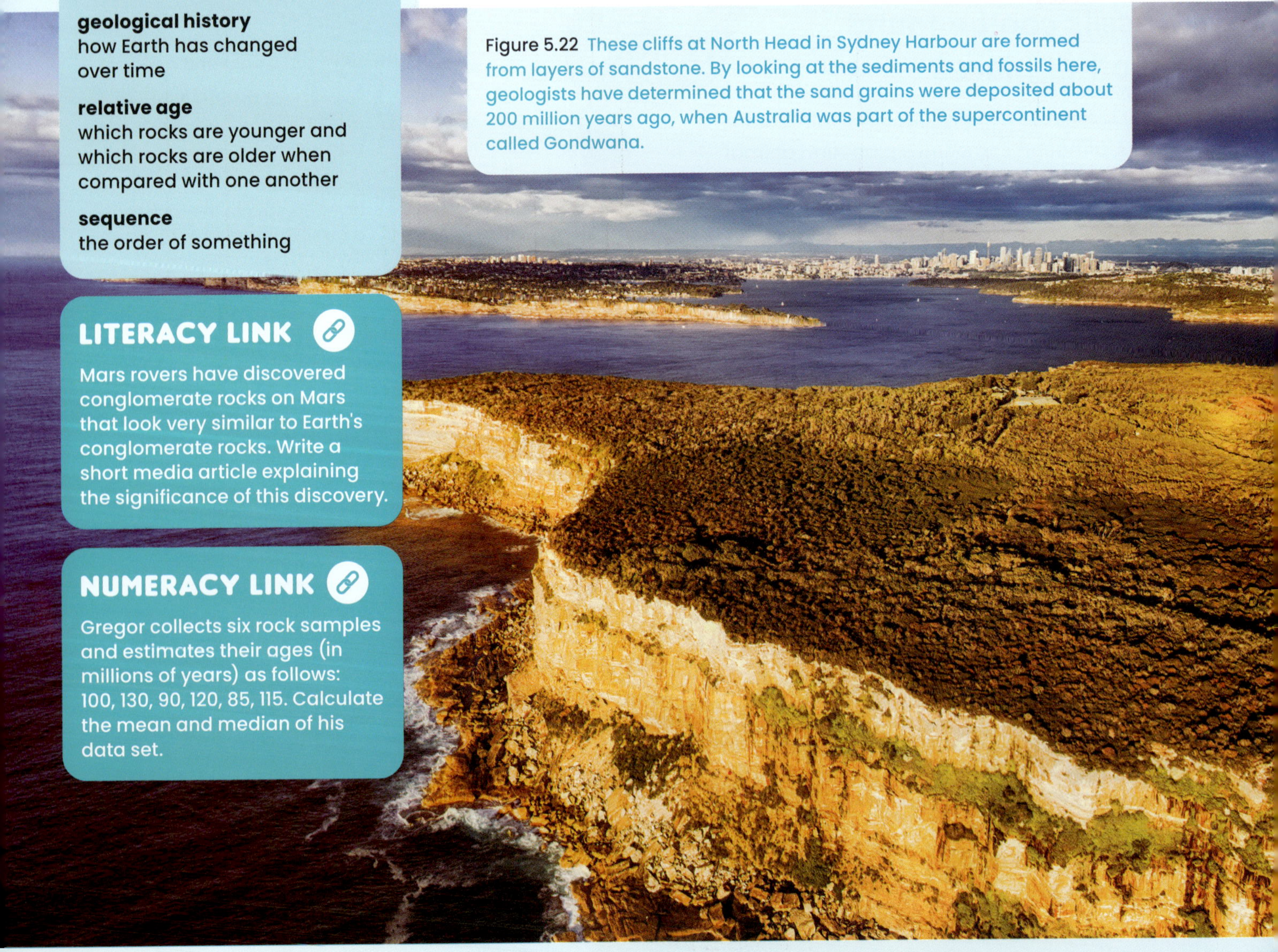

LITERACY LINK 

Mars rovers have discovered conglomerate rocks on Mars that look very similar to Earth's conglomerate rocks. Write a short media article explaining the significance of this discovery.

NUMERACY LINK 

Gregor collects six rock samples and estimates their ages (in millions of years) as follows: 100, 130, 90, 120, 85, 115. Calculate the mean and median of his data set.

It is possible that the place where you are right now was underwater or near active volcanoes millions of years ago.

We can't go back in time, but geologists can piece together the **geological history** of different locations by investigating the **sequence** of rocks found in the area.

1 The youngest rocks are at the top

When sedimentary layers are deposited, younger ones are deposited on top of older ones. Unless something unusual has happened, the oldest rocks in a cliff face will be at the bottom and the youngest rocks will be at the top. Knowing this, geologists are able to work out the relative age of the layers. Finding the same layers in different areas means they must have been deposited at the same time.

Where would you expect to find the oldest rocks in a cliff face?

Figure 5.22 These cliffs at North Head in Sydney Harbour are formed from layers of sandstone. By looking at the sediments and fossils here, geologists have determined that the sand grains were deposited about 200 million years ago, when Australia was part of the supercontinent called Gondwana.

2 Different rocks are deposited in different environments

If you are looking at a cliff face and observe different types of sedimentary rock, you can work out its geological history. This is because different rocks are deposited in different environments. Fossils in the rocks can also be used to provide more information, and can be used to work out how old the rocks are. The presence of igneous rocks might mean that a volcanic eruption happened.

What can the presence of different rock types tell us?

Table 5.2 Sedimentary rock types and environments of formation

Rock type	Sediment size	Environment when formed
Limestone	Very small Often fossils of reef organisms	Warm shallow seas
Siltstone and mudstone	Extremely small	Deep, calm water off the continental shelf
Sandstone	Small	Beaches and just offshore from land Deserts
Conglomerate	Large	Rivers

Figure 5.23 NASA has found **conglomerate rocks** on Mars that look very similar to Earth's conglomerate rocks. This may be evidence that there was once a riverbed on Mars!

CHECKPOINT 5.9 

1 Copy and complete this sentence.
Generally, the youngest layers of rocks are found on the __________ and the oldest rock layers are found on the __________.

2 You observe a rock that contains limestone with fossils of coral. What must the environment have been like when that limestone formed?

3 You observe a cliff face that has a layer of limestone at the bottom and a layer of mudstone on top. Has the sea level risen or fallen during the period that the rocks were deposited?

4 You observe a cliff face that has a layer of limestone at the bottom and a layer of sandstone on top. Has the sea level risen or fallen during the period that the rocks were deposited?

5 You observe a layer of conglomerate in a cliff face. What can you tell about the environment at the time that this layer was deposited?

CHALLENGE

6 Geologists use fossils called index fossils to help them determine the age of rocks. Use the internet to find out three features that a fossil must have to be used as an index fossil.

7 Geologists can use several techniques to date rocks. Use the internet to find out what evidence geologists have used to determine the age of Earth.

SKILLS CHECK

- I can explain how different layers of sediment can show geological history.
- I can identify in a sequence of sedimentary rocks where the youngest and oldest rocks can be found.

5.10 SCIENTISTS INVOLVED IN MINING

At the end of this lesson I will be able to:

- **describe** some examples of how different scientists contribute to the mining process in Australia.

KEY TERMS 

exploration
processes undertaken to find rocks that contain minerals

mineral deposit
rocks that contain a particular mineral

ore body
a mineral deposit that is profitable to mine

rehabilitation
processes that return the environment to close to how it was before mining

LITERACY LINK 

Write a paragraph that explains the difference between the following terms: mineral, mineral deposit, ore body, mineral ore.

NUMERACY LINK 

Wiremu, a geochemist, draws two parallel lines on a map while surveying an area and labels them A and B. He then draws a third line that crosses line A at a 56° angle. What angle does it cross line B at?

Figure 5.24 A geologist who works in a mine will make sure the best locations are being mined.

Mining allows us to access resources from Earth that we use in our everyday lives. The process of locating, removing and processing a resource requires expertise from scientists in many different areas.

1 Geophysicists and geochemists search for resources

To meet the demand for resources, mining companies conduct **exploration** to find areas that have high concentrations of the **mineral deposits** that contain the resource that they want to mine. If they are granted permission from the government and landholders to explore in a particular area they use many different exploration techniques to find these deposits and to work out if they will be profitable to mine.

Geophysicists use equipment that can tell them what types of rock might be underneath the surface. They can compare the densities of the rocks, and figure out if they are magnetic or radioactive. They can even measure how long sound takes to travel through rocks.

Geochemists analyse the chemical make-up of soil samples, water and even plants on the surface to search for clues about the minerals in the rocks under Earth's surface.

Three-dimensional modellers process all of the data that has been collected to build up a picture of what is happening underground.

What scientists are involved in the search for mineral deposits?

❷ Geologists and geotechnical engineers work in mining

An **ore body** is a mineral deposit that is profitable to mine. After an ore body has been located, mining companies need to work out the most cost effective, safe and environmentally friendly way of removing the ore from the surrounding rock.

Geotechnical engineers will gather information on the site and create a plan to construct the mine safely.

Mining geologists analyse the ore and decide if they should keep mining a particular area or move to a new one.

What do mining companies need to consider before they begin mining?

❸ Metallurgical engineers extract metals from ore

After ore has been removed, it needs to be processed to extract (take out) the resource inside. The main processes involve crushing the ore and using chemical reactions to separate the resource from the other elements in the mineral.

Metallurgical engineers specialise in extracting metals from ore. They analyse the ore mineral and determine the most effective method for extraction of the metal. They monitor the extraction process to make sure that it is efficient and safe for the environment.

Why must ore be processed after it is removed from the ground?

❹ Environmental scientists rehabilitate mine sites

At all times throughout the mining process, a mining company must take care to protect the natural environment. When mining is finished, any land that has been impacted by the mine must undergo **rehabilitation** to return it to the way it was before the mining began. This is to make sure that no damage is done, biodiversity is maintained and the site is safe.

Environmental scientists identify ways to protect and restore the natural environment before, during and after mining.

Why is rehabilitation of a mine site important?

Figure 5.25 Iron ore is removed from the ground and processed to extract resources.

EXPERIMENT 5.10
Mining for chocolate chips

CHECKPOINT 5.10 ✓

1 Identify the four main stages of the mining process.

2 Identify the types of scientists and engineers involved in each stage of mining.

3 Companies exploring for gold often focus on areas where there are gold mines. Propose a reason why they might do this.

4 Propose an impact on the finished product if the metallurgical engineer did not pay careful attention to the extraction process.

5 Propose what impacts there may be to the environment if mining companies did not employ environmental scientists.

CHALLENGE

6 Many different resources are mined in Australia. Research one and present the following information:
- location of a major mine
- types of exploration techniques used to find the deposit
- type of mining method used
- methods used to process the ore and extract the resource
- methods used or planned to rehabilitate the mine site.

SKILLS CHECK

- I can describe the role of geologists, metallurgical engineers, geophysicists and environmental scientists in the mining process.

CHAPTER SUMMARY

What mineral is that?

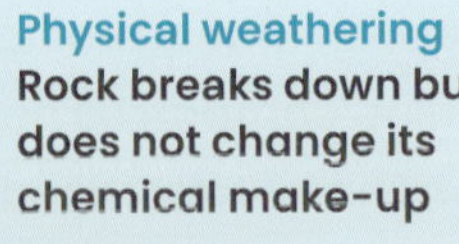

- Check the hardness of a mineral and score it using Moh's scale.
- Check a mineral's lustre – is it dull or shiny?
- Scratch the mineral to check the colour of the streak.
- Check the structure of its crystals.
- Check the colour, but this is not very reliable.

Physical weathering ▶ Rock breaks down but does not change its chemical make-up

◀ Chemical weathering Rock breaks down through chemical reactions that change the minerals in the rocks.

▲ Geologist analyses ore and suggests where to mine

Metallurgical engineer works to take metal out of ore

Environmental scientist ensures the natural environment is protected and restored

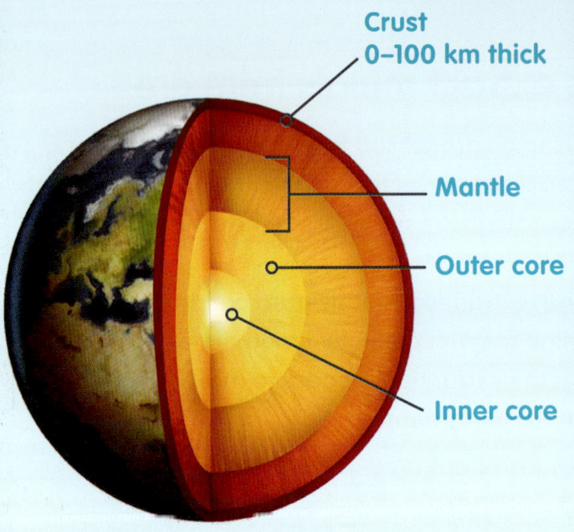

Fossils ▶ form when a dead organism is quickly covered by sediment. Many layers are added, compact and then cement into sedimentary rock.

The layers of sedimentary rock give us clues to the history of the area.

- Did it used to be underwater?
- Did ancient animals live here?
- Was it part of a vast mountain range?

We can study the type and size of the rocks to answer many of these questions.

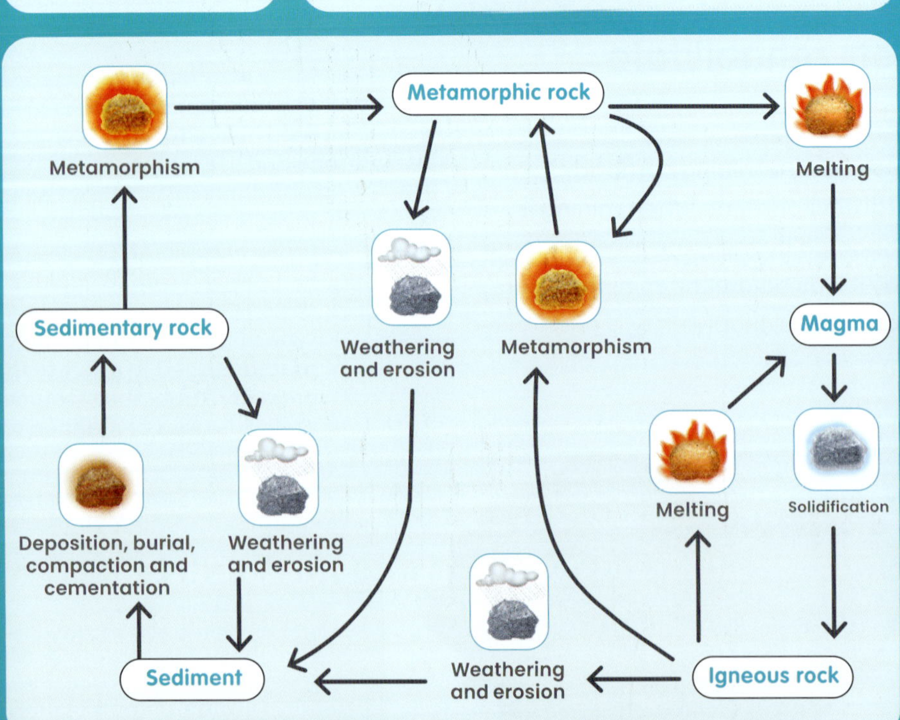

Igneous rocks form when molten rocks cool and solidify.

Sedimentary rocks form when sediment such as clay, sand and shells are buried and eventually compacted, joining together

Metamorphic rocks form from other rocks that have been changed by heat and/or pressure

★ FINAL CHALLENGE ★

1. Describe each of Earth's three main layers.
2. Which type of rock forms when molten rock cools and solidifies?
3. Explain why fossils are so rare.

LEVEL 1

50xp

LEVEL UP!

4. This list shows scientists involved in the mining process in Australia. Draw an arrow between their title and the description of what they do.

Geologist	Uses equipment to identify the types of rocks under the surface
Geophysicist	Specialises in extracting metal from ore
Geotechnical engineer	Analyses the chemical make-up of samples from the mining site
Geochemist	Analyses the ore that is being mined
Metallurgical engineer	Identifies ways to protect and restore the natural environment
Environmental scientist	Gathers information on the site and helps plan to make the mine safe

5. Explain how you would classify an unknown rock based on observable properties.
6. If you were asked to grow a very large crystal, how would you go about doing this? Give evidence to support your answer.
7. Streak is a better way to identify minerals than colour – suggest why.
8. Explain how the sedimentary layers within a cliff face can be used to learn about geological history.

9. You are presented with two samples of different white minerals. Outline what you would do to tell the difference between the two.
10. Explain how sedimentary, metamorphic and igneous rocks are related using an annotated flow chart.
11. Metamorphism can be likened to baking a cake – suggest why.

12. Using your understanding of the processes of deposition, compaction and cementation, explain the steps required for a skeleton to become a fossil.
13. Create a Venn diagram to identify the similarities and differences between the three rock types – igneous, sedimentary and metamorphic.

6 EARTH, SUN AND THE MOON

Many ancient cultures thought that Earth was at the centre of the universe. They believed that the Sun and the other planets revolved around Earth, and they thought they had proof – these bodies seemed to move around the sky during the day and the night.

We now know that Earth revolves around the Sun, the largest body in our solar system, with the largest gravitational pull. For the same reason, our Moon revolves around Earth. We also understand that the tilt of Earth causes the seasons, and that the rotation of Earth causes day and night.

1 LEARNING LINKS

What do you already know about Earth, Sun and the Moon?

What revolves around the Sun?

What revolves around Earth?

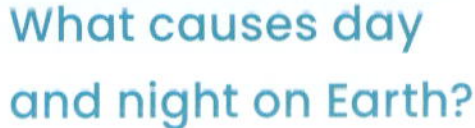

What causes day and night on Earth?

What would it be like in the Northern Hemisphere right now?

2 SEE-KNOW-WONDER

List three things you can **see**, three things you **know** and three things you **wonder** about this image.

3 CRITICAL + CREATIVE THINKING

What if ... Earth suddenly began spinning backwards?

Commonality: List three things that the Moon and winter have in common.

Five questions: Think of five questions that have *eclipse* as the answer.

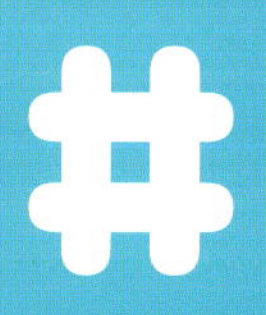

THE DARKEST!

Direct sunlight is not seen for up to six months of the year at the poles of Earth. For about six months at the south and north poles, the Sun is above the horizon (half a year of daylight) and for the other six months it is below the horizon (half a year of night or twilight). Imagine six months without a sunrise or sunset!

This also means that the poles have only two seasons: summer and winter. The seasons are caused by the tilt of Earth towards the Sun – when it's winter, Earth is tilted away.

6.1 DAY AND NIGHT

At the end of this lesson I will be able to:

- **explain** that predictable phenomena on Earth, including day and night, are caused by the relative positions of the Sun, Earth and Moon.

KEY TERMS 

axis
a real or imaginary line through the centre of an object

orbit
the curved path a smaller object takes around another object

revolve
move in a circular path around another object

rotate
spin on an axis

LITERACY LINK 

Summarise this section into a single paragraph. When you're finished, give yourself one point (for each point below) if your paragraph:

- is easy to understand
- is written in your own words
- includes at least one example
- is written formally and scientifically
- includes a simple diagram or image.

NUMERACY LINK 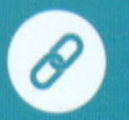

There are seven days in a week, and the Moon takes 28 days to orbit the Earth.

$28 \times \frac{1}{7} =$ ______ weeks

Humans have always wondered about the stars. The Sun and Moon stood out in the sky and people wondered about them as well.

Little by little our understanding increased. By observing and exploring, scientists worked out what causes day and night.

1 The Sun is at the centre of our solar system

The Sun is the star at the centre of our solar system. It is an enormous, dense ball of gas that produces light and heat. The nuclear reactions inside the core of the Sun are so powerful that the energy produced can light and heat Earth. This light takes about eight minutes to reach Earth.

Earth is one of the planets that **revolves** around the Sun. Earth's **orbit** is roughly the shape of a squashed circle. Earth takes one year to revolve around the Sun.

What star is at the centre of our solar system?

Figure 6.1 The planets in our solar system revolve around the Sun (not to scale).

2 Earth rotates on its axis

Earth is the third planet from the Sun. Earth is roughly a sphere, and it **rotates** on its own **axis**. A full day is 24 hours because Earth takes 24 hours to complete one full rotation on its axis.

Viewed from the north pole, Earth rotates in an anticlockwise direction. So the Sun appears to rise in the east and set in the west. At any one time, half of Earth is in daylight and half is in darkness. The side of Earth facing the Sun has day, and the side facing away from the Sun has night.

How long does Earth take to rotate once?

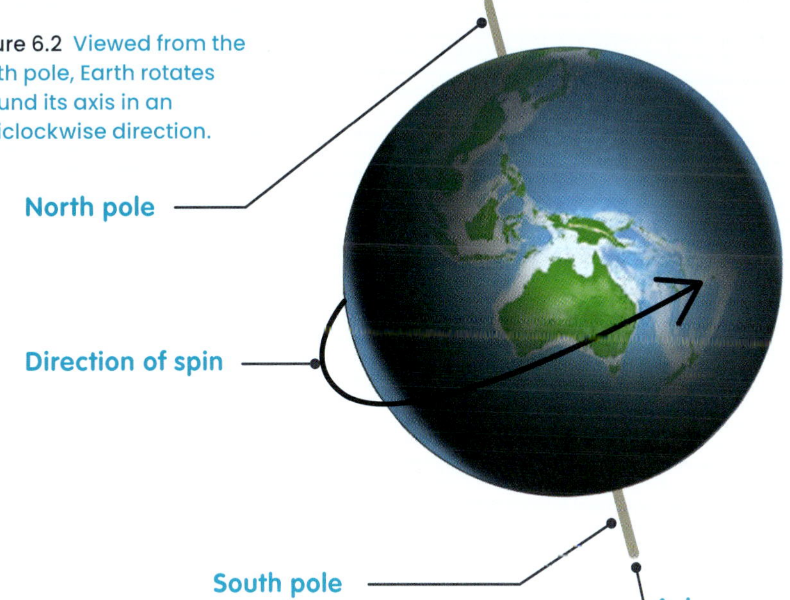

Figure 6.2 Viewed from the north pole, Earth rotates around its axis in an anticlockwise direction.

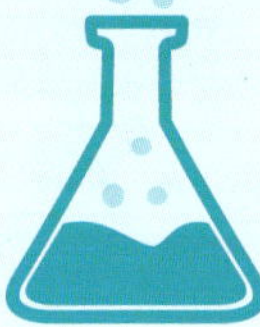

INVESTIGATION 6.1
Modelling day and night

CHECKPOINT 6.1

1 Write definitions for these words.
 a day-time
 b night-time
 c a day
 d a year

2 Copy and complete these sentences.
 a The Sun is a ______.
 b Earth rotates in a/an ____________ direction.
 c Earth revolves around the ______ and the ______ revolves around Earth.

3 Demonstrate how it's possible to see the Moon during the day. You can use objects around you, draw a diagram or any other method of your choice.

4 Investigate why we can't see most stars during the day.

5 Explain the difference between the words revolve and rotate by writing two sentences.

6 Demonstrate the difference between the words rotate and revolve by moving your body.

CHALLENGE

7 Research the length of a day for the other planets in our solar system.

 Mercury, Venus, Mars, Jupiter, Saturn, Uranus, Neptune

3 The Moon revolves around Earth

Moons are small bodies that orbit planets. Some planets have more than one moon.

Earth's Moon takes about 28 days to orbit Earth. It also takes 28 days to make one full rotation on its axis. This is why the same side of the Moon is always facing Earth.

Whenever the Moon passes over the side of Earth that is in daylight, it can be seen at the same time as the Sun. When the Moon is on the daylight side of Earth, the night side of Earth has night-time with no visible Moon.

How long does the Moon take to orbit Earth?

Figure 6.3 The same side of the Moon always faces Earth.

SKILLS CHECK

- I can explain what causes day and night.
- I can use a simple model or diagram to show what causes day and night.

6.2 SEASONS

At the end of this lesson I will be able to:

- **explain** that predictable phenomena on Earth, including seasons, are caused by the relative positions of the Sun, Earth and Moon.

KEY TERMS

equinox
the two times each year when night and day are about the same length

solstice
the two times each year when night and day are the most different in length

tilt
a sloping position or lean

LITERACY LINK

Write a short creative story that mentions a solstice and equinox.

NUMERACY LINK

There are four seasons, and each season is 13 weeks long.

If $x = 13w$, find x when $w = 4$.

Ice cream and the beach. Crunchy leaves and warm jackets. Hot drinks and beanies. Beautiful flowers and the first warm weather after the cold. Which of the seasons is your favourite? Did you know that Earth's tilt is why we have seasons?

1 Most places have four different seasons

Most areas around the world experience four separate seasons. Throughout the year, the northern and southern hemispheres experience opposite seasons. So when it's summer in Australia, which is in the southern hemisphere, it's winter in northern hemisphere countries such as the USA and England. The closer you are to the north and south poles, the greater the difference between winter and summer.

In some parts of the world, such as the far northern parts of Australia, there's not much difference between the winter and summer average temperatures. The year in these tropical areas in the southern hemisphere has a wet season from October to March and dry season from April to September.

What are the four seasons in most parts of the world?

Table 6.1 Earth's non-tropical seasons in the northern and southern hemispheres

Season	Temperature	Time of year	
		Southern hemisphere	Northern hemisphere
Winter	Cool/cold	June – August	December – February
Spring	Warming	September – November	March – May
Summer	Warm/hot	December – February	June – August
Autumn	Cooling	March – May	September – November

2 Earth's tilt causes the seasons

Earth's seasons are due to the **tilt** of its axis. Earth is tilted about 23.5° compared to the path Earth takes around the Sun, which you can see in yellow in Figure 6.5.

The seasons are caused by the intensity of sunlight – how much it spreads on Earth's surface. For example, in January, Australia is directly facing the Sun and the weather is warmer. In June, Australia is not directly facing the Sun, so the sunlight is spread over a larger area, making it cooler.

What causes seasons on Earth?

Figure 6.4 When it's summer in Australia, our part of the world is tilted towards the Sun and getting more intense sunlight.

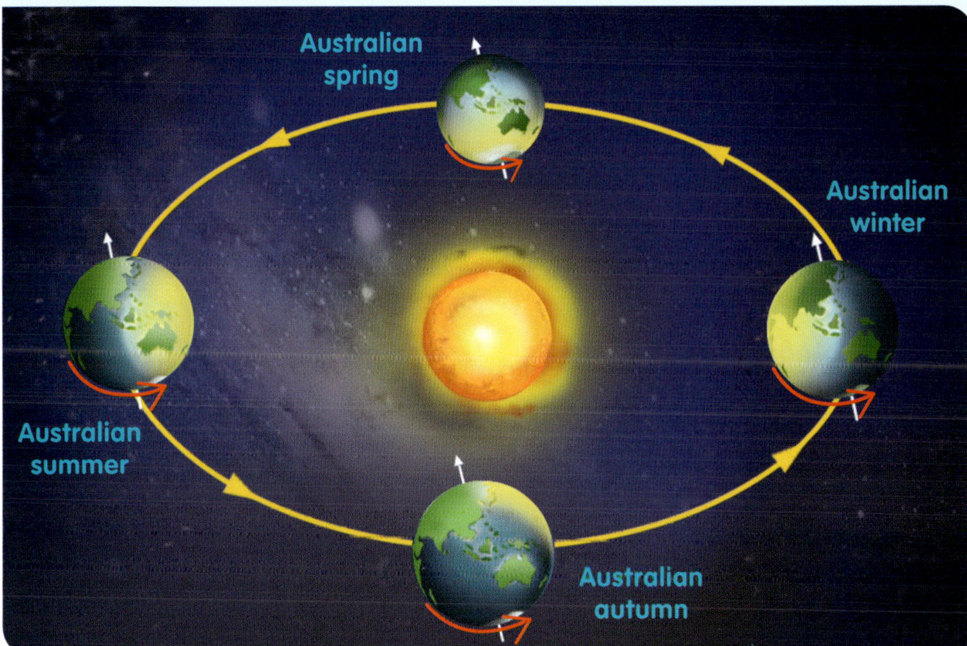

Figure 6.5 Earth's tilt means that different countries have more direct or less direct sunlight, depending on the time of year.

INVESTIGATION 6.2
Modelling the seasons

❸ Equinoxes and solstices – equal and different lengths of day and night

The **equinox** is the period when the day and night are the same length – 12 hours each. Equinoxes happen twice a year, once in spring and once in autumn, and they happen at the same time all around the world. When one hemisphere has spring equinox, the other hemisphere has autumn equinox, and vice versa.

The **solstice** is the time when the day and night are most different. The summer solstice is the day when the Sun reaches the highest point in the sky. It has the longest daylight of the year and the shortest night. The winter solstice has the shortest daylight and the longest night.

Table 6.2 Equinox and solstice dates around the world

Event	Approximate date
Equinox	*23 September* and *20 March* worldwide
Solstice	*21 June*: winter solstice in southern hemisphere and summer solstice in northern hemisphere *21 December*: summer solstice in southern hemisphere and winter solstice in northern hemisphere

In which month is the Australian summer solstice?

CHECKPOINT 6.2

1 Copy and complete these sentences.
  a When the southern hemisphere has summer, the northern hemisphere has ______________.
  b The seasons are caused by how much the sunlight __________ on Earth's surface.
  c The time of year when the length of the day and night are roughly equal is called the ______________.

2 If the summer solstice is when the Sun reaches the highest point in the sky for the whole year, where do you think the Sun will reach during the winter solstice? Why?

3 Some cultures have recognised more than four seasons. Find out what they are and why.

CHALLENGE

4 Research how Earth's tilt is stabilised by the Moon's gravitational pull on Earth. Suggest what life on Earth might be like if there was no Moon.

SKILLS CHECK

- I can explain what causes the seasons.
- I can use a simple model or diagram to show what causes the seasons.

6.3 ECLIPSES

At the end of this lesson I will be able to:

- **explain** that predictable phenomena on Earth, including eclipses, are caused by the relative positions of the Sun, Earth and Moon.

KEY TERMS 

annular
ring-shaped

eclipse
the blocking of the Sun's light from Earth

penumbra
the outer part of the Moon's shadow on Earth

umbra
the inner part of the Moon's shadow on Earth

LITERACY LINK 

Copy and complete this sentence: 'The alignment of the Sun, Moon and Earth is called ...'

NUMERACY LINK 

One lunar eclipse lasts for 13 minutes and 47 seconds. How many seconds did it last in total?

NEVER LOOK DIRECTLY AT THE SUN, INCLUDING DURING A SOLAR ECLIPSE. LIGHT FROM THE SUN CAN DAMAGE YOUR EYES.

ALWAYS USE AN INDIRECT METHOD OF OBSERVATION, SUCH AS A PINHOLE PROJECTED ONTO ANOTHER SURFACE.

The Sun is about 400 times wider than the Moon, and the Sun is about 400 times further away from Earth. This means that the Sun and Moon appear nearly the same size as seen from Earth.

As Earth and the Moon move, there are times when they and the Sun line up, in events called **eclipses**.

1 The Moon blocks the light of the Sun during a solar eclipse

A solar eclipse happens when the Moon passes between the Sun and Earth. The Moon blocks sunlight and casts a shadow on Earth. The centre of this shadow is the **umbra**, and the outer ring of shadow is the **penumbra**.

There are three types of solar eclipses, and they differ in how much the Moon blocks out the light of the Sun, as seen by a viewer on Earth. A total solar eclipse can be seen when the Moon completely lines up with both the Sun and Earth. People within the umbra see the Moon completely block the light of the Sun, but those within the penumbra only see it block part of the Sun.

An **annular** solar eclipse happens when the Moon is further away from Earth, making it appear smaller than usual. Because of this, the Moon only covers the centre of the Sun. During an annular eclipse, we can see the outer edges of the Sun. This is called an annulus or 'ring of fire'.

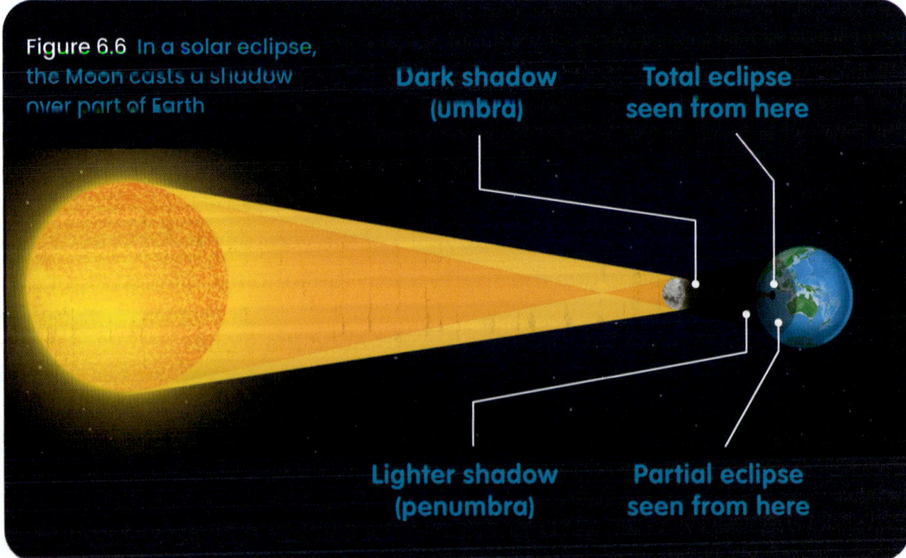

Figure 6.6 In a solar eclipse, the Moon casts a shadow over part of Earth

Partial solar eclipses occur when the Sun, Moon and Earth aren't completely lined up. The Moon only blocks part of the Sun when this happens. There's no umbra in a partial eclipse – everyone who sees it is in the penumbra.

What are the three types of solar eclipse?

INVESTIGATION 6.3A
Modelling a solar eclipse

INVESTIGATION 6.3B
Modelling a lunar eclipse

❷ Earth blocks the light of the Sun during a lunar eclipse

A lunar eclipse happens when Earth passes directly between the Sun and the Moon. Depending where you are on Earth, Earth blocks some or all sunlight and casts a shadow on the Moon.

Lunar eclipses always happen when the Moon is full. Full moons happen once every 29.5 days, which is the time the Moon takes to make one full revolution around Earth. During most months, the full Moon happens when the Moon does not line up with the Sun and Earth. During a lunar eclipse, the Moon is aligned with the shadow cast by Earth.

There are three different types of lunar eclipse, much like there are three types of solar eclipse. Each type is named according to how much Earth blocks the light of the Sun, as seen by a viewer on Earth.

A total lunar eclipse is seen when the Sun, Earth and Moon are perfectly in line, so Earth's shadow completely blocks the Moon. Total lunar eclipses are known as 'blood moons' because the Moon changes to a striking red colour.

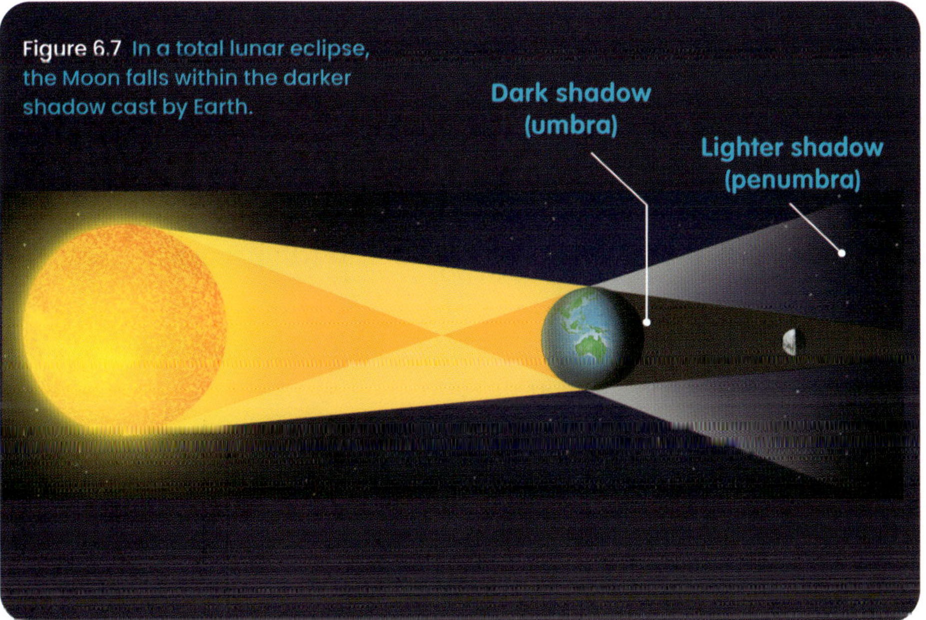

Figure 6.7 In a total lunar eclipse, the Moon falls within the darker shadow cast by Earth.

Partial lunar eclipses are seen when the Sun, Moon and Earth aren't completely in line. Only part of the Moon is covered by Earth's shadow. During a partial eclipse, you can see the curved shape of Earth's shadow on the Moon.

Penumbral lunar eclipses happen when the Moon only passes through the penumbra, which is the outer edge of Earth's shadow. These eclipses are often not noticed because the Moon appears only slightly dimmer than a regular full Moon.

When do lunar eclipses happen?

CHECKPOINT 6.3 

1 Copy and complete these sentences.
   a During a solar eclipse, the Moon casts a __________ on Earth.
   b An annulus is also known as a ______________.
   c Partial solar eclipses happen when the Sun, Earth and Moon ______________.
2 Copy and complete these sentences.
   a Lunar eclipses always happen during a ______________.
   b Penumbral eclipses make the Moon appear ____________.
3 In what ways is a total lunar eclipse different to a penumbral lunar eclipse?
4 When is the next solar eclipse? Which type is it? Where is the best place to observe it from?

CHALLENGE

5 Write two truths and one lie about lunar eclipses. Trade your answers with another student to test your knowledge.

SKILLS CHECK

- I can explain what a solar eclipse is, including how they occur.
- I can explain the difference between a partial, annular and total solar eclipse.
- I can explain what a lunar eclipse is, including how they occur.
- I can explain the difference between a total, partial and penumbral lunar eclipse.

6.4 MODELS OF THE SOLAR SYSTEM

At the end of this lesson I will be able to:

- **demonstrate**, using examples, how ideas by people from different cultures have contributed to the current understanding of the solar system
- **compare** historical and current models of the solar system to show how models are modified or rejected as a result of new scientific evidence.

KEY TERMS 

evidence
facts and observations that can be used to support or oppose a theory

model
a simplified way of explaining something complex and real based on evidence

LITERACY LINK 

Choose one ancient culture and write a report on how they explained natural phenomena such as the behaviour of the Sun, Moon and stars.

NUMERACY LINK 

Earth has a radius of 6378 km, while the planet Neptune has a radius of 24 776 km.

Create a ratio in its simplest form to compare Earth's radius to Neptune's.

The Sun has guided all living things on Earth since life began. When looking at the sky, we can understand how early civilisations thought the Sun and stars revolved around Earth.

The Ancient Greeks were the first people known to make **models** of nature to explain patterns that they observed. These models allowed them to try to make sense of the world around them using reasoning.

1 The geocentric model put Earth at the centre of the universe

Perhaps the first model of the solar system was the geocentric model, which stated that Earth was at the centre of the universe. The word *geocentric* comes from 'geo' (Earth) and 'centric' (centred).

Aristotle (384–323 BCE) was one of the earliest known writers on astronomy. His observations were made with only the naked eye. He argued that:

- Earth was a sphere at the centre of the universe
- the planets and the Sun orbited on many perfect and unchanging spheres
- these spheres revolved around the unmoving Earth.

Claudius Ptolemy (100–175 CE) was another early philosopher and scientist. He supported the idea that Earth was at the centre of the universe, but he suggested that the planets and the Sun revolved around a point outside Earth.

What are the main features of the geocentric model?

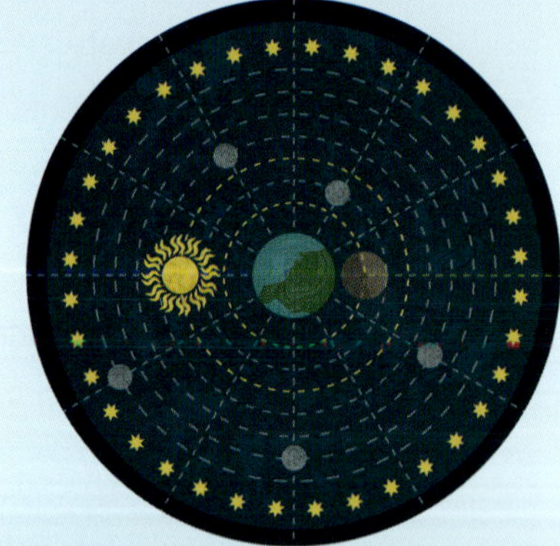

Figure 6.9 In the geocentric model, Earth was at the centre of the universe and the planets and the Sun revolved around it.

Figure 6.8 Stonehenge is a circle of standing stones in England, built about 5000 years ago. It was probably used as a calendar and a way to predict solstices.

❷ The heliocentric model put the Sun at the centre of the universe

The heliocentric model replaced the geocentric model. 'Helios' was the Greek god of the Sun – this model got its name because it stated that the Sun was at the centre of the universe.

The heliocentric model was created by Nicolaus Copernicus (1473–1543 CE), who used mathematics (rather than a telescope) to explain the motion of objects in the heavens. His model proposed that:

- the Sun, rather than Earth, was at the centre of the universe
- Earth and the planets moved in circular orbits around the Sun.

Many people didn't agree with the heliocentric model. Some thought it was dangerous, because it suggested that humans were not the most important beings in the universe. In particular, the early Catholic Church refused to accept this model.

Galileo Galilei (1564–1642 CE) was one of the first astronomers to use a telescope. His observations of the solar system provided **evidence** that overwhelmingly supported the heliocentric model. He faced punishment by the Catholic Church for his work and was placed under house arrest.

Eventually the heliocentric model was accepted around the world. But, as telescopes developed, people noticed that stars didn't orbit the Sun. By the early 19th century, scientists realised that although the Sun is the centre of our solar system, it isn't the centre of the universe.

What are the main features of the heliocentric model?

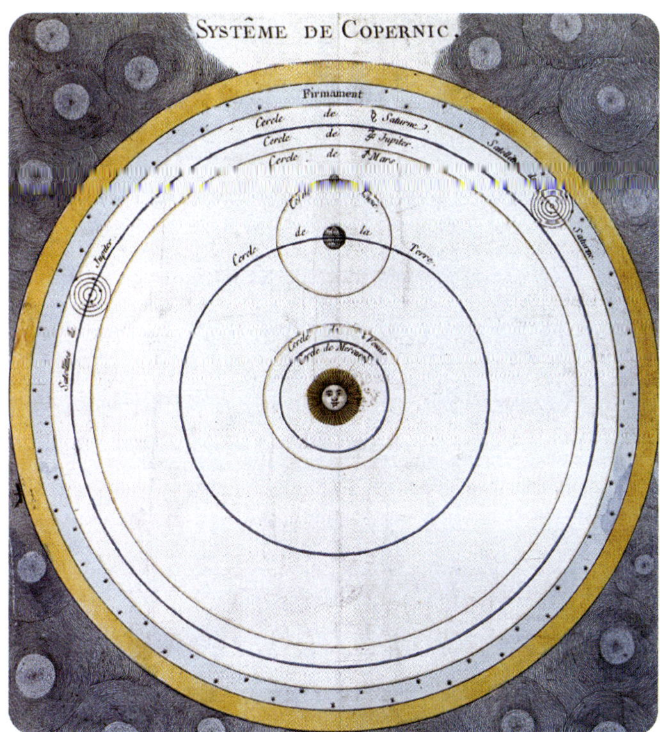

Figure 6.10 In the heliocentric model, the Sun was at the centre of the universe and the planets had circular orbits.

CHECKPOINT 6.4 

1 True or false?

a *Geo* means 'planet'.

b The Ancient Greeks believed that Earth was round.

c Copernicus made observations using a telescope.

d The orbits of the planets are elliptical.

2 Create a Venn diagram to compare and contrast the geocentric and heliocentric models of the solar system.

3 Describe the contributions towards our current understanding of the solar system made by

a Ptolemy

b Aristotle

c Galileo

4 Copernicus first presented the heliocentric model, which was thought to be dangerous. Explain why.

CHALLENGE

5 Research the scientists who have contributed to increasing our knowledge of the solar system through time. Organise their discoveries on a timeline to demonstrate your learning.

SKILLS CHECK

- I can explain how historical models of the solar system changed over time.

6.5 THE TECHNOLOGY OF DISCOVERY

At the end of this lesson I will be able to:

- **describe** some examples of how technological advances have led to discoveries and increased scientific understanding of the solar system.

KEY TERMS 

astronomer
a scientist who studies space and the objects within it

galaxy
a large system of stars

LITERACY LINK 

Write a newspaper article detailing the first Moon walk in 1969. Include quotes and pictures.

NUMERACY LINK 

Venus has a radius of approximately 6000 km.

Calculate the circumference of Venus.

Formula: $C = 2\pi r$

Have you ever looked through a pair of binoculars, or perhaps a telescope? What did you see? How did you feel about being able to observe objects and places that were so far away?

People throughout history have always been curious. Technology has allowed us to discover distant objects and events and wonder about more things than ever before.

1 Edwin Hubble first calculated the true size of the universe

Until the early 20th century, **astronomers** thought that the entire universe consisted of just one group of stars. They thought that the universe was only a single **galaxy**, with the stars in it relatively nearby. In the 1920s, astronomer Edwin Hubble was studying a star in what was called the Andromeda Nebula. He worked with a 2.5-metre telescope at Mt Wilson in California. His observations led him to conclude that:

- Andromeda was not nearby, but very distant
- it was not a cloud of gas or nebula but a galaxy like ours (It's now known as the Andromeda Galaxy)
- there are an incredible number of galaxies in the universe
- each of these galaxies contains tens of millions of stars.

This completely changed our understanding of the universe!

Edwin Hubble went on to create a system to classify galaxies according to how they look. Also, he proved that the entire universe is expanding at the same rate everywhere. He is now considered to be one of the most important astronomers in history, and the Hubble Space Telescope is named after him.

What did astronomers think the universe looked like before Edwin Hubble's work?

2 The Hubble Space Telescope revolves around Earth

The orbit of the Hubble Space Telescope is about 600 km above Earth. The space shuttle *Discovery* placed it in orbit in 1990. Because it's outside Earth's atmosphere, it can be used to observe light and radiation filtered out by Earth's atmosphere.

Figure 6.11 The Hubble Space Telescope has provided us with information about deep space since 1990.

INVESTIGATION 6.5
Making a simple telescope

Hubble researchers collect information to answer questions such as these:

- What is the size of the universe?
- How are stars formed?
- How fast is the universe growing?

The Hubble Space Telescope has special features that allow it to perform this mission:

- a 2.4-metre mirror to collect light from deep space
- cameras that detect different types of light
- instruments that separate and analyse the collected light
- systems that control where the telescope is pointed and keep it in orbit.

How is the Hubble Space Telescope different to telescopes on Earth's surface?

3 Humans first walked on the Moon in 1969

The *Apollo 11* mission was the first human-piloted effort to land on the Moon. Neil Armstrong, Michael Collins and Edwin 'Buzz' Aldrin, Jr formed the crew on this historic mission. The US space agency, NASA, succeeded in their attempt on 20 July 1969. Since then, humans have landed on the Moon four more times.

One of the main aims of these missions was to collect rock samples from the Moon's surface for study back on Earth. The missions collected over 382 kilograms of lunar samples. These samples allowed scientists to develop theories about the origin of the Moon.

Lunar samples show that some of the Moon's matter comes from Earth, and some comes from another source. The current theory for the Moon's origin is that a very young Earth was hit by a stray body about half its size. This collision threw debris out around Earth, which eventually collected together to form the Moon.

What was the first human-piloted effort to land on the Moon?

Figure 6.12 Astronauts collected lunar soil samples during the *Apollo 12* mission, the second human-piloted Moon landing.

CHECKPOINT 6.5

1 Copy and complete these sentences.
  a The Andromeda Galaxy was once thought to be a ______________.
  b Edwin Hubble used a ______________ to study space.
  c Each ______________ contains tens of millions of stars.
  d The universe is ______________ at the same rate everywhere.
  e The Hubble Space Telescope can collect ______________, ______________ and ______________ light.

2 What does an astronomer do?

3 How has the Hubble Space Telescope increased scientific understanding of the solar system?

4 Research the different technologies produced by NASA during the 'space race'. Choose one to create a fact sheet about.

CHALLENGE

5 Imagine you are part of a mission to Mars. List the five most important pieces of technology for this mission and justify why you have included each of them.

SKILLS CHECK

- I can describe some examples of how technological advances have increased scientific understanding of the solar system.

CHAPTER SUMMARY

The Sun is the star at the centre of our solar system

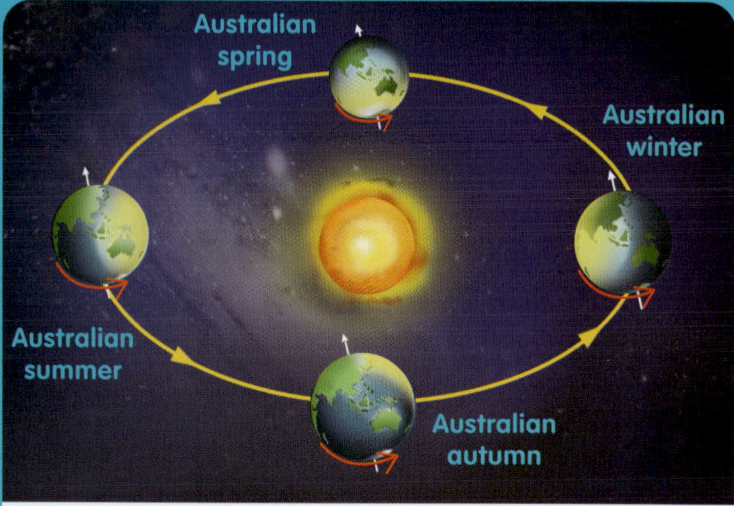

▲ Earth is tilted on its axis, which is what causes seasons. Its orbit around the Sun is an oval shape, which causes days and nights to become longer and shorter.

Earth facts
- third planet from the Sun
- roughly a sphere
- rotates on its own **axis** once every 24 hours

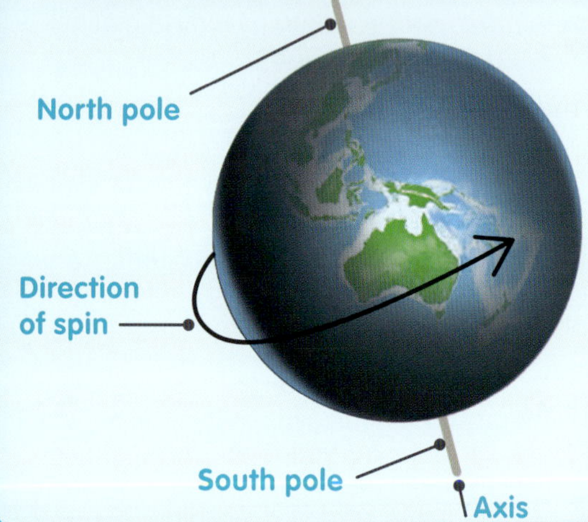

▼ During a lunar eclipse, Earth casts a shadow that covers the Moon.

The Hubble Space Telescope has provided us with information about deep space since 1990.

▼ In a solar eclipse, the Moon casts a shadow over part of Earth.

Geocentric model
- Aristotle argued that Earth was at the centre of the universe
- Ptolemy suggested that the planets revolved around a point outside Earth

The heliocentric model
- Nicolaus Copernicus proposed that the Sun, rather than Earth, was the centre of the universe
  - Galileo Galilei provided mathematical **evidence** that supported the heliocentric model

★ FINAL CHALLENGE ★

1. In your own words, explain what causes a solar eclipse.
2. How did Galileo provide evidence to support Copernicus' heliocentric model of the solar system?
3. When Australia, which is in the Southern Hemisphere, is experiencing winter, what season is France experiencing in the Northern Hemisphere?

LEVEL 1
50xp
LEVEL UP!

4. Draw a diagram showing the positions of Earth, the Sun and the Moon during:
   a A lunar eclipse
   b A solar eclipse
5. Earth's seasons are caused by the tilt of its rotation – suggest why.

6. Describe the difference between an equinox and a solstice.
7. How long does it take for:
   a Earth to rotate once on its own axis?
   b The Moon to orbit Earth?

8. Describe how human missions to the Moon have led to increased scientific understanding of the solar system.
9. Design a model that could be used to explain how day and night occurs on Earth. Provide an annotated diagram of your model as well as a list of materials.
10. Early astronomers believed that Earth was the centre of the universe. Why do you think this was a popular idea?

11. Do you think the Moon experiences day and night? Justify your response.
12. Predict how the seasons would change if Earth's rotation was not tilted.

7 RESOURCES

Resources can be described as the things that living things need to survive. We use many different resources in our daily lives, for food, shelter and to make our lives easier. Some of these resources are found naturally, while others are made.

We need to think carefully about how we use Earth's resources. Managing our resources well will make sure that they are still able to be used by future generations.

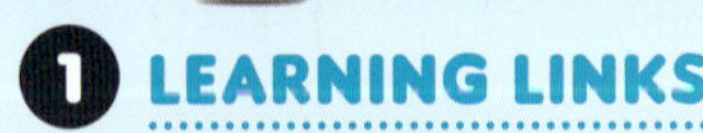

1 LEARNING LINKS

What do you already know about resources?

What are some common resources that we obtain from nature?

How do we use materials obtained from Earth?

What can we do in our daily life to use resources more sustainably?

What are some changes to your local environments caused by human activities?

2 SEE-KNOW-WONDER

List three things you can **see**, three things you **know** and three things you **wonder** about this image.

3 CRITICAL + CREATIVE THINKING

Variations: How many ways can you reuse a plastic bag?

Disadvantages: List as many disadvantages of recycling as you can. Now list ways of eliminating these disadvantages.

Mismatches: If you had a giant bucket, the world's largest diamond and a windmill, how could you solve the problem of plastic that ends up in the ocean?

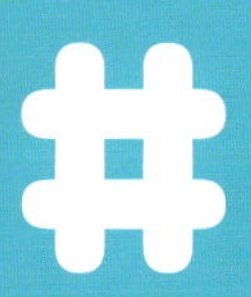

4 THE MOST WASTEFUL!

We're good at many things in Australia, but recycling and reusing materials is not one of them. Australians throw out a lot of stuff. Every year we generate hundreds, even thousands, of kilograms of waste per person.

Waste, including plastic waste, can end up in the oceans. There is so much plastic in oceans that scientists predict that by 2050 there will be more plastic than fish in the ocean! Sea life can get tangled up in plastic, eat it and starve, or even suffocate.

7.1 WHAT IS A RESOURCE?

At the end of this lesson I will be able to:

- **classify** a range of Earth's resources as renewable or non-renewable.

KEY TERMS 

made resource
a resource that is manufactured from natural resources

natural resource
a resource that is valuable in its natural form

non-renewable resource
a resource that can run out, or one that takes longer than a human life span to be restored

renewable resource
a resource that cannot run out, or can be restored in a human life span

resource
a source of something that is useful

LITERACY LINK 

An antonym is a word that is the opposite of another word. Think of an antonym for the words *finite* and *natural*.

NUMERACY LINK 

Anika estimates that an area of soil will take x^4 years to form. If $x = 6$, calculate how many years it will take.

Anything that a person can use is a **resource**.

Natural resources, such as vegetables, oxygen, timber and wool, don't need to be changed much or at all before we can use them. Some natural resources are processed to become something that's very different and also useful. These are **made resources**, such as plastic, concrete and some fabrics.

Figure 7.1 These wind turbines create electricity from the motion of the wind, which is a renewable resource.

❶ Natural resources are useful in their natural form

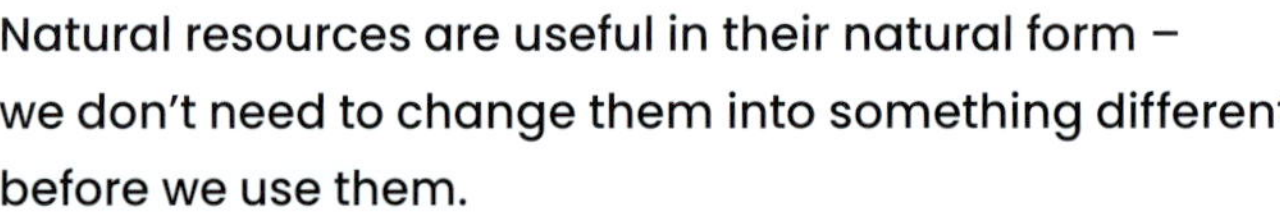

Natural resources are useful in their natural form – we don't need to change them into something different before we use them.

Fresh food and water are two common natural resources. Water is collected from rivers and lakes to supply towns and farms. Fish are taken from the oceans to provide food. Fruit and vegetables are grown and harvested for food.

What is a natural resource?

❷ There are renewable and non-renewable resources

Some natural resources replenish themselves, and the length of time this takes depends on the type of resource.

A **renewable resource** can be infinite (can never run out), such as wind or sunlight, or will replenish itself within the average human life span (about 80 years). Crops are renewable because they can grow back within a human life span.

A **non-renewable resource** is finite (can run out), or only replenishes itself over a much longer period than a human life span. Fossil fuels such as coal, oil and gas are

Figure 7.2 Cotton is a natural resource because the fibre from the cotton plant is not changed a lot to make clothes – it is the same basic material.

non-renewable resources. Coal (the compressed remains of ancient plants) takes hundreds of millions of years to form, and it only forms under certain conditions.

Some trees can grow back in a life span, but some are many hundreds of years old. This means that trees can be renewable or non-renewable.

What is the difference between a renewable and non-renewable resource?

INVESTIGATION 7.1
Classifying resources used in the classroom

3 Made resources are made from natural resources

Not every resource can be used without changing it. Made resources are manufactured from natural resources, and they are very different to the original resource.

There are an incredible variety of made resources. Plastic packaging and polyester fabrics are made from oil extracted from the ground. Medicines are made from many different natural products. Steel is made by mixing iron metal with carbon and other substances. Another term for made resource is synthetic resource.

What is a made resource?

Figure 7.3 The coal extracted from this mine took millions of years to form.

CHECKPOINT 7.1

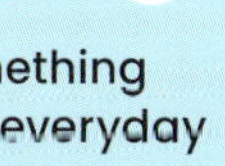

1 A __________ is something that societies use in everyday life __________. A __________ resource can be restored within a __________ life span. A non-renewable __________ is finite or takes much __________ to restore.

2 List three examples of natural resources and three examples of made resources.

3 List three examples of renewable resources and three examples of non-renewable resources.

4 Identify if these resources are renewable or non-renewable.
a banana
b cotton
c coal
d copper
e paper
f plastic
g wind
h water

CHALLENGE

5 Use the internet to find out the natural resources used to make a smartphone. Select one of these materials, find out how it is obtained and what it is used for in the phone.

SKILLS CHECK

- I can describe the difference between non-renewable and renewable resources.
- I can give three examples of each type of resource.

7.2 NON-RENEWABLE RESOURCES

At the end of this lesson I will be able to:

- **outline** features of some non-renewable resources, including metal ores and fossil fuels.
- describe uses of a variety of natural and made resources extracted from the lithosphere.

KEY TERMS 

fossil fuel
a natural fuel formed over millions of years from the remains of living things

non-renewable resource
a resource that can run out, or one that takes longer than a human life span to be restored

sediment
small particles of rocks, such as clay, sand and pebbles

LITERACY LINK 

Quickly scan this section, reading the headings and considering the images. What can you predict about this text? What do you think it will be about and include? Also consider what you may already know about this topic.

NUMERACY LINK 

In some regions it takes 1000 years to produce 1 cm of soil from natural processes. How long would a natural soil depth of 6 cm take to develop?

Non-renewable resources run out or are not replenished in a human life span.

Rocks, minerals, fossil fuels and soil are some of the non-renewable resources we use that come from Earth's lithosphere. These resources are used to produce materials, generate energy and develop agriculture.

Figure 7.4 Huge amounts of stone, such as this marble, are quarried for use as building materials.

1 Rocks and minerals provide metals

Rocks and minerals are non-renewable resources because there are only limited amounts close to Earth's surface that humans can access. Rocks are useful resources as building materials. They can be cut to create building stone or crushed to add strength to roads and concrete.

Rocks can contain minerals that hold useful metals such as aluminium, copper and iron. These minerals are known as metal ores, and when they are found in large amounts they are called mineral deposits. Mining and extraction processes remove and purify metals from their ores for use to manufacture many items. Australia has many important mineral deposits that produce a variety of useful metals.

Table 7.1 Common metals mined in Australia

Metal	Mineral ore/s	States mined	Useful property	Use
Iron	Hematite, magnetite	WA	strong	Steel
Aluminium	Bauxite	Qld, NT	lightweight	Aluminium cans, aircraft
Copper	Chalcopyrite	Qld, SA, NSW, WA	good conductor of electricity	electrical wiring, computers, coins
Zinc	Sphalerite	Qld, NSW, NT, Tas, WA	resistant to corrosion	galvanising steel, zinc creams

Why are rocks and minerals non-renewable resources?

2 Fossil fuels include coal, oil and gas

Fossil fuels include coal, crude oil and natural gas. These resources were produced from the remains of ancient plants and animals, in processes that take hundreds of millions of years. While fossil fuels are still being formed today, they are being removed and used at a rate that is much faster than they are being restored.

Coal is formed from the remains of ancient swamps. This plant matter builds up over time and doesn't decay. As the plant matter is buried, heated and squashed under tonnes of soil, the water and impurities are squeezed out and it slowly changes into coal.

Oil and natural gas are formed from the remains of tiny marine organisms, such as algae and plankton, that die and sink to the bottom of the ocean. Conditions at the bottom of the ocean stop them from breaking down, so the remains build up in the **sediment**. This happens over millions of years. As the sediments are buried, heated and squashed, the remains undergo chemical reactions that produce oil and natural gas.

Fossil fuels are often burnt to provide heat energy. In power plants, this heat energy is used to change water into steam, which then spins large, fan-like machines called turbines. Crude oil is processed to make petrol, diesel, motor oil and bitumen, as well as chemicals that have uses such as in making medicine, cosmetics and plastics.

What are the three main fossil fuels?

3 Soil takes hundreds of years to form

Soils are vital for plant growth and are important habitats for living things. They are a mixture of sediments, minerals, plant and animal material, air and water. A healthy soil is important for agriculture to produce crops, pasture for livestock, and forests. Soils are produced when rocks are weathered into sediment that is then mixed with the broken down remains of dead plants and animals. Because these processes take a long time, one centimetre of soil can take hundreds, even thousands, of years to form. Because soils take so long to form, they are considered a non-renewable resource.

What is soil a mixture of?

Figure 7.5 Overgrazing can cause erosion and the loss of precious soil.

INVESTIGATION 7.2
Investigating soil erosion

CHECKPOINT 7.2

1 List three resources that can be obtained from rocks.
2 Explain how metals such as iron come from rocks.
3 Explain how fossil fuels are used as a resource.
4 What are some of the problems with using fossil fuels as a resource?
5 Why are minerals, fossil fuels and soil all non-renewable resources?
6 Justify why recycling is important to manage a non-renewable resource.
7 Of the three non-renewable resources discussed in this section, identify which one you think is the most important. Justify your response.
8 Fossils fuels are considered finite. Explain why.

CHALLENGE

9 Research the process of how one of the fossil fuels in this section forms. Create an annotated flow chart that summarises this process.

SKILLS CHECK

- I can explain what a non-renewable resource is.
- I can give examples of some non-renewable resources and their features.
- I can describe what a fossil fuel is.

7.3 RENEWABLE RESOURCES

At the end of this lesson I will be able to:

- **describe** uses of a variety of natural and made resources extracted from the biosphere, atmosphere, and hydrosphere.

KEY TERMS 

atmosphere
the mixture of gases surrounding Earth

biofuel
a substance produced from living things that can be burnt to create energy

biosphere
the parts of Earth where living things are found

hydrosphere
all the water on Earth's surface

renewable resource
a resource that cannot run out, or can be restored in a human life span

solar energy
energy made by the Sun

LITERACY LINK 

Consider the first paragraph in this section. Rewrite it to make it sound more technical and scientific. Your goal is to make the language more formal.

NUMERACY LINK 

Arjun builds a chair out of renewable wood and bio-materials. If it cost him $40 to make the chair, and he sells it for $65, calculate the profit as a percentage.

We use **renewable resources** for food, materials, and energy. Some renewable resources can replenish within a human life span, and some will never run out. Renewable resources can be from living things, such as plants and animals, and non-living things such as the Sun, wind and water.

The living parts of the world contain many different renewable resources, including plants and animals. **Solar energy**, wind and water are non-living renewable resources. We can use these and other resources for food, fuel and power.

Figure 7.6 The oils from some plants can be made into biodiesel.

1 The biosphere has many renewable resources

The **biosphere** is the living parts of the world, including in parts of the atmosphere, underground and underwater. Those living things are an important resource that provide us with food, materials and even energy. If we manage our use carefully, living resources can restore within a human life span.

Trees are grown and harvested to provide materials such as timber and paper. Other plants provide fruit and vegetable crops. Animals provide meat and materials such as wool, leather and honey.

Biofuels are fuels produced from plants and animal waste. Wood is a biofuel used by humans for hundreds of thousands of years. Modern biofuels include motor fuels such as biodiesel, which is made from vegetable oils.

What are some examples of living things as resources?

2 The atmosphere is a resource

The air in our **atmosphere** is an infinite resource. Air provides living things with oxygen for respiration, and plants with carbon dioxide for photosynthesis.

When air moves, it creates wind. Wind can be harnessed by windmills to pump water, and by turbines to create electricity.

How is the wind useful as a resource?

INVESTIGATION 7.3A
Designing a windmill to lift a weight

INVESTIGATION 7.3B
Making bioplastic

3 The hydrosphere is a resource

All living things need water to survive. Water resources are sources of water that can be useful, such as rivers, lakes and dams. These resources can be used for drinking, for farming, and by factories. Most of the water that is used by humans is fresh and these sources must be managed carefully to make sure they do not run out or become polluted.

The energy from moving water, such as waves or in dams, can be harnessed to create electricity.

What is a water resource?

4 The Sun is the source of all energy on Earth

The energy from the Sun is **solar energy**. It is the starting point for most of the processes on Earth. Without the Sun's heat and light, nothing would live on Earth – our planet would be a cold, airless rock.

Plants use the energy from the Sun for photosynthesis, which enables them to grow. This energy is then passed on to the organisms that eat the plants. The Sun's heat warms the air unevenly, causing air of different temperatures to meet and move as wind.

Solar power panels can change sunlight into electricity called solar power. Solar thermal systems use sunlight to heat water or air in buildings.

What can solar power panels change solar energy into?

Figure 7.7 Dams throughout the Snowy Mountains are used to generate hydro-electricity. The movement of falling water is used to spin turbines.

CHECKPOINT 7.3 

1 Identify two examples of resources provided by:
  a living things
  b air
  c water
  d the Sun.

2 Identify four examples of renewable energy resources.

3 Explain why the time taken to restore a resource is important for determining if it is renewable or not.

4 Trees and other plants are versatile resources as they can be used to produce many things. Give some examples.

5 Hydro-electric schemes use the movement of water to spin big turbines to generate electricity. Can you think of any impacts on the environment that could be caused by hydro-electric schemes?

CHALLENGE

6 As a class, research and debate the use of renewable energy resources in Australia.

SKILLS CHECK

- I can explain what a renewable resource is.
- I can give examples of some renewable resources from the biosphere, atmosphere and hydrosphere and their features.

7.4 CONSERVING EARTH'S RESOURCES

At the end of this lesson I will be able to:

- **investigate** some strategies used by people to conserve and manage non-renewable resources, e.g. recycling and the alternative use of natural and made resources.

KEY TERMS

population
all the living things of one species in a particular area

recycle
change something into something else that is useful

sustainable
able to be maintained at a certain rate or level

LITERACY LINK

Write a pamphlet (or similar) explaining the benefits of recycling as well as how to properly recycle (things that can and can't be recycled) for your local neighbourhood.

NUMERACY LINK

A recycling centre calculated that it can obtain x kilograms of plastics from donations.

$4(x-2) = 40$ kg

What is the value of x?

Using a resource in a **sustainable** way means using it in a way that allows others to use it in future. This might mean limiting the amount of the resource that is used, using another resource, or reusing that resource again.

1 Reducing resource use

Some resources can be restored in a human lifetime, but this may only happen if they are used sustainably.

Overfishing is an example of a renewable resource being used unsustainably. If too many fish are taken from wild **populations**, there won't be enough wild fish left to breed, leaving no fish left for the future. Many natural fish populations are much smaller than they were 100 years ago. To combat this, governments place size and catch limits for people who fish individually or for their business. They do this so that young fish have time to grow and breed, and that enough fish are left to maintain the population.

What may happen to a renewable resource if it is used unsustainably?

2 Non-renewable resources should be used carefully

It's important that the world manages its use of non-renewable resources, particularly fossil fuels and minerals, so that the environment is healthy and there will be enough resources for future generations. There are three major ways to manage these resources.

One way is to reduce the amount of a resource that is used. Governments can influence how much a resource costs. If costs are high, people will use less. Some groups encourage citizens to change their habits so that they use less resources, such as fuel.

Another way is **recycling**. Materials such as plastics and metals can be collected and recycled into new items. This means that the resources can be used more than once.

The third way is to find other resources. Scientists and engineers are working to find and develop other ways of using different energy sources. In Australia, more and more of our energy supplies come from hydro-electric, solar and wind sources.

What are three ways to conserve a non-renewable resource?

Figure 7.9 Petrol for transport is the main use for oil worldwide.

INVESTIGATION 7.4
The sustainability game

❸ Non-renewable resources need to be reused and recycled

Some items are designed and made to be reused again and again, such as reusable coffee cups and glass milk bottles. Items can also be repurposed – used in a different way than they were originally intended. For example, old car tyres being used to make the walls of a house.

If an item has come to the end of its useful life, the materials in it should be recycled. Recycling is where an item is broken down into raw materials that can be made into new products. Aluminium is a resource that is easily and cheaply recycled – in fact, recycling aluminium uses only 5% of the energy and produces only 5% of greenhouse gas emissions that obtaining new aluminium from mining requires.

What is the difference between reusing and recycling?

Figure 7.10 Some of the walls of this house have been made from old car tyres filled with compacted dirt.

CHECKPOINT 7.4

1 Identify what could happen if a resource was not managed sustainably.

2 Explain how overfishing can be prevented.

3 List four other renewable resources that need to be managed sustainably.

4 Explain, using examples, how the following strategies can be used to conserve resources:
a reducing consumption
b reusing items
c recycling materials.

5 Explain why it is beneficial to recycle aluminium cans.

6 Is it more important to conserve renewable or non-renewable resources? Justify your response.

CHALLENGE

7 Find out how plastic is recycled. Summarise the steps and present them creatively.

8 Find out how your school manages its resources, such as water, electricity and paper. Can you suggest ways that the resources can be used more sustainably?

SKILLS CHECK

- I can explain why it's important to sustainably manage resources.
- I can give examples of how to conserve and manage resources.

7.5 RESOURCE USE AND THE ENVIRONMENT

At the end of this lesson I will be able to:

- **outline** the choices that need to be made when considering whether to use scientific and technological advances to obtain a resource from Earth's spheres.
- **discuss** different viewpoints people may use to weight criteria in making decisions about the use of a major non-renewable resource found in Australia.

KEY TERMS 

biodiversity
the number of different types of organisms in an area

habitat
the place where an animal or plant naturally lives

LITERACY LINK 

Prepare speaking cue cards for a debate that argues for and against single-use plastic items. Try your arguments against a classmate.

NUMERACY LINK 

It costs Earth Solutions $11 to make each solar cell that they then sell for $20.

Write an equation to calculate their profit if they sell 10 solar cells.

Figure 7.11 The Super Pit is the biggest open-pit gold mine in Australia. Extracting this important resource affects the environment.

Using and obtaining resources from Earth can damage the natural environment.

It can be hard to keep a balance between the need for a resource and conserving the environment. When making these decisions it is important to understand how the resource will benefit society, how obtaining it can damage the environment, how any damage can be minimised as well as the opinions of different groups in the community.

1 How will we benefit from this resource?

Resources that we obtain from Earth can provide us with food, shelter and other ways to make our lives easier. We not only need to consider the benefits of using a resource to make various products, but we also need to consider other ways that the resource can be of benefit to society. For example, mining makes a significant contribution to Australia's economy, providing many people with jobs.

What are some ways that society can benefit from a resource?

2 Will obtaining or using this resource damage the environment?

Obtaining a resource may change, pollute or destroy native **habitats** or reduce the **biodiversity** of an area. Sometimes it is the use of the resource that can be damaging to the environment. The burning of fossil fuels results in the release of carbon dioxide into the atmosphere at levels well above what would naturally occur. Carbon dioxide traps heat in the atmosphere that would normally escape into space. This has caused average global temperatures to rise and is contributing to climate change. If we are to limit the amount of carbon dioxide being emitted, we need to use alternatives to fossil fuels to produce energy.

What are three ways that obtaining or using a resource can damage the environment?

❸ Can we reduce the environmental impact?

Sometimes an environmental 'trade off' needs to be made if society can't do without a resource. We have to accept that some environmental damage will occur in return for obtaining or using the resource. In some cases, technologies and strategies can be used to lessen the impact on the environment.

Mines are required to rehabilitate the land that they use so that native plants and animals will return, but it will never be exactly the way it was before the mine started.

What is an environmental trade off?

❹ Conflicting interests and making choices

Before a resource is extracted and used, information and opinions are gathered about the resource and the local environment so that choices and decisions can be made.

There are always groups of people with different opinions and this can make the final decision difficult to make. Coal is one of these resources that is frequently debated in Australia. Different groups will be affected in different ways depending on the choices that are made.

Table 7.2 Different points of view about mining and using coal

Group	Point of view
Mining company	Wants to extract as much coal as possible, spending as little as possible to maximise profits.
Energy company	Wants to purchase coal cheaply to burn to make electricity.
Coal industry workers	Wants the industry to continue to provide them with work.
Environmentally conscious people	Want the mining and burning of coal to stop so that carbon dioxide emissions are greatly reduced.
Renewable energy companies	Want governments to support renewable energy technologies so that renewable energy becomes cheaper than fossil fuels.
People living near mines and power plants	Are not happy with the damage to the local environment and have concerns that pollution is harming their health.
Scientists	Have undertaken many studies showing that the burning of fossil fuels such as coal is causing climate change.

Why can making decisions about using resources be difficult?

CHECKPOINT 7.5 

1. Describe how obtaining and using a resource can be of benefit to society.
2. Describe how the natural environment can be affected by obtaining or using resources.
3. What gas is released into the atmosphere when a fossil fuel is burnt? Describe the environmental impact of this gas.
4. What are mines required to do to reduce their impact on the environment?
5. Describe how farms can reduce their impact on the environment.
6. Stopping coal mining in Australia would reduce a significant portion of income coming into the country's economy, but would stop the considerable environmental damage from the mining and use of the coal. Does the benefit of conserving the environment outweigh the cost to the economy? Use evidence to support your response.

CHALLENGE

7. Research your local energy supplier. Find out where your energy is sourced from. What percentage is renewable and what percentage is non-renewable? Does your household have the option to purchase renewable energy from this supplier?

SKILLS CHECK

- I can state the impact of obtaining and using resources on the environment and the economy.
- I can describe some different points of view people may have on the use of non-renewable resources in Australia.

CHAPTER SUMMARY

Non-renewable resource
a resource that can run out, or one that takes longer than a human life span to be restored

Fossil fuels are non-renewable resources
- coal
- oil
- gas.

Renewable resource
a resource that cannot run out, or can be restored in a human life span

The biosphere has many renewable resources
- wind
- water
- sun.

Recycling is one way to conserve and manage non-renewable resources.

Using resources sustainably means the demand for resources needs to be balanced with maintaining the environment.

Key stakeholders
- government
- local communities
- mining companies
- environmentalists
- clean energy providers
- technology companies

Biofuels ▶
resources made from plant and animal waste

★ FINAL CHALLENGE ★

1. What is the difference between a renewable and a non-renewable resource?
2. Explain what fossil fuels are in your own words and give three examples.
3. List one resource that comes from each of the biosphere, lithosphere, atmosphere and hydrosphere.

4. Match the resource with some of its uses

Oil	Drinking, habitats, electricity generation
Metal	Building materials, paper products, food
Trees	Fuel products, plastics, cosmetics
Water	Fuel, electricity generation
Coal	Clothing products, food, transport
Animals	Building materials, electrical wiring, technology

5. Explain why soil is considered to be a non-renewable resource.
6. Identify and describe the uses of three different resources obtained from rocks.
7. Explain what is meant by the sustainable use of resources.
8. A wind turbine's output varies over 6 months. Create a column graph using this data.

Month	Jan	Feb	Mar	April	May	June
Percentage of possible output	80%	70%	40%	60%	40%	50%

9. 'Reduce, Reuse and Recycle' is a common slogan. Use your knowledge of resources to explain why it is important.
10. Identify two non-renewable resources and suggest a renewable alternative for each of them.

11. There are limited amounts of fossil fuels, and if their use continues, we will run out of them in less than 100 years. Discuss the impact this will have on the resources that we have available.

8 WATER AS A RESOURCE

About 70% of Earth's surface is covered with water, and the oceans contain more than 96% of all the water on Earth. Where is the rest? In lakes, rivers, ice, underground, in the air ... even in living things. Scientists call this the hydrosphere (Earth's water sphere).

The water cycle model is used to explain how water constantly moves through all of these places. Only about 2.5% of all of the water on Earth is fresh water, and even less is located where humans can easily access it. Hydrologists are scientists who study the water cycle and help make sure that we use this precious resource sustainably.

1 LEARNING LINKS

What do you already know about water as a resource?

How many ways do you use water everyday? List as many ways as you can.

What action do you take at home or school to use water sustainably?

How has scientific knowledge and skill been used to manage water supplies in Australia?

How has human use of water changed an environment near your home or school?

2 SEE-KNOW-WONDER

List three things you can **see**, three things you **know** and three things you **wonder** about this image.

3 CRITICAL + CREATIVE THINKING

Brainstorm: Brainstorm a list of solutions to the problem of 'running out of fresh water in Australia'.

Music smart: Write a jingle to encourage people to reduce their water use. Use the tune of a song you know already.

Variations: How many ways can you source fresh water?

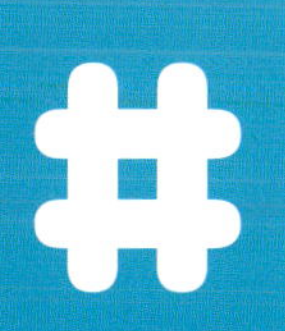

4 THE DRIEST!

Deserts are regions that receive on average of less than 200 mm of precipitation (rain, hail, sleet or snow) per year. One of the driest places on Earth are the McMurdo Dry Valleys in Antarctica. These valleys are unlike other places in Antarctica, because weather patterns prevent moisture laden air from passing over the valleys. Scientists think these valleys might be the closest thing on Earth to the planet Mars, and so they are being investigated to help the search for extraterrestrial life.

8.1 WATER IN THE WORLD

At the end of this lesson I will be able to:

- **identify** that water is an important resource that cycles through the environment.

Figure 8.1 Earth may be called the blue planet, but only about 2.5% of its water is fresh water, and most of that is frozen or underground.

Most of the water on Earth is salt water. Only a very small percentage is fresh water, and even less of this water is easily accessible by humans and other living things. Fresh water is an important resource – we use it to drink, to grow our food, and in industry. Water is also a very important part of living systems.

KEY TERMS

adhesion
the 'sticking' of molecules to other substances

cohesion
the 'sticking' of molecules to each other

ecosystem
a system of living things and their environment

molecule
a group of atoms chemically bonded together

LITERACY LINK

Use your existing knowledge, as well as any research you need to do, to describe the hydrosphere. Include an annotated diagram.

NUMERACY LINK

The human body is about 65% water.

Draw a pie chart to demonstrate how much 65% is.

1 The chemical formula for water is H_2O

Water is the only substance found naturally on Earth in the solid, liquid and gas states. A water **molecule** has the chemical formula H_2O. This means it is a compound made up of two hydrogen atoms and one oxygen atom, chemically bonded together.

The shape of the water molecule gives it special properties that are important for life. Water molecules are attracted to each other and will 'stick' together – this is called **cohesion**. Water molecules can also 'stick' to other substances – this is called **adhesion**. These properties help water travel from the roots of trees up to the leaves, and through the human body.

What is the chemical formula for water?

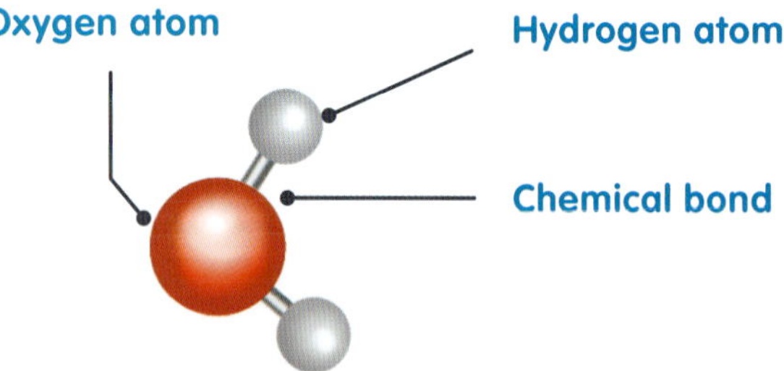

Figure 8.2 A water molecule is made up of two hydrogen atoms bonded to an oxygen atom.

2 The human body is made mostly of water

The human body is about 65% water. Water can be found throughout the tissues and cells of your body, as well as in liquids such as blood.

The water inside your body is important for giving your cells structure. It allows all of the important chemical reactions that keep you alive to take place. Humans can only survive about three days without fresh water.

What percentage of the human body is water?

INVESTIGATION 8.1A
Investigating cohesion of water

INVESTIGATION 8.1B
Observing capillary action

❸ Water is vital to agriculture and ecosystems

Fresh water is an important agricultural resource. It's used to water crops and to support livestock. Without enough water supplies, farmers are unable to produce enough food, or other resources such as cotton and wool.

Fresh water is also an essential part of natural **ecosystems** – without it, they wouldn't exist. Not only does water allow plants to grow, but the collection of rain water in waterways and wetlands creates habitats for other organisms. Even deserts rely on water, which may come from occasional rains or underground reservoirs.

How is water an important part of an ecosystem?

Figure 8.3 Wetlands are important for many of the functions of an ecosystem. They allow water to be naturally cleaned and stored.

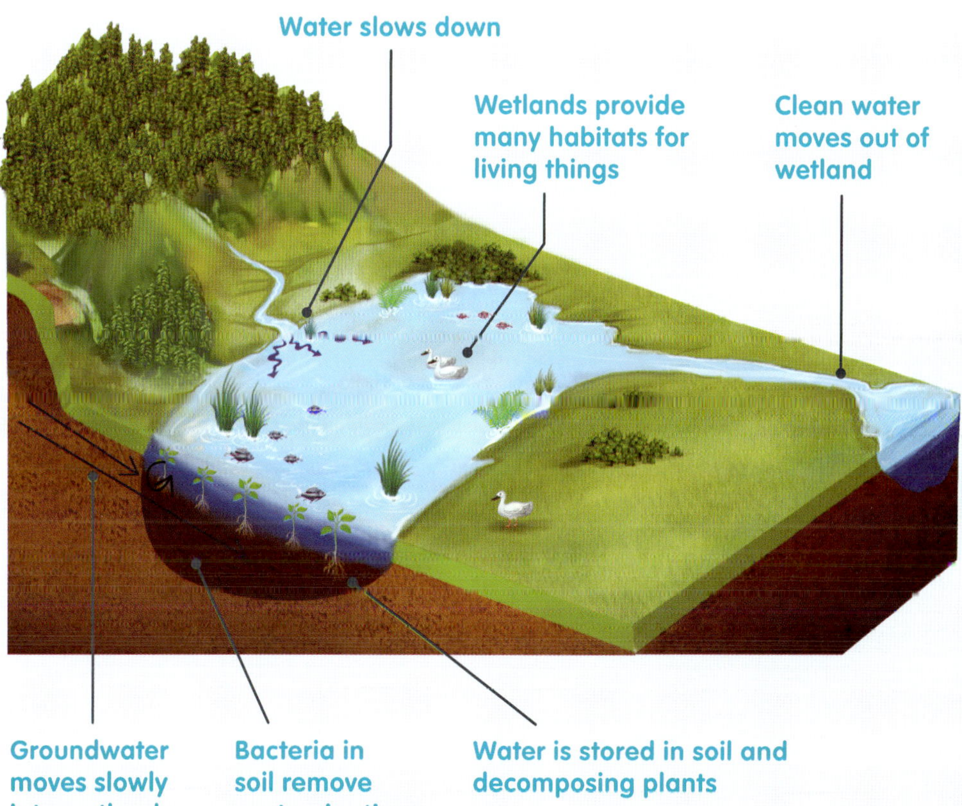

CHECKPOINT 8.1 

1 Identify the chemical formula for water.

2 Why is water important for farmers?

3 What is the difference between cohesion and adhesion?

4 Provide an example of how water can be found on Earth in each state of matter.

5 Water is very important for living things in ecosystems. Suggest why.

6 Using Figure 8.3, predict what would happen to a wetland in a drought.

7 Suggest the impact on living things and ecosystems in the case of:
a a drought
b a flood.

CHALLENGE

8 Undertake some research to find out what crops or livestock are grown by farmers in your region or a nearby region. How much water is required to grow them? Where do the farmers get their water from?

SKILLS CHECK

- I can explain why water is an important resource.
- I can describe various uses of water.

8.2 THE WATER CYCLE

At the end of this lesson I will be able to:

- **explain** the water cycle in terms of the physical processes involved.

KEY TERMS 

condensation
changing from a gas to a liquid

evaporation
changing from a liquid to a gas

groundwater
water within soil and rocks in the ground

precipitation
water falling from clouds as a solid (e.g. hail) or liquid (e.g. rain)

transpiration
how water in the soil travels through plants

water vapour
the gas state of water

LITERACY LINK 

Summarise the water cycle in a postcard to your teacher. Include all the most important aspects.

NUMERACY LINK 

Rainwater fills a tank at a rate of 3 litres per hour. If the tank holds 120 litres, how long will it take to fill completely?

Every molecule of water on Earth cycles through the atmosphere, rocks of the crust and living things as part of the water cycle.

The water cycle is driven by energy from the Sun, which enables liquid water to evaporate in the atmosphere. Once in the atmosphere, the wind moves it around the planet. It then condenses back to water and falls again.

1 Liquid water evaporates into the atmosphere

Water in oceans, lakes, rivers and on the surface of the land gains energy from the Sun. This energy causes the liquid water to **evaporate** and move into the atmosphere as **water vapour**.

Plants can also help in the process of evaporation through a process called **transpiration**. Liquid water moves from the soil into plant roots up to the leaves. The water in the leaves then evaporates.

How does water in the ocean move into the atmosphere?

2 Water vapour condenses into clouds

As water vapour rises in the atmosphere it starts to cool. As it cools it **condenses** around tiny solid particles, such as ash and dust that are small enough to stay up in the atmosphere. These form liquid water droplets. These gather together to form clouds. Movement of air in the atmosphere moves the clouds around.

What happens to water vapour in the atmosphere?

3 Water falls to Earth as precipitation

When liquid water droplets in clouds become too heavy to stay in the atmosphere, they fall to Earth as **precipitation**. Rain is when the water falls as a liquid. If the atmosphere is cold enough, water can fall as a solid such as snow or hail.

What is precipitation?

INVESTIGATION 8.2
Modelling the water cycle

❹ Water runs over or into the land

When liquid water reaches the land, it runs over the surface to waterways, or it moves through the soil. If it isn't absorbed by plants, it keeps moving through the soil to the rock below and becomes part of the water underground.

Groundwater is contained within soil and in cracks and pores in rocks underground. Groundwater can slowly move back to the surface through springs, or make its way into oceans and lakes.

What is groundwater?

Figure 8.4 During the water cycle, water moves through the environment and it also changes between solid, liquid and gas.

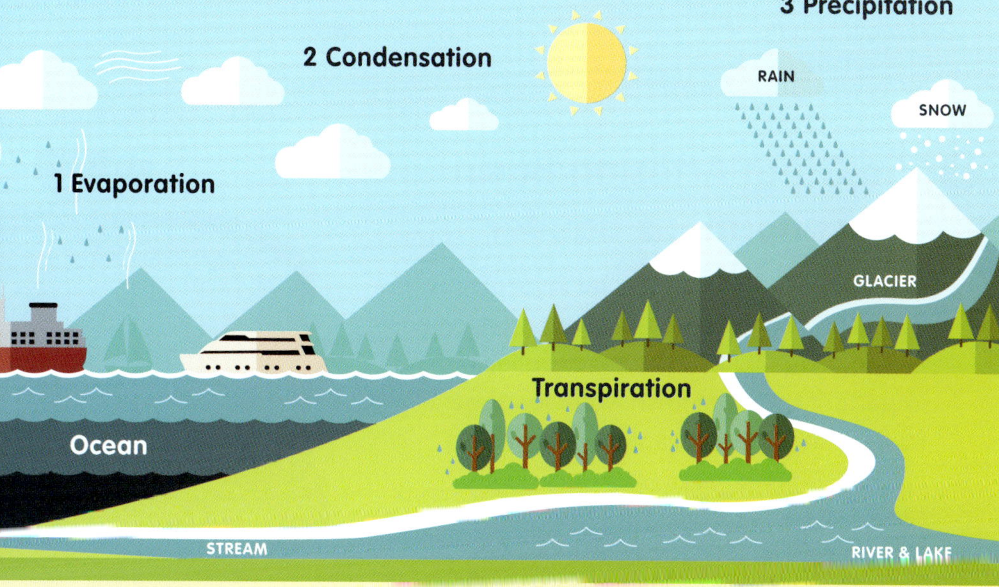

❺ Ice melts slowly over time

Water can be stored for long periods of time as ice. This ice can be found in glaciers on mountains and ice sheets at Earth's poles. It can also be found in frozen soil called permafrost. The ice melts slowly over time, adding liquid water to waterways or groundwater.

How does ice affect the water cycle?

CHECKPOINT 8.2 

1 Identify where in the water cycle these events take place.
   a Water changes from a liquid to a gas.
   b Water changes from a gas to a liquid.
   c Water changes from a liquid to a solid.
   d Water changes from a solid to a liquid.

2 In what state is most of the water in clouds?

3 What is the difference between evaporation and transpiration?

4 Would you expect to find a higher rate of transpiration in a desert or a rainforest? Justify your response.

5 Would you expect to have a higher rate of evaporation on a cloudy day or a sunny day? Justify your response.

6 Describe what could happen to rainwater that falls onto a grassy hill.

CHALLENGE

7 Dams are often built to provide water sources for major towns and cities. Consider how the construction of a dam may alter the water cycle in a local area. Draw diagrams to illustrate this change.

SKILLS CHECK

- I can identify when water changes state in the water cycle.
- I can describe how water moves through the water cycle.

8.3 MANAGING WATER USE

At the end of this lesson I will be able to:

- **demonstrate** how scientific knowledge of the water cycle has influenced the development of household, industrial and agricultural water management practices.

KEY TERMS 

aquifer
a layer of rock that contains water

bore
a pipe or well dug into an aquifer to access water

runoff
water that runs off the surface of the land into waterways such as rivers

LITERACY LINK

Write a poem that could be used to teach young children about using water in a sustainable way.

NUMERACY LINK 

If there is a linear relationship between temperature and evaporation, what happens to the rate of evaporation when the temperature increases?

Fresh water is a precious resource that must be managed and used sustainably.

Scientists use their knowledge of the water cycle and ecosystems to find the best ways to manage water supplies and to limit the impact of human activities on the water cycle.

1 Dams supply towns and farms with water

A dam is a wall built across a waterway to allow water to collect behind it. Dams are built to store water to supply water to cities, towns and farms.

To work out the best place to build a dam, scientists and engineers calculate the amount of **runoff** in the area. The dam is placed at the spot in the waterway where the runoff will fill it most effectively.

A dam on a river limits the natural flow of water downstream. Water is often released from dams to make sure that the downstream waterways remain healthy.

Why are dams built?

2 The Great Artesian Basin is an enormous aquifer

An **aquifer** is an underground layer of rock that contains groundwater. The Great Artesian Basin is a huge aquifer in eastern Australia, and an important source of fresh water for Queensland, New South Wales and South Australia.

Figure 8.5 The Great Artesian Basin, shaded on this map, spreads over almost a third of Australia.

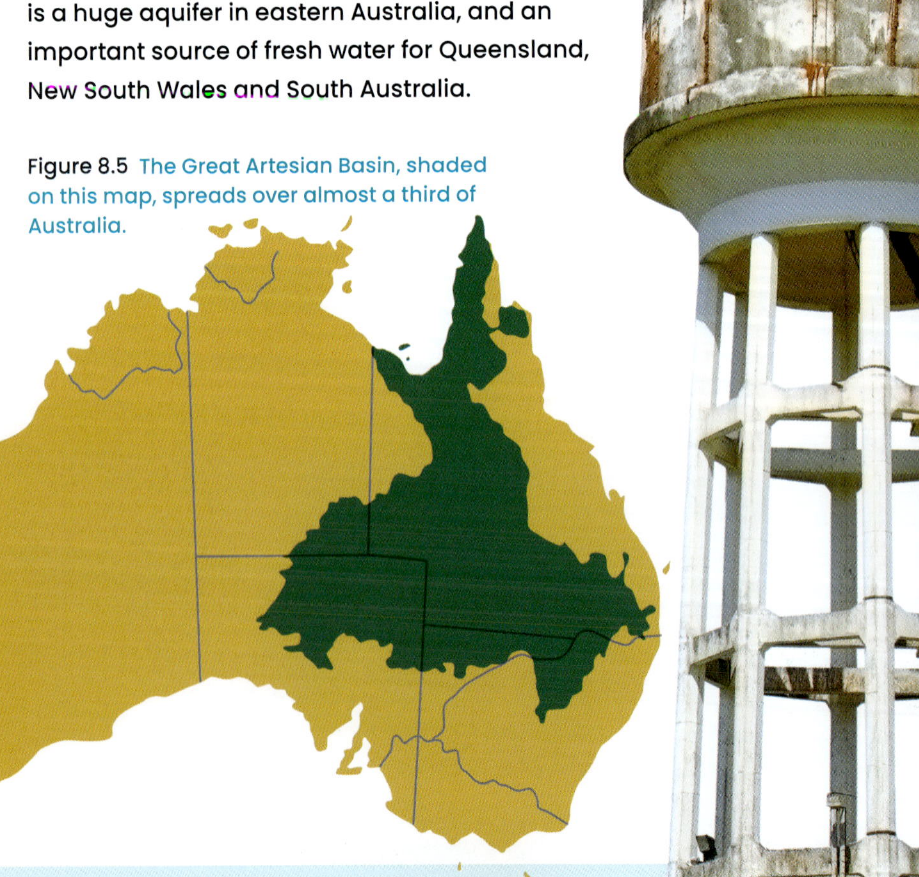

Farmers and remote communities use **bores** to access this water for humans, livestock and crops.

Over the past 50 years, some communities have noticed that their bores no longer work, and that natural springs have dried up. Scientists worked out that the amount of water flowing into the Great Artesian Basin from the water cycle is less than the amount of water being removed.

Governments now limit how much water can be removed from the Great Artesian Basin, to allow the water levels to increase again. Some natural springs that once ran dry have refilled again.

What is an aquifer?

3 Stormwater systems link cities to natural waterways

Many materials, such as concrete and road bitumen, prevent rainwater from soaking into the ground. This water runs off into stormwater systems, which eventually link to natural waterways such as rivers and the ocean.

When stormwater runs over hard surfaces, it can pick up pollutants and may take these into natural waterways. To prevent this, some city councils now make wetlands in stormwater systems. These constructed wetlands allow natural processes to take place, including the removal of pollutants, evaporation and the formation of new wetland ecosystems. This means less pollutants are deposited into natural waterways.

Scientists regularly test the quality of water in urban areas. This helps them to identify sources of pollution and to make sure natural waterways remain healthy.

What happens to rainwater that falls in urban areas?

Figure 8.8 Urban wetlands, like these in Sydney Park, help prevent pollutants from entering natural waterways.

INVESTIGATION 8.3
Testing water quality

CHECKPOINT 8.3

1 Suggest how a dam may impact the river system downstream from it.

2 Describe what the Great Artesian Basin is and how it is used as a water source.

3 What is stormwater?

4 Where does the water in stormwater systems flow to?

5 Explain why it is important to know about how the water cycle operates in the local area before a dam is constructed.

6 Identify what aspects of the water cycle influence the amount of groundwater in aquifers. Propose how these might affect a bore or spring drying up.

7 How do hard surfaces in urban areas affect the water cycle?

CHALLENGE

8 Investigate the organisation that manages the water supply in your local area. What is the storage capacity, and what size of population does the supply support? What are the current storage levels? What can be done if the storage levels drop?

SKILLS CHECK

- I can describe three different ways water is managed in Australia.
- I can demonstrate how knowledge about the water cycle has informed these water management practices.

8.4 INDIGENOUS AUSTRALIAN WATER MANAGEMENT

At the end of this lesson I will be able to:

- **research** how Aboriginal and Torres Strait Islander peoples' knowledge is being used in decisions to care for Country and place, with regards to aquatic resource management.

KEY TERMS 

collaboration
working cooperatively together

cultural significance
importance to a particular culture

LITERACY LINK 

Find out more about one of these Dreamtime stories.

- Tiddalik the Frog
- the Rainbow Serpent
- Baiame and the Fish Traps

Summarise the story and identify how the importance of managing water resources is included in the story.

NUMERACY LINK 

A fish trap on the Barwon River contains 7 bream and 3 mulloway. If you pick a fish out of the trap at random, what is the probability that it is a bream?

For tens of thousands of years, Australian Aboriginal and Torres Strait Islander peoples have used their knowledge of water sources to survive in times of drought, during travel and to find food. Water is also an important part of Aboriginal and Torres Strait Islander cultures and traditions.

Governments and science organisations are starting to learn how Indigenous knowledge can improve the way they manage water.

1 Observing the land can lead you to water

Imagine you're standing in a desert, with no visible water. If you look carefully, you might see a group of ghost gums in the distance, tracks from different animals all leading in the same direction, or a trail of ants coming from one direction and suddenly disappearing into the sand. These are all signs that water is nearby.

Indigenous Australians have closely observed their environment over many thousands of years, including working out ways to find water sources.

What are some clues that could suggest a water source nearby?

2 Water is a resource and is part of Indigenous culture

Water and waterways are highly valued by Indigenous Australians. They provide drinking water, habitats for food plants and animals, and can also be places of **cultural significance**. For these reasons, part of Indigenous Australian culture and tradition is careful use of water.

Figure 8.7 These fish traps on the Barwon River in Victoria may be thousands of years old. They were made and used by the Ngunnhu people to harvest fish and eels.

Indigenous Australians have used their knowledge of the seasons and weather observations to predict when the rainy season will arrive and end, so that they could prepare. Traditionally, this might mean moving to a different place, or being ready to harvest plants and animals that flourish after the seasonal rains. They covered wells that they dug into aquifers to prevent water evaporating or being polluted by animals.

Indigenous Australian cultural knowledge about water has been passed on through many generations in stories, songs, dance and art. Many places related to water have been and are still protected.

Why is water important in Indigenous Australian culture and tradition?

3 Scientists are collaborating with Indigenous communities

Scientists from universities and other organisations have partnered with Indigenous Australian scientists, rangers and communities to share knowledge. This **collaboration** helps to make sure that water resources are valued, managed and maintained for the future.

Indigenous and non-Indigenous scientists gather data about water quality and meet with local Indigenous Australians who are skilled at making detailed observations. These may be of changes to the landscape, waterways and wildlife each season or over several years.

By using all of the information, scientists and others can predict how human activities may affect water supplies, ecosystems and culture. They can then propose ways to prevent or minimise harmful effects.

How is information from Indigenous Australians being used to manage waterways?

Figure 8.8 Indigenous scientists and rangers collaborate with universities and other organisations to gather and understand environmental data.

CHECKPOINT 8.4 

1 Identify three observations that you could use in a desert to find hidden water.

2 How have Indigenous Australians protected wells dug into aquifers?

3 Why is it traditionally important for Indigenous Australians to be able to predict when rain would arrive?

4 What types of observation can support scientific data on water quality?

5 Explain why ghost gums are an important indicator that there is water in the desert. Where is this water located?

6 Explain why linking observations of landscapes, waterways and wildlife to water quality data is important for managing Australian water resources.

CHALLENGE

7 The D'harawal people from the Sydney and Shoalhaven region traditionally recognise six seasons throughout the year. They have related observations of their environment to weather patterns and resource availability.

Use the internet to research the six seasons and their features.

8 Internet research is an important skill. As a class, brainstorm a set of techniques you can use to ensure a website provides reliable, scientifically accurate and safe information.

SKILLS CHECK

- I can describe at least two water management practices carried out by Aboriginal and Torres Strait Islander peoples.

CHAPTER SUMMARY

1 Evaporation

2 Condensation

3 Precipitation

RAIN

SNOW

GLACIER

Transpiration

Ocean

STREAM

RIVER & LAKE

4 Groundwater

The chemical formula for water is

H_2O

Oxygen

Hydrogen

3% Fresh water

97% Salt water in oceans

69% Locked away in glaciers

30% Underground

1% Easily available for human use

▲ The Great Artesian Basin is an important source of fresh water for Queensland, New South Wales and South Australia

Indigenous and non-indigenous scientists, rangers and community leaders collaborate to share water management strategies. ▶

The careful management and use of water is a community and scientific priority.

Collaboration is key to successful water management.

★ FINAL CHALLENGE ★

1. State the chemical formula for water.
2. Identify the physical changes that water goes through as part of the water cycle.
3. Identify why water is important:
   a. for the human body
   b. for a farmer
   c. for an ecosystem.

LEVEL 1
50xp
LEVEL UP!

4. Create a flow chart to show how the major processes of the water cycle link together.
5. Explain what stormwater is in your own words.
6. How is Aboriginal and Torres Strait Islander peoples' traditional knowledge being used in decisions about water management?

7. Explain why it is important to manage the water in aquifers, such as the Great Artesian Basin.
8. Explain why it would be beneficial to include information from local Indigenous people in a study of rainfall and river flows in your local area.

9. Identify and explain three ways that human action can have a negative impact on the water cycle.
10. Identify and explain a possible negative consequence of stormwater running directly into natural waterways.

11. Design a water audit at your home or your school using these steps:
    a. Find out how much water your home or school uses in total over a year or a month.
    b. Identify how this water is used in your home/school.
    c. Propose ways that the water use could be decreased or used more sustainably.

FORCES

Have you ever watched Formula One racing? The cars fly around the track, sometimes at over 200 kilometres per hour (km/h).

Every aspect of each car is designed to keep them safely on the track, while not slowing them down too much. Their shape is low and streamlined so that the air flows over them. Their tyres propel the car forward but also provide enough friction to stop it sliding off the track. The hard metal cage of the inside of the car absorbs the force of the car in a crash, helping protect the driver.

1 LEARNING LINKS

What do you already know about forces?

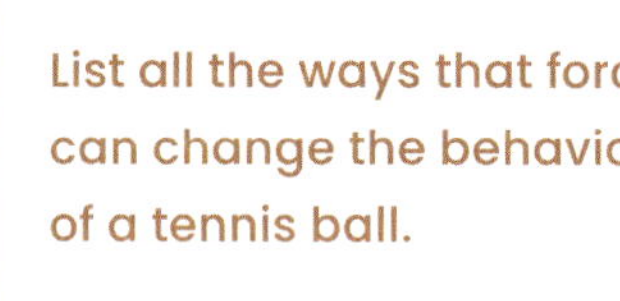

List all the ways that forces can change the behaviour of a tennis ball.

How does gravity affect you in your daily life?

How many types of forces can you think of?

2 SEE-KNOW-WONDER

List three things you can **see**, three things you **know** and three things you **wonder** about this image.

3 CRITICAL + CREATIVE THINKING

What if ... The wheel had never been invented?

B-A-R: Consider an everyday family car. What would you make *bigger*, what would you *add* and what would you *replace*?

Ridiculous! Justify this statement: 'All speed limits should be changed to be speed minimums'.

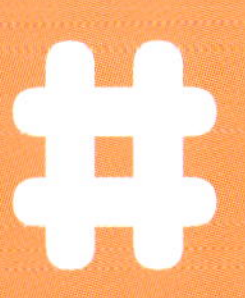

4 THE FASTEST!

In 2009, athlete Usain Bolt officially became the fastest person on Earth. He still holds the record for the fastest human speed – 44.72 km/h – which he ran at the World Championships in Berlin.

Bolt was born in a small town in Jamaica. As a child, all he thought about was sport, and he played cricket and soccer before being identified as an extremely fast sprinter.

Bolt went on to win nine Olympic gold medals and break many world records – even his own. All of this is even more incredible considering he has a medical condition that has made one of his legs more than a centimetre shorter than the other.

9.1 FORCES AT WORK

At the end of this lesson I will be able to:

- **identify** changes that take place when particular forces are acting.

KEY TERMS 

field
an area of influence that exerts a force on an object

force
a push, pull or twist on an object when it interacts with another object

friction
a force opposite to the motion of surfaces in contact

interact
act on each other

motion
the change in position of an object over time

LITERACY LINK 

Create a story including a scenario where no (or nearly zero) forces are acting upon an object.

NUMERACY LINK 

An unbalanced object is moving at 6 metres per second. If its speed triples, what is its new speed?

Every object on Earth is always being acted upon by **forces**, such as gravity and **friction**. When all the forces on an object are balanced, the object will continue doing whatever it's been doing, whether that's remaining still or continuing to move.

You can look at the forces on an object to predict what will change about the object or its **motion**.

1 Forces happen when objects interact

Forces happen when objects **interact** with each other or with an area of influence called a **field**. The interaction could make the objects move, stop moving, change direction or change shape.

A ball sitting out on an oval won't move until it's forced to. If you kick the ball, several things happen. The ball changes shape for a moment, because the kick has applied a force to the ball. The ball moves in the direction of the force from the kick. As it moves, it gets its shape back. Finally, it hits the ground and stops, rolls or bounces.

Each of these events is an applied force – a force on one object by a person or another object. You can describe each effect as pushing, pulling or twisting an object. The force from the kick, for example, is a push.

What are the three possible effects of a force?

2 A stationary object is affected by balanced forces

Even objects that are stationary (not moving) have forces acting on them. Take a book and put it on a table. The book doesn't move, but it's still interacting with the table, and that means forces are acting on it.

Figure 9.1 The forces acting on the book are equal but opposite, so they balance out.

Force of table on book (normal force)

Force of gravity

What would happen if the table suddenly vanished? The book would fall to the floor due to gravity, the force on all objects in Earth's gravitational field.

The table applies a force on the book, which is called the normal force. The normal force is the same size as the force of gravity but it acts in the opposite direction, preventing the book from falling. The forces acting on the book are balanced, so it remains stationary.

What are balanced forces?

3 A moving object might also be affected by balanced forces

If a stationary object is affected by balanced forces, does that mean the forces affecting a moving object are not balanced? Not always.

When forces on an object are balanced, the object keeps doing what it's been doing. If a moving object keeps going in the same direction and at the same speed, then all the forces acting on it are balanced. When a car moves at a constant speed, the force from the car's engine is balanced by forces acting in the opposite direction, such as friction on the tyres.

Sometimes these forces are almost balanced, such as when an ice hockey puck slides across an ice rink. After the person slides the puck, they can't apply any more force. The surface of the rink is slippery, so the puck keeps sliding at *almost* the same speed until it hits something.

How can objects in motion have balanced forces?

Figure 9.2 An ice hockey puck slides across the slippery surface of an ice rink. It moves, but the forces on it are balanced.

INVESTIGATION 9.1
Push, pull or twist

CHECKPOINT 9.1

1 Explain what forces are, using an example.
2 You can apply a force in three ways: a push, a pull and a twist. Give an everyday example of each.
3 Describe what it means for the forces on an object to be balanced.
4 Describe two forces that could be acting on a stationary object.
5 Using your understanding of balanced forces, suggest the characteristics of unbalanced forces.
6 A motorbike is travelling down the road at a constant speed of 100 km/h. Are the forces on the motorbike balanced? Explain your answer.

CHALLENGE

7 A soccer ball has a rubber bladder inside and a synthetic outer covering. These prevent it from getting waterlogged in rainy conditions, which would change the way the ball reacted to the force of a kick. Identify another piece of sporting equipment that has been designed to optimise the forces applied to it during use, and explain how it achieves this.

SKILLS CHECK

- I can define forces.
- I can explain what balanced forces are, using an example.

9.2 UNBALANCED FORCES

At the end of this lesson I will be able to:

- **predict** the effect of unbalanced forces acting in everyday situations.

KEY TERMS 

force
a push, pull or twist on an object when it interacts with another object

friction
a force opposite to the motion of surfaces in contact

motion
the change in position of an object over time

LITERACY LINK 

Use these words to create a mind map: *force, balanced, unbalanced, motion, opposite, gravity*. Link the words using lines and write an explanation for each link along each line.

NUMERACY LINK 

An object sits at position (1,1) on a graph, but then unbalanced forces translate it 4 units down and 2 units to the right. What is its new position?

Every day, you push, pull and twist objects to change the way they move. When this happens, the **forces** on these objects become unbalanced.

The effect of unbalanced forces can be seen anytime an object starts or stops moving, speeds up or slows down, or changes direction.

1 Adding or removing a force can unbalance the forces on an object

There are several ways that the forces acting on objects, including us, become unbalanced. Here are two examples.

A force can be added.

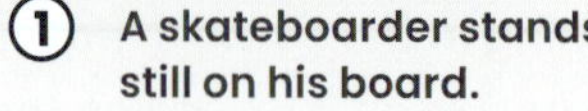

① A skateboarder stands still on his board.

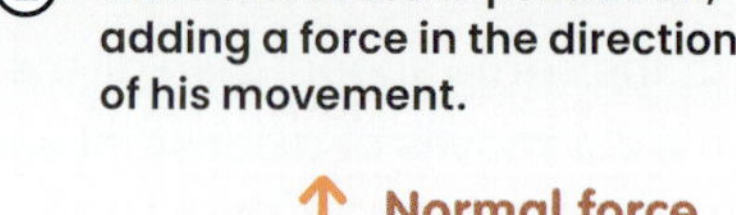

② The skateboarder pushes off, adding a force in the direction of his movement.

A force can be removed.

① A diver stands on a diving board.

② The diver jumps off the board, removing the normal force opposing the force of gravity.

What can cause forces to become unbalanced?

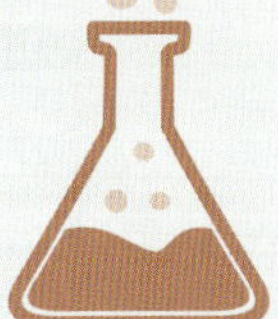

❷ We can predict how forces will affect motion

When the forces acting on an object are unbalanced, that object's motion will change. If you look at how the forces are unbalanced, you can describe how the **motion** of the object will change.

It's easy to predict how a stationary object such as a basketball will move when you apply a force to it, such as by throwing. A force acting in the opposite direction to an object's movement is also fairly easy to predict. For example, a rolling ball on a flat floor will slow down and eventually stop due to **friction**.

How can you explain the change in motion of an object?

INVESTIGATION 9.2
Blowball

❸ Forces can be added up to work out their effect

Adding up the forces on a ball isn't necessary to predict the direction it will go. But when many forces act on an object, adding them up is the best way of working out how it will behave.

If you push a ball away from you to pass it, you're applying a third force to the ball. The forces acting on it become unbalanced, and the ball moves away from you.

In what direction does the ball move? You can work this out by adding up the forces acting on the ball. The force of gravity and the force from your hands holding the ball cancel each other out, leaving only the force from the push.

What happens when you unbalance the forces on an object?

Figure 9.3 The applied force of a throw unbalances the forces acting on a ball.

CHECKPOINT 9.2

1 What are the possible results of unbalanced forces acting on an object?

2 What are the three ways that unbalanced forces can change an object's motion?

3 What are two ways that the overall forces acting on an object can become unbalanced?

4 An object with a force acting on it in the opposite direction to its motion will slow down – suggest why.

5 If a person is sitting on a bike at the top of a hill, then slowly rolls down the hill without pedalling, describe the forces acting on the bike and why the rider doesn't need to pedal.

6 Explain how you think the size of the force applied to an object affects its motion.

CHALLENGE

7 Rockets require an extremely large unbalanced force in order to be launched into space. Research how this force is created.

SKILLS CHECK

- I can explain what unbalanced forces are.
- I can draw a simple force diagram with arrows showing unbalanced forces.

9.3 THE SIZE AND DIRECTION OF FORCES

At the end of this lesson I will be able to:

- **identify** characteristics of specific forces in terms of size and direction.

KEY TERMS 

magnitude
the size or power of an object, energy or force

net force
the sum of all forces acting on an object

LITERACY LINK 

Outline a situation in which you have had to think about both the size and direction of a force that you've applied to an object.

NUMERACY LINK 

The formula to convert kilograms to Newtons is:

1 kg = 9.807 N

Convert these to Newtons:

a 1 kg
b 5 kg
c 120 kg

NUMERACY LINK 

A person pushes a trolley with a force of 500 N to the right. A second person pushes the trolley in the opposite direction with a force of 300 N. Which direction will the trolley move and with what force?

Every moment of the day, you're either applying forces to objects or having forces applied to you.

Each force has two key characteristics. The direction of the force affects the direction that an object moves, or whether it slows down or speeds up. The size, or **magnitude**, of the force affects how much the object might move, or how quickly it speeds up or slows down.

1 Every force acts in a certain direction

A force is shown in a diagram as an arrow that starts at the centre of the object and points outwards. The direction that the arrow is pointing indicates which way the force is acting on an object.

Imagine you and your family are at the supermarket, and you're handling a shopping trolley. In what directions could you apply force?

1. You could push the trolley. This applies force in a forwards direction and makes the trolley move that way.

2. You could pull the trolley to make it move backwards.

3. You could try swinging the handle to the side to make it spin. The force is directed to the side of the trolley, but the motion is a turn, rather than a sideways push.

4. If you started running while pushing the trolley ahead of you, you could keep pushing on it to make it go faster.

5. The only way to stop the trolley quickly would be to apply a force in the opposite direction.

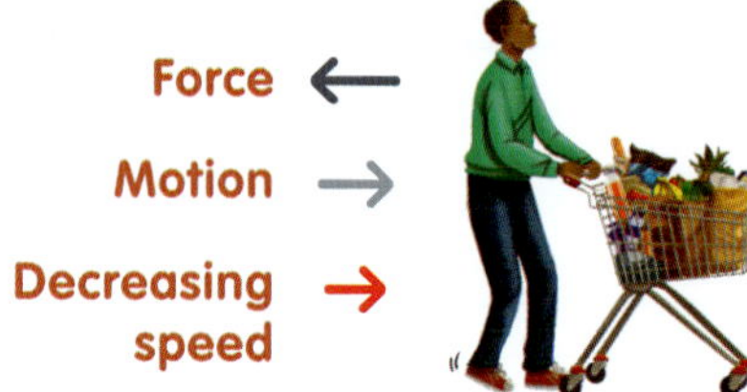

The direction of the applied force determines the direction an object will move. It can also determine if the object speeds up, slows down or changes direction.

How does the direction of a force affect how an object moves?

2 Every force has a magnitude (size)

A force can be strong or weak, great or small. As with direction, the size of the force should be shown when drawing diagrams. A larger force arrow means a larger force. Force is measured in Newtons (N). The **net force** is the sum of all forces acting on an object.

After your family finishes at the supermarket, it's time to head home and make dinner. At home, you need to set the table. How much force should you apply when you shut the cutlery drawer?

① If you slam the cutlery drawer with a large force, it will close quickly.

② The drawer frame applies a force of the same size and opposite direction to stop the drawer. This is what causes the loud slamming noise.

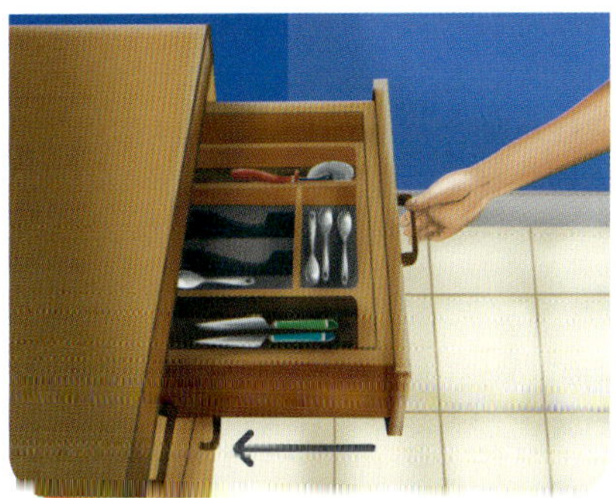

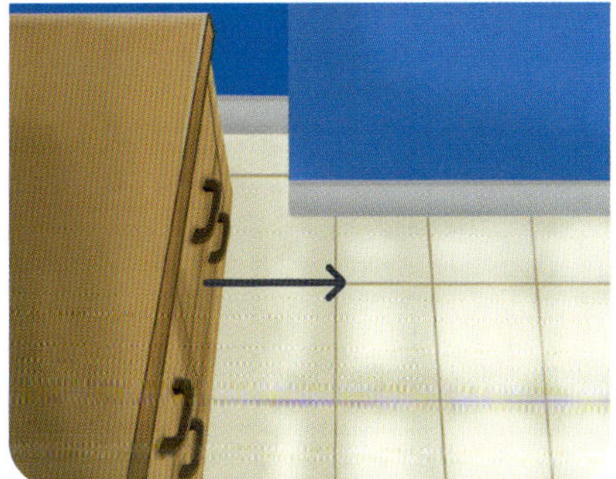

③ If you close the drawer more gently (using a smaller force), it will take longer to close or may not close at all.

④ The drawer frame applies a force of the same size and opposite direction to stop the drawer. Because the force is much smaller, there's no slamming sound.

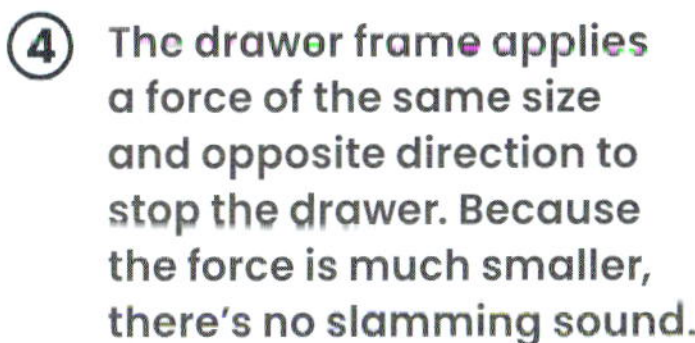

How does the size of a force affect how an object moves?

CHECKPOINT 9.3 

1 What are two key characteristics of forces?

2 What words are used to describe the direction that a force is applied in?

3 What happens differently when you apply a force in the same direction as an object's motion as compared to applying a force in the opposite direction to an object's motion?

4 How can you make a ball go further when throwing it?

5 Draw a simple force diagram showing a ball rolling off a table.

6 Vehicles must be able to both speed up and slow down, and to do so they can apply forces onto the wheels in both directions. What main part of a car applies a force in the forward direction? Which parts of the car apply forces in the opposite direction?

7 Why is it important for vehicles to be able to apply very large and very small forces?

CHALLENGE

8 Whenever you play a sport, make a piece of art or play a musical instrument, you apply a variety of forces during each activity. Choose an example from the list above. Explain how and why you change either the magnitude or direction of the forces you apply to an object (sports equipment, the medium of your art or the musical instrument) during this activity.

SKILLS CHECK

- I can describe the result of a change in the direction of a force.
- I can describe the result of a change in the size of a force.

9.4 REDUCING THE IMPACT OF FORCES

At the end of this lesson I will be able to:

- **describe** some examples of technological developments that have contributed to finding solutions to reduce the impact of forces in everyday life.

KEY TERMS 

deform
change shape

impact
the effect of a force

LITERACY LINK 

Write a brochure for the 'safest car on Earth'. Include information on all its safety features.

Figure 9.4 The soles of walking shoes deform around rocks, spreading out the contact force.

Many activities involve forces that can be uncomfortable, painful or even harmful.

Engineers and scientists have developed products that can reduce the **impact** of these forces. These products decrease the forces we feel by absorbing them or spreading them over a larger area.

1 Shoes can reduce forces on your feet

When was the last time you ran barefoot across dirt or gravel? All it takes is landing on a pointy rock to remember to wear footwear next time!

Most footwear has a sole made out of some combination of rubber, foam or other synthetic material. The design of these soles is meant to:

- provide grip, or traction, on a variety of surfaces
- reduce the force of impact from walking or running on a variety of surfaces.

When you step on a pointy rock wearing shoes, the sole briefly changes its shape – it **deforms** around the rock, spreading the contact force over a greater area.

When you're walking or running, the impact of the ground can be evenly spread out across your foot, if your shoes are designed well and fit you properly. Next time you put on your sports shoes, take a look at how they're designed. You should be able to see how the design minimises the forces you feel when running.

How do shoes reduce the force of objects such as rocks?

2 Crumple zones absorb some impact in car accidents

Vehicles let us travel much faster than we can on our own. This increased speed comes with more risk. Why would you rather bump into a wall while walking than when running? It's because the forces you experience are much greater at higher speeds than at lower ones.

If a modern car crashes into another object, its design means it will deform. This reduces the forces on the driver and any passengers. The front and rear of the car each have a crumple zone.

How do crumple zones protect us? Think of it like running into a wall, but this time you're holding a cardboard box in front of you. The box will be crushed first, absorbing some of the impact before it reaches you.

How does a crumple zone reduce the forces of a collision on people in a car?

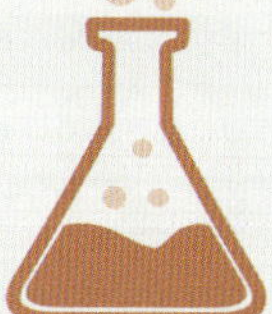

INVESTIGATION 9.4
Crash cushions

Figure 9.5 The front and back sections of modern cars act as crumple zones if there is a collision.

3 Airbags absorb impact for individual car passengers

Almost all modern cars have airbags built into surfaces. Airbags work in a similar way to crumple zones, but instead of being overall protection for people inside the vehicle, they protect the individual passengers in certain ways.

How do airbags reduce the forces of a collision on people in a car?

① The bag inflates upon collision.

② The passenger keeps moving forward until they contact the airbag.

③ The airbag slows the passenger's movement and begins to deflate upon contact.

④ The passenger slows to a stop, but takes longer to do so because of the airbag. This means they endure a force that is less than if there was no airbag.

CHECKPOINT 9.4

1 Other than grip or traction, what else does a shoe's sole provide to the wearer?
2 Why do cars need specially designed safety features for people?
3 How do the sole of a shoe and the crumple zones and airbags of a car all behave in response to forces?
4 Why are straps on a large backpack often several centimetres wide and padded?
5 Bicycle helmets are often made out of a stiff foam. Give two reasons why this material is suitable for its purpose.
6 Explain why a collision involving a cyclist or motorcyclist can be more hazardous to the riders, compared to a collision with only cars involved.

CHALLENGE

7 Design your own wearable safety equipment for cyclists, highlighting its key safety features and how it would increase the safety of a rider during a collision.

SKILLS CHECK

- I can describe some everyday examples of technology that reduce the impact of forces.

9.5 SIMPLE MACHINES

At the end of this lesson I will be able to:

- **investigate** some simple machines, such as levers, pulleys, gears or inclined planes.

KEY TERMS 

fulcrum
the point on which a lever turns when moving an object

inclined
tilted up at an angle from horizontal

lever
a bar acted upon at different points by two forces

pulley
a wheel with a grooved rim for carrying a cable

LITERACY LINK 

Research and provide a written report on how the Ancient Egyptians used simple machines in their culture, including the building of the great pyramids.

NUMERACY LINK 

A pulley lets Graeme exert force at a 60° from the horizontal, rather that pulling straight up. What is the difference between 60° and the vertical?

Early machines built by humans were designed to make work easier. They are simple machines such as **levers**, **pulleys** and **inclined** planes.

Simple machines make work easier by increasing the size of the force on an object, or by decreasing the amount of force needed to move an object.

1 Levers increase the size of a force

Levers come in a range of shapes and sizes, but the essential parts are always the same: a long arm or beam that rests or turns around a **fulcrum**.

When you were younger, you probably played on a seesaw, which is a lever resting on a fulcrum. You might have noticed that the closer to the centre/fulcrum you sat, the easier it was for someone on the other end to move you up and down. This makes levers helpful machines – the closer to the fulcrum that an object is, the less force is needed to move it.

When you were on the seesaw, did you also notice that when you sat near the centre, the height you moved up and down was less? This is the other aspect of a lever – the closer an object is to the fulcrum, the less distance it can be moved.

① If you sit on the end of a seesaw, you need a greater mass or force on the other end to lift you all the way up.

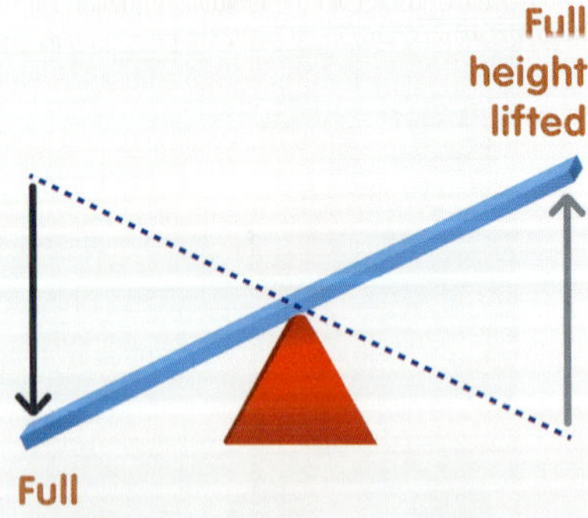

② If you sit halfway towards the fulcrum, you only need half the applied force, but you are only lifted half as high as before.

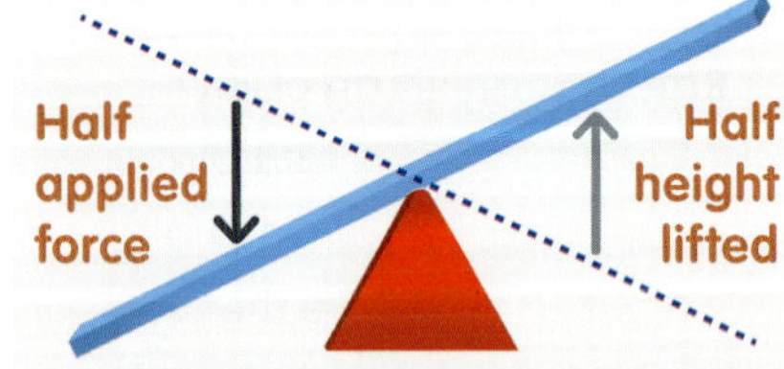

The height that a lever can lift an object depends on the length of the arms on either side of the fulcrum, and on the applied force. If you use a claw hammer to pull out a nail, the handle moves a lot further down than the nail moves up, but less force is needed to pull out the nail.

When using a lever, what affects the height an object is lifted and the force needed?

INVESTIGATION 9.5A
Investigating levers

INVESTIGATION 9.5B
Investigating pulleys

2 Pulleys redirect a force being applied

A pulley is another simple machine. It's a wheel on an axle (which allows the wheel to rotate), with a rope or wire resting on the rim of the wheel so it can be pulled back and forth.

Similar to levers, pulleys work by reducing the force required to lift an object. When they do, they also reduce the height that an object is lifted.

① When using one pulley, the force to lift an object is equal to the force due to gravity on that object. For every metre you move when pulling the object up, it moves up by a metre.

② When using two pulleys, the force to lift an object is half the amount of force due to gravity on that object. For every metre you move when pulling the object up, it only moves up by half a metre.

In the same way as a lever, and like the seesaw opposite, while the force required to lift an object using two pulleys is halved, the distance it moves is also halved. The more pulleys you use, the less force you need but the less distance the object will move. For example, three pulleys only require a third of the original force, but only move the object a third of the distance.

What percentage of force could lift an object with five pulleys, compared to one pulley?

3 Inclined planes change the direction of force

Rather than lift an object directly up, you could slide it up a ramp, also known as an inclined plane. When you do this, the force needed to move the object up the ramp is much less than if you pick the object straight up. This is because the force you are working against isn't the full force going straight down, but a reduced force at an angle along the surface of the inclined plane.

How does a ramp help walkers if it has to be longer?

① More force is required for steeper ramps, but there is less distance to get to the required height.

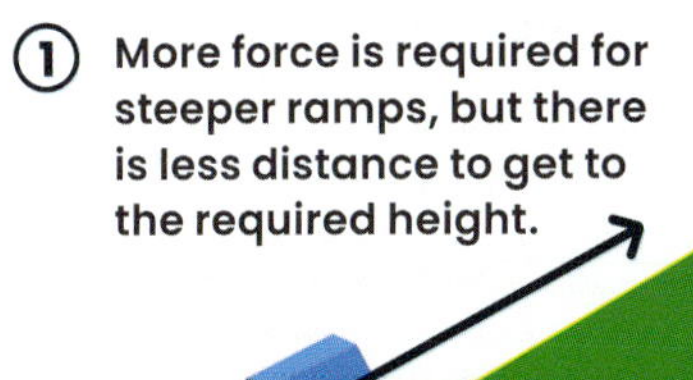

② Less force is required for shallower ramps, but there is more distance to get to the required height.

CHECKPOINT 9.5

1 What is a simple machine?
2 What are simple machines designed to do?
3 Give three examples of simple machines.
4 For your answers to question 3, identify a real-world situation where each one is used.
5 If a lever had a force applied a third of the way on the arm from the fulcrum and the arm on the other side of the fulcrum moved an object 10 cm, how far would the object move if the same force was applied at the end of the arm?
6 If you can only generate a quarter of the force required to lift an object, what is the fewest number of pulleys you would need to lift it?

CHALLENGE

7 A screw can be described as 'an inclined plane wrapped around a nail'. Explain how this means that a screw is easier than a nail to put into a piece of wood, and why it takes longer to insert than a nail.

SKILLS CHECK

- I can explain what a simple machine is and outline the benefits of them.
- I can give at least four examples of simple machines.

9.6 FRICTION

At the end of this lesson I will be able to:

- **analyse** some everyday common situations where friction operates to oppose motion and produce heat.

KEY TERMS 

friction
a force opposite to the motion of surfaces in contact

static friction
a friction force that keeps an object in place on a surface

LITERACY LINK 

Design an experiment that tests the effect of different surfaces on the force required to move an object. Write an aim and hypothesis for your experiment.

NUMERACY LINK 

Shoshana measures the force required to push an object up an incline. She carries out the experiment 4 times and obtains the following results: 52 N, 48 N, 57 N and 51 N. Calculate the average force required.

Friction is a force that is opposite to the motion of an object. It depends on the mass of an object and the types of surfaces involved.

Even if an object isn't moving, a force of friction may be stopping it from slipping or sliding. Air resistance and 'drag' are common names for the friction on objects as they interact with the air around them.

❶ Static friction prevents objects from moving

Objects don't have to be moving for friction to act upon them. **Static friction** can act on objects, keeping them in place on another surface.

Any time you've taken a step without slipping, that was due to static friction. Imagine trying to run with no static friction. It would be like trying to run on ice! Your feet would slip out from under you with every step because they couldn't get traction on the ground.

Static friction also applies to objects on an incline. You can stand on a ramp or slope without sliding down because the interaction between the slope and the soles of your shoes produces a friction force large enough to oppose the forces acting downhill.

If you're standing still and not sliding,the static friction force must be equal to the downhill force. If the downhill force is greater than the maximum friction force of your shoe, you'll begin to slip. This maximum depends on the materials of the slope and of your shoes.

What stops objects from sliding down slopes?

Figure 9.6 The downhill force on a slope depends on the normal force and the force due to gravity.

❷ Kinetic friction makes moving objects slow down or stop

When an object starts to move or slide, that doesn't mean friction has gone away. Whatever the object is moving along or through – whether it's a surface, the water or just air – a kinetic friction force is acting on it.

Have you ever ice-skated, or slid across a polished floor in socks? If so, you know that sometimes the kinetic friction forces can be pretty small, and you can keep moving for a long time. But you come to a stop eventually, even without running into something. This is because of the continual action of kinetic friction against your movement.

The kinetic friction acting on a moving object is less than the static friction needed to keep it still. If you were pushing a box up a ramp, getting it moving requires slightly more force than keeping it moving.

Which is larger: static friction force or kinetic friction force?

① When the box is stationary, static friction force = applied force.

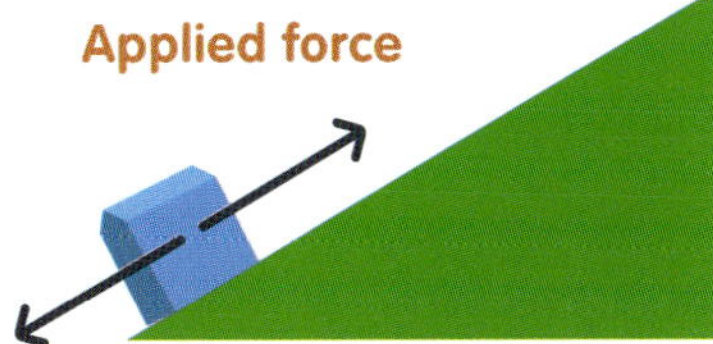

The box won't slide until the applied force is more than the static friction force.

② When the box is moving, applied force only needs to overcome the kinetic friction force, which is less than the static friction force.

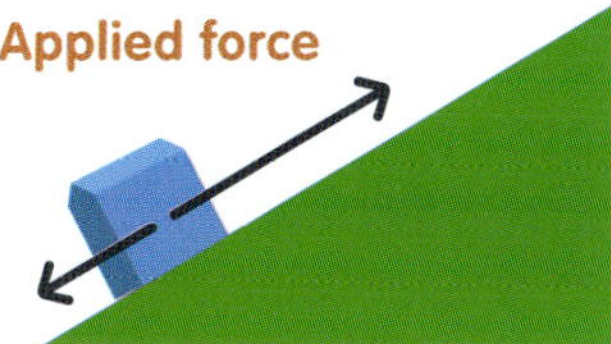

The kinetic friction force depends on the speed of the objects and the size and materials of the surfaces in contact.

❸ Heat is a product of friction

Place your hands together, apply a little bit of pressure and then quickly rub them back and forth. It shouldn't take long for a bit of heat to build up. This heat is caused by the friction between the particles of your hands as they rub against each other.

If you have ever seen a video or picture of the Space Shuttle re-entering Earth's atmosphere, then you've seen how much heat that friction can generate at high speeds. This is because of the high number of collisions between particles in the air and the shuttle. The air gets hot enough to form a glowing, fiery cloud around the shuttle.

What causes surfaces to produce heat when they interact?

INVESTIGATION 9.6
Heat from friction

CHECKPOINT 9.6

1 What does friction oppose?
2 What are two types of friction?
3 Which friction force is larger: the maximum static friction force or the kinetic friction force?
4 What type of energy is the main by-product of friction?
5 Outline three factors that affect the kinetic friction force.
6 A lot of machines use special materials and substances to minimise the friction between moving parts. Explain why this is done.

CHALLENGE

7 Compare the soles of a pair of sneakers to those of a pair of school shoes. Identify the key differences and explain why they might be different with regard to friction.

SKILLS CHECK

- I can explain the difference between kinetic and static friction forces.
- I can explain why friction can produce heat.

9.7 FACTORS AFFECTING FRICTION

At the end of this lesson I will be able to:

- **investigate** factors that influence the size and effect of frictional forces.

KEY TERMS 

coefficient of friction
a value which indicates how easily an object moves when interacting with another material.

lubrication
a substance that makes a surface slippery or smooth

LITERACY LINK 

Summarise what can create more or less friction, in a short tweet (140 characters or less).

NUMERACY LINK 

A football boot has a rubber sole. On dry concrete its coefficient of friction is 0.6, while on wet concrete it is 0.45. Express the coefficient of the boot on wet concrete as a percentage of its coefficient on dry concrete.

Friction is an essential part of movement. It is always there when objects start, stop and continue moving.

The moving parts of machines and vehicles, such as engines and tyres, are carefully designed to minimise or make the best use of this force.

1 Different materials apply different amounts of friction

Why aren't the bottoms of your sports shoes made out of metal? They would be much more durable and easier to clean. It would also be very hard to get traction when running in them – but why?

Rubber, even without shoe patterns, is much less slippery than a sheet of metal. This is because the particles of the materials interact. It is called a material's **coefficient of friction**. This number is a measure of how easily an object moves when interacting with another material.

Rubber has a high coefficient of friction, so rubber soles grip the ground more strongly and get more traction. Metals have a low coefficient of friction, so they don't get as much traction on the ground. This makes metals the right materials for ice skates, which need to glide rather than grip.

You can place water, oil or grease between two surfaces to reduce the friction between them. This **lubrication** means that the surfaces are interacting with the slippery material of the lubricant rather than with each other.

Why do different materials generate different friction forces when they interact?

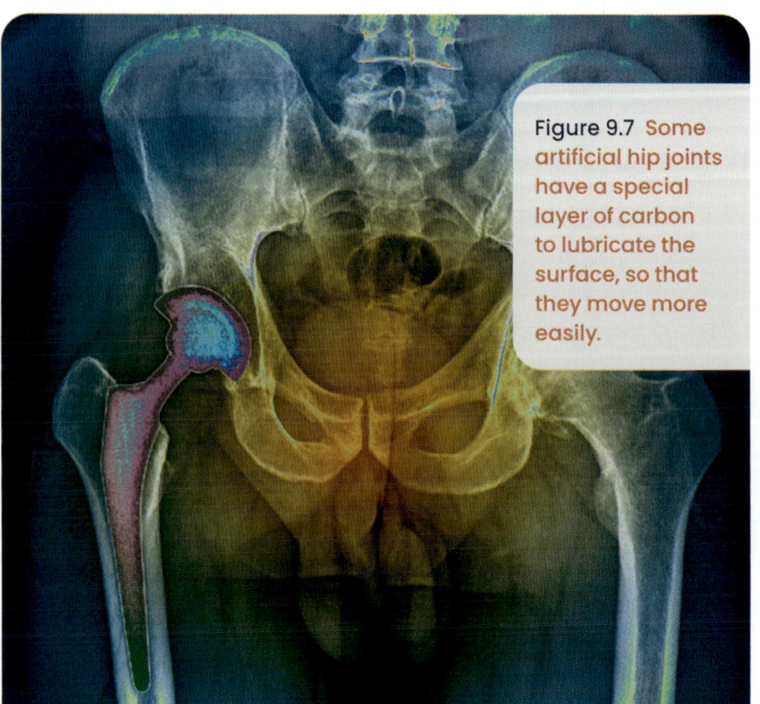

Figure 9.7 Some artificial hip joints have a special layer of carbon to lubricate the surface, so that they move more easily.

2 Other forces in an interaction can change a friction force

Imagine you're packing equipment into a sled on a ramp. Which do you think would be more likely to slide away: a sled with a little equipment on it, or a fully packed sled?

The fully packed sled would be more likely to slide off. This is because friction is due to the material's coefficient of friction *and* how much the objects are being pushed together by other forces.

At the start of this chapter you learnt about the forces acting on a stationary object. A book on a table is pulled down by gravitational forces, but it stays in place because the 'normal force' applied by the table pushes it back up.

Figure 9.8 shows how the friction force keeps the stationary sled in balance.

Figure 9.8 A sled on a ramp is affected by gravity, the normal force and the friction force. If these forces balance, the sled will not move.

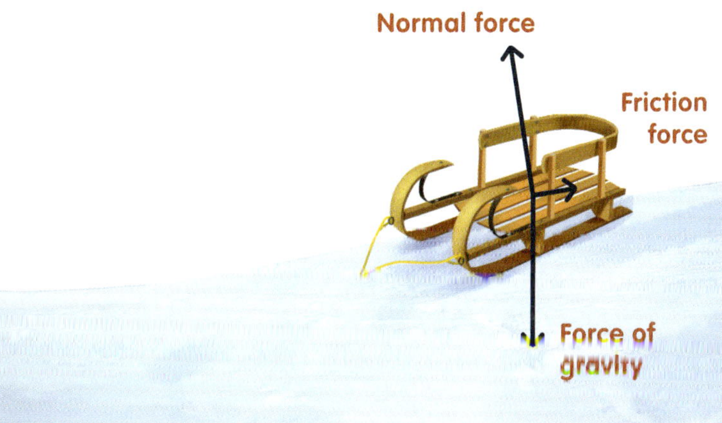

If you loaded the sled with more equipment, the force of gravity and the normal force would increase. This means the friction force would also increase to keep the sled still. But there's an upper limit to the friction force, depending on the materials of the sled and the ramp. So eventually the friction force wouldn't be strong enough to balance the other forces, and the sled would slide away.

Let's look at this another way – how do you use your brakes when riding a bike? When you need to slow down or stop quickly, what do you do with the brakes? You apply a larger force on them. Why? Because the larger force applies more friction on your wheels from the brake pads, causing them to slow down more quickly.

How does the normal force affect the friction force?

INVESTIGATION 9.7
Friction of materials

CHECKPOINT 9.7

1 What two aspects of materials determine the level of friction they have when they interact?
2 What does the coefficient of friction measure?
3 How can you make an object have less friction when interacting with another?
4 Why does more mass generally result in a greater friction force?
5 On bicycles, brakes are generally made out of two rubber pads that apply pressure on the rim of the tyre to slow it down. Why are they made out of rubber?
6 Explain why people need to drive more carefully when it is raining.
7 Why do tyres and brake pads have to be replaced occasionally?

CHALLENGE

8 Curling is a sport where a large stone is pushed down an icy track. Players use brooms to manipulate the friction between the ice and the stone, speeding it up or slowing it down. Research and describe another activity that involves manipulating friction, and explain how the participants do this.

SKILLS CHECK

- I can describe some causes of differences in friction.

CHAPTER SUMMARY

Forces are the result of objects interacting with each other or with a field.

A stationary object is affected by balanced forces.

Force of table on book (normal force)

Force of gravity

Adding or removing a force can unbalance an object.

Normal force

Force applied by skateboarder

Force of friction

Force of gravity

Every force acts in a specific direction and has a size or magnitude.

We can use simple machines to change how a force is applied.

- levers increase the size of a force
- pulleys redirect the force being applied
- inclines change the direction of a force

Engineers and scientists have developed products that can reduce the impact of forces, such as crumple zones in cars or the treads of shoes.

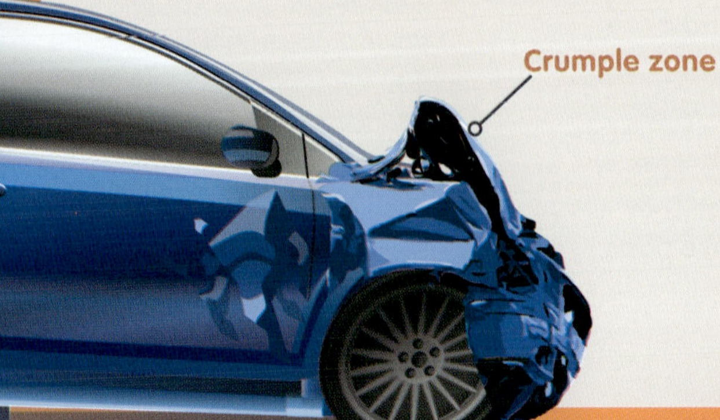

Friction is a force that can be described as the resistance of motion when one object rubs against another.

★ FINAL CHALLENGE ★

1. List the three ways you can apply a force to an object.
2. Recall the two things that every force has.
3. Define friction in your own words.

4. Examine surfaces that are designed to either increase or decrease friction, for example the sole of a running shoe or tyre tread or the hinge on a door. What can you observe about the differences in these surfaces?
5. Make a list of ways you use simple machines like pulleys or ramps in your everyday activities to make them easier.

6. From the list below, select the examples of unbalanced forces:
   - a phone sitting on a table
   - a bike picking up speed as it rolls down a hill
   - a car braking
   - a duck sitting still on the surface of a dam
   - a ball bouncing off a wall.
7. Compare the motion of balanced versus unbalanced forces when acting on the same object.

8. In two minutes, brainstorm as many ways as you can to use a simple machine to lift a heavy crate of books from the ground floor to the first floor of your school.
   a. Which simple machine from your list would make the task easiest for you?
   b. Draw a plan of your simple machine that you could share with an engineer.
9. From personal experience, evaluate how well different types of shoes reduce the force on your feet when running.

LEVEL 4

200xp

LEVEL UP!

10. Write a short paragraph about what your life would be like if you could turn friction on and off.
11. Create a sketch of a shoe, helmet or car that would be perfect at reducing the impact of forces. Label features on your sketch that demonstrate this.

10 FIELDS

When you think of forces, you might think of physical actions such as dragging a box along a floor, or the thrust from a car's engine. These are called contact forces because objects are physically interacting (coming into contact) with each other.

Some forces, called non-contact forces, have no physical interactions. Gravity can pull things towards Earth without touching them, and magnets attract pieces of metal without contact. A place where a non-contact force is acting is called a field. Right now, you're being acted on by gravity from Earth – you're in Earth's gravitational field.

1 LEARNING LINKS

What do you already know about fields?

How many ways can you think of to move something without touching it?

Why do some planets have less gravity than others?

Think of as many uses for magnets as you can.

2 SEE-KNOW-WONDER

List three things you can **see**, three things you **know** and three things you **wonder** about this image.

3 CRITICAL + CREATIVE THINKING

What if ... Earth suddenly swapped gravitational fields with the Moon?

Mismatches: How could you lift a piano using a coathanger, a magnet and a box of matches?

B-A-R: What would you make *bigger*, *add* and *replace* in a rocket ship?

THE PRETTIEST!

Auroras are beautiful displays of light seen in the night sky in areas near or at Earth's poles (such as in the Arctic or Antarctica). They happen when charged particles drawn in to Earth's magnetic field excite atoms in the air, causing patterns of green, red and blue light.

Due to their rarity in some parts of the world, auroras are part of several myths and legends about Greek, Roman and Norse gods, and even Chinese dragons.

10.1 INTRODUCTION TO FIELDS

At the end of this lesson I will be able to:

- **use** the term 'field' in describing forces acting at a distance.

KEY TERMS 

electrostatic force
a non-contact force between any objects with an electric charge

field
an area of space where objects are affected by a non-contact force

field line
a line used to show the direction of a force within a field

gravitational force
a non-contact force that affects all matter

magnetic force
a non-contact force that affects any object made of certain metals, such as iron

LITERACY LINK 

In fiction, force fields are invisible barriers or shields. Create a comic strip that includes a fictional force field.

NUMERACY LINK 

Three points exist within a field: *A* (3, 2), *B* (5, 5) and *C* (7, 9).

Plot all three points on a Cartesian plane.

Most forces are contact forces. This means that objects need to touch for one object to exert force on another. You can't lift a box just by pointing at it or thinking about it – you need to use the force generated by your muscles.

A non-contact force can act on an object without having any physical contact. This happens when the object is within an area of influence, called a **field**, and has properties affected by that field.

1 Non-contact forces can be gravitational, magnetic or electrostatic

Non-contact forces can work at a distance, without contact. Only a few of these forces exist in the universe, and they only affect certain objects in their fields:

- **Gravitational forces** are the most common, and they affect any objects containing matter.
- **Magnetic forces** affect any object made of certain metals, such as iron, nickel or cobalt.
- **Electrostatic forces** affect any object that has an electrical charge, even very weak charges.

Why are gravitational, magnetic and electrostatic forces called non-contact forces?

2 Fields are the areas where non-contact forces act

Any area where non-contact forces act on objects is a field. For example, any object in space that is affected by a gravitational force from Earth is said to be within Earth's gravitational field. This is the same for magnetic and electrostatic fields.

If something is outside the field of a non-contact force, it's not affected by that force. The limits of a field can be hard to measure.

How can you tell if an object is within a field?

Figure 10.1 The nail within the magnetic field is attracted to the magnet, but the one outside the field is not.

❸ The strength of a force varies within a field

Have you ever seen an image of iron shavings around a bar magnet, or observed them in class? One thing you may have noticed is that the iron filings are thickest at the ends of the bar magnet, but are less thick in other areas. This is evidence that the non-contact forces within fields are not always the same.

Field lines can help to locate the source of a force. For example, the field lines around a bar magnet show that the ends of the magnet generate the force, not the whole magnet.

An object that is close to the source of a field will be under a larger force than an object that is further away. This is shown by drawing field lines that are closer together.

How can you tell where a non-contact force is the strongest?

Figure 10.2 Iron filings gather in the areas around a magnet where the magnetic field is strongest.

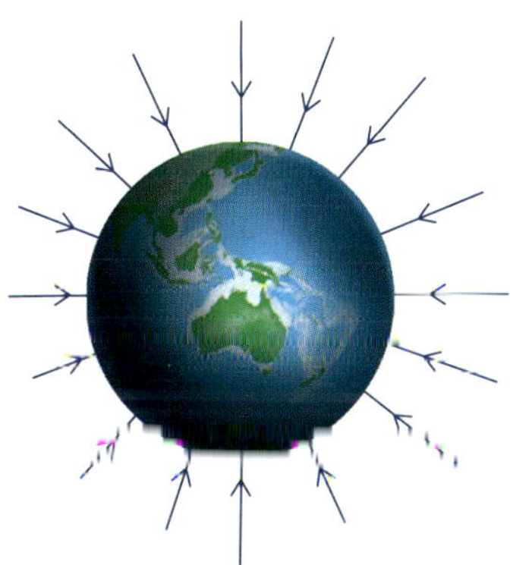

Gravitational field lines

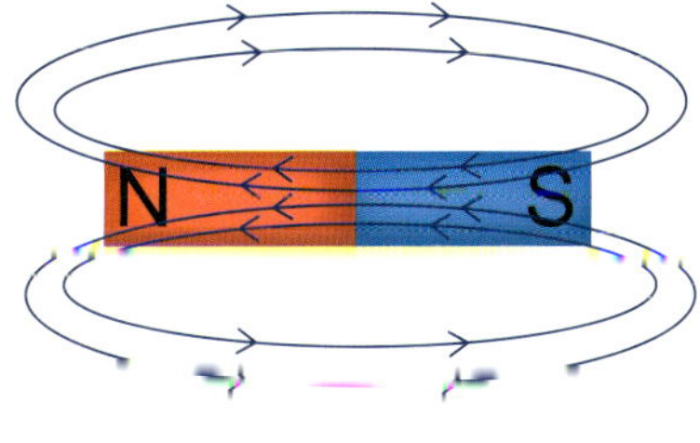

Magnetic field lines

Figure 10.3 Field lines show how the different non-contact forces operate within fields.

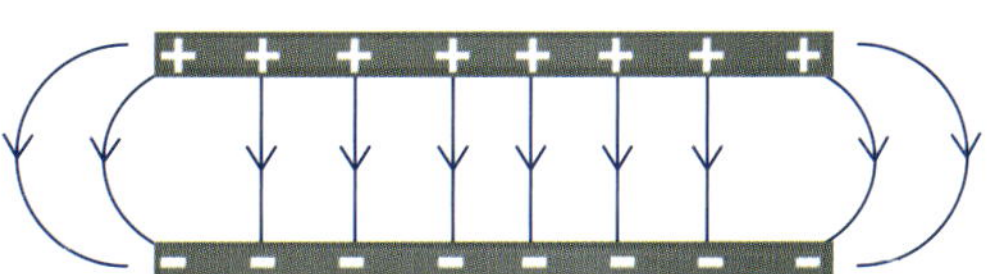

Electrostatic field lines

CHECKPOINT 10.1  

1 What is a non-contact force?

2 What are three examples of non-contact forces?

3 What is the name of the area that non-contact forces act in?

4 Outline the evidence that suggests that fields generate forces of different sizes in certain areas.

5 What is something that reduces the force on an object and is common between the three types of non-contact forces?

6 Using the diagram of the field lines around the bar magnet in Figure 10.3 to help you, draw how you think field lines would look around a horseshoe magnet.

CHALLENGE

7 Suggest how you could use a compass to find the magnetic field direction at different points around a magnet.

SKILLS CHECK

- I can explain what fields are and how they work.
- I can describe the difference between contact and non-contact forces and give examples of each.

10.2 HOW MAGNETS BEHAVE

At the end of this lesson I will be able to:

- **describe** the behaviour of magnetic poles when they are brought close together.

KEY TERMS 

attractive force
a force that pulls objects towards each other

like poles
magnetic poles that are the same (both north or both south); they repel each other

magnetic pole
one of the two ends of a magnet

repulsive force
a force that pushes objects away from each other

unlike poles
magnetic poles that are different (south-north or north-south); they attract each other

LITERACY LINK 

Write a two-part statement to describe the general rules of behaviour between two magnets.

NUMERACY LINK 

A magnet exerts a force of 0.07 N on a piece of iron. If that force becomes 100 times stronger, what will it be now?

A magnet is made of a material that has two **magnetic poles**, north and south. The direction of magnetic force in a magnet is from the north pole to the south pole.

When poles are brought together, the lines of force interact, pulling them together or pushing them apart. **Like poles** repel each other and **unlike poles** attract.

1 Magnetic field lines show the direction of magnetic forces

Each field line around a magnet shows the magnetic forces of the magnet's poles at that place. These lines can also show the direction of the magnetic force, and how that affects objects within the magnetic field.

One way to examine the field lines around a magnet is to use a compass. When you put a compass in a magnetic field, the compass needle lines up with the direction of the field lines around the magnet. The needle will always point from the magnet's north pole to its south pole. This shows that the direction of a magnetic force around a magnet is from north to south.

Figure 10.4 The direction of magnetic forces in a bar magnet is in a loop from the north pole to the south pole.

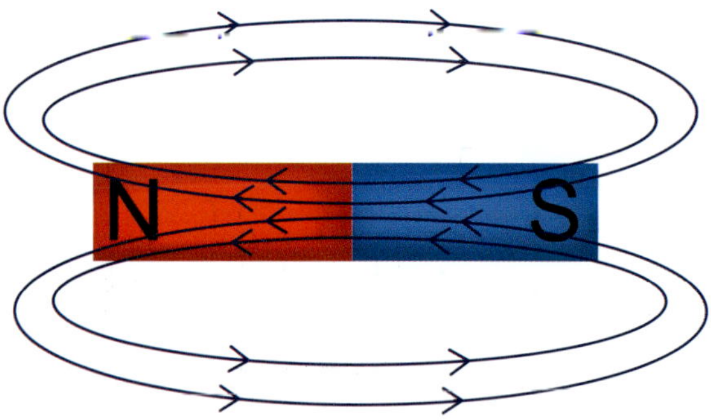

Here are some other things to remember about magnetic field lines:

- The distance between field lines represents the strength of a field in any given area.
- Field lines can never cross one another.
- Field lines continue through the magnets.

In which direction do the arrows point on magnetic field lines?

2 Like magnetic poles repel each other

When the north poles of two magnets are moved closer together, a **repulsive force** is made and pushes them away from each other. This is because the field from one north pole cannot enter or cross the field of the other north pole. Instead, the fields move away from each other and the field lines become more tightly packed.

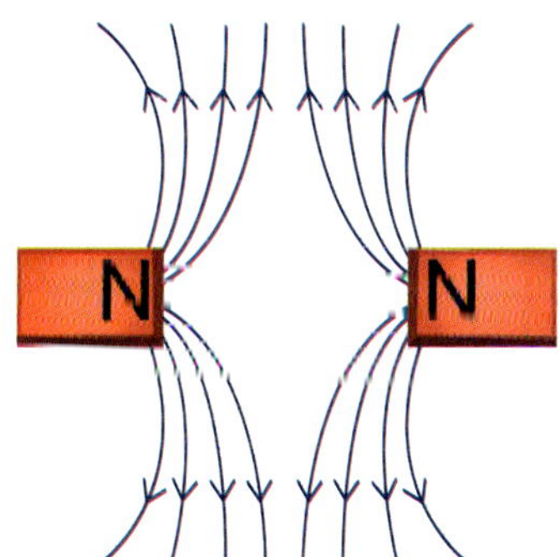

Figure 10.5 When two like poles are put together, the repulsive force can be strong enough to cause one object to hover over the other.

Although the fields between two north poles have been squeezed into a smaller area, the distance between field lines still shows the strength of the field. The force pushing the two poles apart becomes stronger.

A repulsive force also happens between the south poles of two magnets. This means we can say that, for all magnets, like poles repel.

When two like poles are moved closer together, what happens to their magnetic fields?

3 Unlike poles attract each other

When the opposite poles of two magnets are brought close together, an **attractive force** is made. This is because the field lines from the north pole of one magnet can flow easily into the south pole of the other magnet. In other words, unlike poles attract.

When two magnets contact one another, they act as one big magnet rather than two joined magnets.

What type of force happens between the north pole of one magnet and the south pole of another?

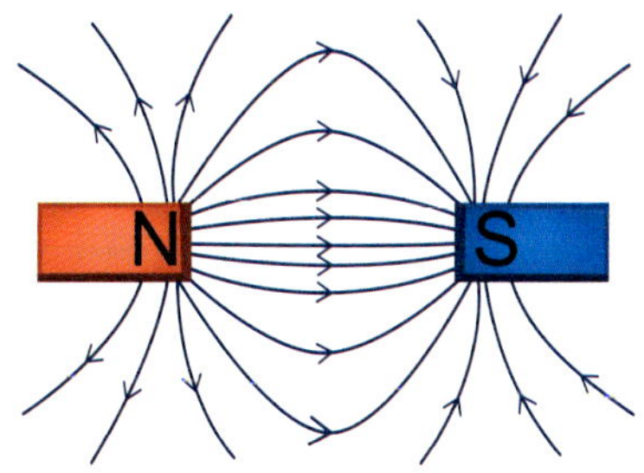

Figure 10.6 When two unlike poles are put together, the forces pull the north pole of one magnet towards the south pole of the other.

INVESTIGATION 10.2
Magnets and Matchbox cars

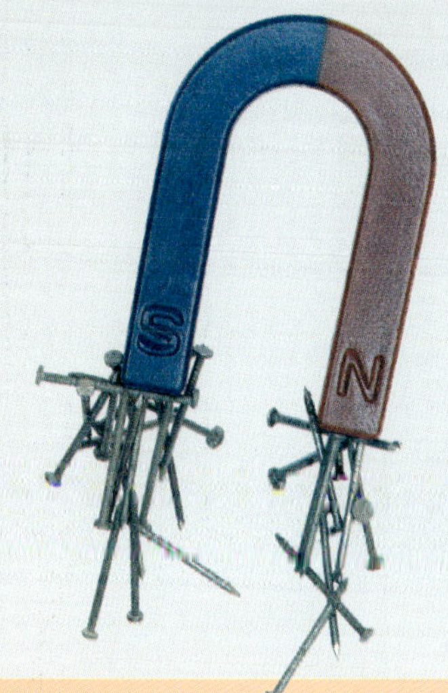

CHECKPOINT 10.2

1 Which pole on a magnet does a compass needle point towards?
2 What happens when the like poles of two magnets are pushed closer together?
3 Draw a diagram that shows the field lines of a magnet that is attracted to another magnet.
4 How could you map the direction of the field lines between two magnets?
5 Write a use for both the attractive force and the repulsive force between two magnets.

CHALLENGE

6 Some trains use magnets to move incredibly quickly. Suggest how you could use both attractive and repulsive forces to move a train, then research 'maglev trains' and see if your suggestion is correct.

SKILLS CHECK

- I can describe the behaviour of magnetic poles when they are brought close together.

10.3 USES OF MAGNETS AND ELECTROMAGNETS

At the end of this lesson I will be able to:

- **investigate** how magnets and electromagnets are used in some everyday devices or technologies used in everyday life.

KEY TERMS 

electromagnet
a device that uses electricity to make a magnetic field

polarity
the direction of a force, such as a magnetic force

LITERACY LINK 

Prepare a formal half-page report on a technological device of your choice that uses either magnets or electromagnets. List any references you use.

NUMERACY LINK 

Electromagnets are also used in the hard drives of computers. Convert 1024 kilobytes into bytes. Formula: 1 kilobyte = 1000 bytes

Ordinary magnets have a magnetic field that is always on – they don't need a supply of electricity. **Electromagnets** make magnetic fields using the flow of electricity.

Electromagnets are very useful because the strength and direction of their field can be easily changed. They are important parts of devices such as speakers, electric motors, medical equipment and computers.

1 Electromagnets use electricity to make a magnetic field

An electromagnet uses the flow of electricity to produce a magnetic field. The movement of electricity through a wire makes a circular magnetic field around the wire. If the wire is coiled into a spiral and electricity flows through it, then the field around the wire builds into one large field that's like the field around a bar magnet.

The field in the spiral can be strengthened by:

- placing an iron core in the middle
- increasing the amount of electricity
- making electricity flow faster
- using more wire to create more coils or loops in the spiral.

What is an electromagnet?

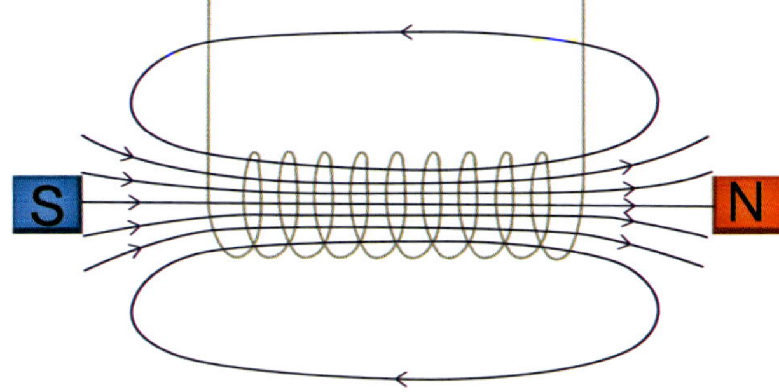

Figure 10.7 The magnetic field around an electromagnet's core has a north and south pole, like a bar magnet.

2 Magnets can hold objects together

Ordinary magnets are sometimes called permanent magnets, because they constantly have magnetic force. The most common use of magnets is to hold objects together. The constant attractive force acts on certain metals or the opposite pole of another magnet.

Magnets used to hold objects are on cabinet and fridge doors, wallet and purse clasps, and even some charger cords. Some tools, such as screwdrivers, have magnetic heads, which allows screws to 'stick' to the tool without needing to be held.

How are magnets commonly used?

Figure 10.8
A magnetic screwdriver head allows you to hold screws in place or pick them up more easily if you drop them.

INVESTIGATION 10.3
Building an electric motor

3 Electromagnets are used in devices such as speakers and medical equipment

Electromagnets are used in an enormous number of devices and technologies. This is because they can change the strength of their magnetic field very precisely. They can also change the direction, or **polarity**, of their magnetic field.

When a speaker is connected to a powered device playing music, the electricity changes the coil in the speaker into an electromagnet. The electromagnet can rapidly flip its field direction. This attracts and then repels a permanent magnet inside the speaker. The electromagnetic coil moves back and forth, pushing sound waves through the 'skin' of the speaker cone.

Computers contain very sensitive electromagnets. They're used in hard drives to read information stored on metallic strips.

Another major use of electromagnets is in medical equipment. Magnetic resonance imaging (MRI) scanners generate magnetic fields that can be used to create images of the inside of the human body.

What is one common use of electromagnets?

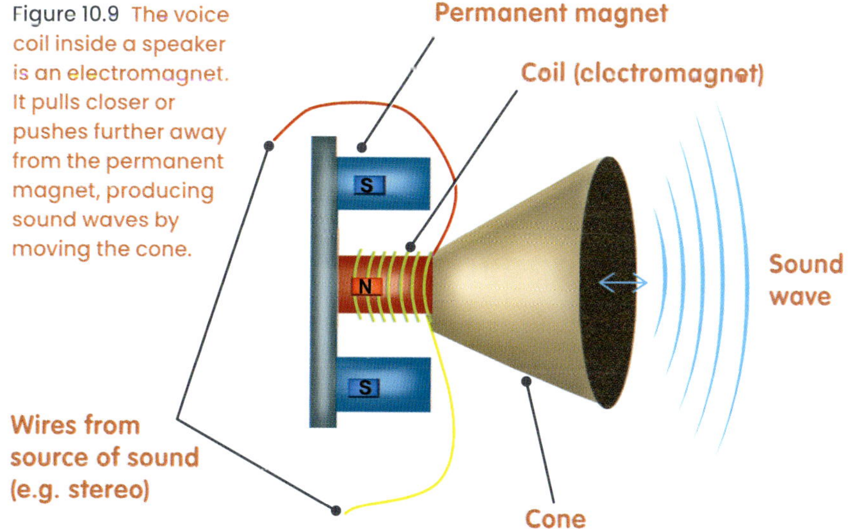

Figure 10.9 The voice coil inside a speaker is an electromagnet. It pulls closer or pushes further away from the permanent magnet, producing sound waves by moving the cone.

CHECKPOINT 10.3

1 What does the flow of electricity make in an electromagnet?

2 How can you increase the strength of an electromagnet?

3 What is the most common way that magnets are used in everyday life?

4 What can electromagnets do that permanent magnets cannot?

5 How does an electromagnet create a magnetic field?

6 List the magnets you can find either around your classroom or around your home. For each one, describe the purpose they serve and why they are used.

CHALLENGE

7 One common use of electromagnets is as a transformer. Research what a transformer is, and give some examples of how it is used in modern devices.

SKILLS CHECK

- I can describe how magnets are used in everyday life.
- I can describe what an electromagnet is and how electromagnets are used in everyday life.

10.4 ELECTROSTATIC CHARGE

At the end of this lesson I will be able to:

- **identify** ways in which objects acquire electrostatic charge.

KEY TERMS 

electron
a negatively charged sub-atomic particle found around the outside of the nucleus of an atom

electrostatic charge
the unmoving electric charge on the surface of an object

neutron
a particle without charge in the nucleus of an atom

nucleus
the centre of an atom, which contains protons and neutrons

proton
a positively charged particle in the nucleus of an atom

LITERACY LINK 

Summarise this section into two sentences. Read your summary to a classmate and ask for two stars (two good things about your summary) and a wish (one thing they wish you could change).

NUMERACY LINK 

An electron leaves a charged plate at a 33° angle. What angle is complementary to 33°?

Magnetic forces only exist in objects that contain certain metals, such as iron. **Electrostatic charge** produces a non-contact force that attracts or repels almost all objects, no matter what they're made of.

Electrostatic force behaves like magnetic force in some ways but is very different in other ways. It is made by the movement of tiny particles called **electrons**.

1 Atoms contain positive, neutral and negative particles

All objects are made of matter, which is made up of atoms. Each atom has a dense core, called a **nucleus**. This is made up of uncharged particles called **neutrons** and positive particles called **protons**. Around each core are one or more negative particles called electrons.

Every proton and electron carries a tiny electric charge. The positive charge of protons is equal but opposite to the negative charge of electrons. So, when the number of protons is the same as the number of electrons in an atom, the atom has no overall charge – it is neutral.

Electrons often move from one atom to another. When an atom has more protons than electrons, it has a positive charge. When it has more electrons than protons, the atom has a negative charge. When this imbalance of charge happens in many of an object's atoms, the object has an electrostatic charge.

What type of charge does an electron have?

Figure 10.10 Atoms have one or more protons, one or more electrons, and (except for hydrogen) one or more neutrons.

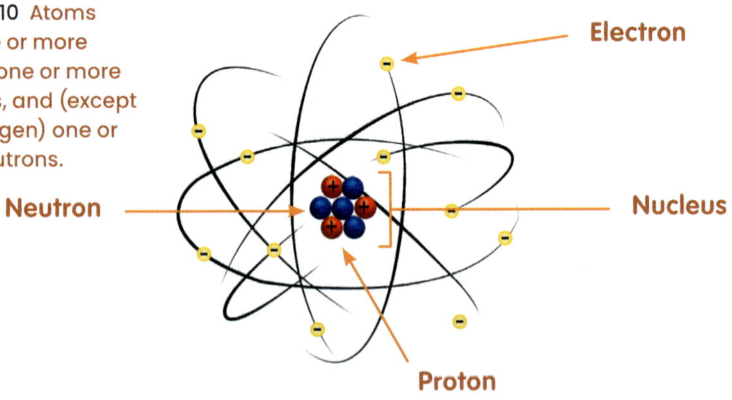

Electron
Neutron
Nucleus
Proton

2 Objects gain and lose electrostatic charge

An object gains electrostatic charge by the movement of electrons. When enough electrons move onto or off the object, it becomes charged.

Some materials have a higher attraction for electrons than others. When these materials come into contact with a material that doesn't attract electrons as strongly, some electrons will move into the electron-attracting material.

Say you're holding a chunk of plastic and a piece of cloth. Both of these objects have no charge at first. If you rub the two objects together, many of the electrons within the cloth will be attracted to the plastic because it attracts the electrons more strongly.

When this happens, both objects become electrostatically charged. When you pull them apart, the plastic will have a negative charge because it now has more electrons than protons. The cloth will have a positive charge because it has more protons than electrons.

Why do electrons move from one material to another?

Figure 10.11 When you rub plastic and cloth together, an electrostatic charge is made on both objects.

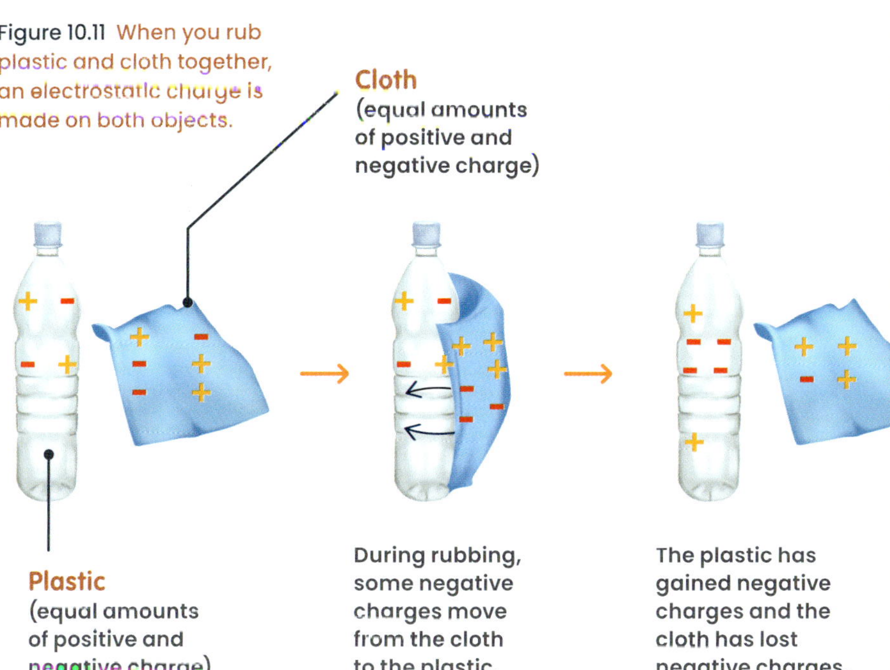

CHECKPOINT 10.4 

1 When an atom has more protons (positively charged particles) than electrons (negatively charged particles) what overall charge does it have?

2 Identify a way in which objects can acquire electrostatic charge.

3 Why do electrons move from one material to another when they come into contact through friction or rubbing?

4 Why do you feel a 'zap' or see a small spark when you touch something that has been statically charged?

5 Provide a reason why electrons, and not another type of particle, move from the atoms of one material to another.

6 If the movement of electrons creates electrostatic charge, what would be the overall charge on an object that loses electrons?

CHALLENGE

7 List all the times you have seen or felt static shocks. For each situation, identify the materials that were interacting to create the static charge. Try to find out if there are any common materials in your own list or other people's lists.

3 Everyday items gain electrostatic charge

Electricity and electrostatic charge are similar, but they're not the same thing. One major difference is that you need a generator (such as a battery) to make an electrical current. It's a lot easier to generate an electrostatic charge – you can do it just by walking around or touching items together. This is often called 'static electricity'.

An easy way to see this is to blow up a balloon and rub it on your hair. The material of the balloon has a higher attraction for the electrons than your hair, so electrons from your hair will join the particles in the material of the balloon, creating a slight negative charge. At the same time, your hair will gain a slight positive charge because it has more protons.

If the electrostatic charge is large enough, it can create a spark as the electrons move quickly back into positively charged material. You might see this if you take off a woollen jumper in the dark on a dry day.

Why do electrons move from one material to another?

SKILLS CHECK

- I can describe what electrostatic charge is.
- I can identify some ways that objects gain electrostatic charge.

10.5 HOW CHARGED OBJECTS BEHAVE

At the end of this lesson I will be able to:

- **describe** the behaviour of charged objects when they are brought close to each other.

KEY TERMS 

attractive force
a force that pulls objects towards each other

electrostatic charge
the unmoving electric charge on the surface of an object

repulsive force
a force that pushes objects away from each other

LITERACY LINK 

Rewrite the first paragraph in this section, simplifying the language and ideas, and adding examples or further explanation so that a younger student can understand it.

NUMERACY LINK 

An object has 4 electrons and 7 protons. What is the net charge?

-4 + 7 = ___________

Objects with **electrostatic charge** can have either a positive or negative charge. When two objects with the same charge are brought close together, the force between them pushes them apart. When two objects with opposite charges are brought close together, the force between them pulls them together.

This behaviour of charged objects is evidence that like charges repel and opposite charges attract.

1 Charged objects can have a positive or negative charge

Protons have a positive charge that is the same size as the negative charge that electrons have. So, when the atoms of an object have more electrons than protons or vice versa, then that object has an overall charge.

An object will have a positive charge if there are fewer electrons than protons. It will have a negative charge if there are more electrons than protons.

If an object has a charge, that charge is spread evenly throughout – there won't be parts with no charge or a different charge. This is one difference between electrostatic force and magnetic force. Magnetic objects always have a north pole and a south pole.

What causes an object to have a negative charge?

Figure 10.12 If an object has a different number of protons than electrons, it has a positive or negative electrostatic charge.

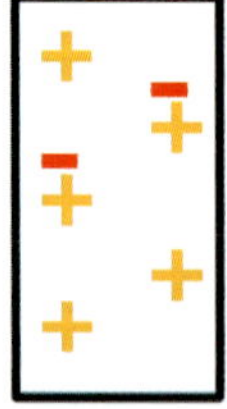

Positively charged = fewer electrons than protons

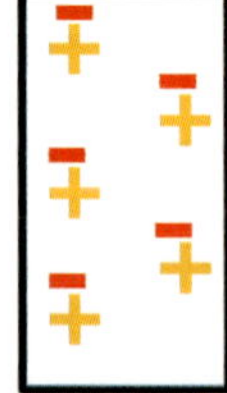

Neutral = same number of electrons and protons

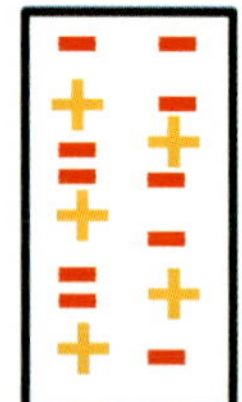

Negatively charged = more electrons than protons

❷ Like charges repel each other

When two objects with the same charge are brought close together, they experience a **repulsive force** that pushes them apart. This is true whether the objects are positively or negatively charged.

You can see this if you hang two balloons near one another. If neither balloon is charged, then they hang straight down. If you charge each one by rubbing them on your hair, or with a piece of woollen fabric, then they push slightly away from each other.

What type of force is made when positively charged objects are brought close together?

Figure 10.13 Two balloons close together and with the same charge experience a repulsive force, pushing them away from each other.

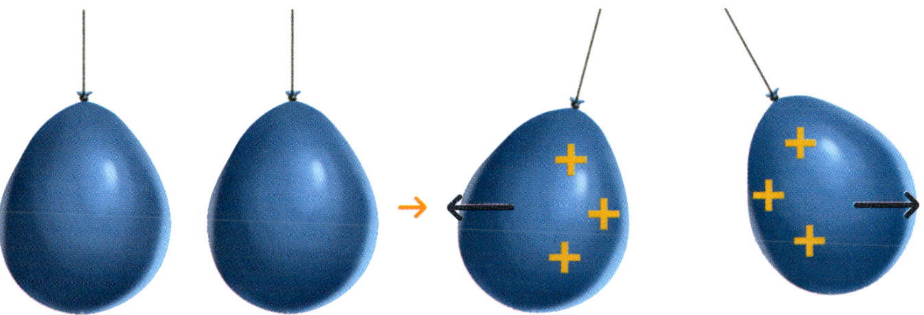

❸ Opposite charges attract each other

When two objects with opposite charges are brought close together, they experience an **attractive force** that pulls them even closer together.

You can see this by hanging two balloons from strings. Rub one of them with wool, and the other one with rubber or a silicone car washing cloth. This will give one balloon a positive charge and the other a negative charge.

This time, the opposite charges of the balloons cause an attractive force, which pulls them slightly together.

What type of force is made between objects with opposite charges?

Figure 10.14 Two balloons close together and with unlike charges experience an attractive force, pulling them towards each other.

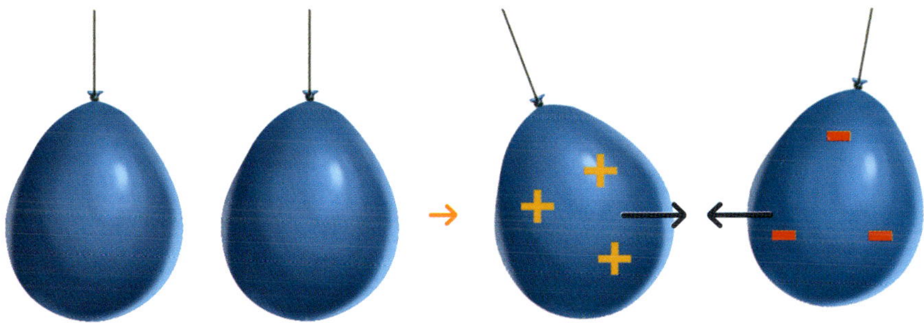

INVESTIGATION 10.5
Charging balloons

CHECKPOINT 10.5

1 What are the three types of charges an object can have?
2 If an object has more electrons than protons and experiences a repulsive force near another object, what charge does the second object have?
3 If there is an attractive force between two objects, what charge(s) do they have?
4 Summarise, in one sentence, the overall rule describing how charged objects behave when they are close together.

CHALLENGE

5 The triboelectric series is a list of objects ordered by their ability to gain or lose charge. Predict which of these materials are most and least likely to gain a positive or negative charge.
skin, wool, aluminium, paper steel, rubber, plastic
Check your predictions by comparing your answers to an online triboelectric series.

SKILLS CHECK

- I can explain what it means for an object to be charged.
- I can describe the behaviour of charged objects when they are brought close to each other.

10.6 ELECTROSTATIC FORCES AND LIGHTNING

At the end of this lesson I will be able to:

- **investigate** everyday situations where the effects of electrostatic forces can be observed, e.g. lightning strikes during severe weather and dust storms.

KEY TERMS 

grounding
providing a pathway to conduct electricity down to the ground

polarise
develop parts with positive and negative charges

resistance
a measure of how difficult it is for electricity to flow through an object

LITERACY LINK 

"A flicker of light, a crash of thunder, a streak of electricity - it's a lightning strike! The clouds far above your head build up electrostatic charge and, when the charge gets large enough... BANG!"

Choose another paragraph from this chapter and rewrite it in this animated style.

NUMERACY LINK 

A lightning bolt can contain one billion joules of power.

Express one billion as a power of 10.

Convert one billion joules to kilojoules.
Formula: 1 joule = 0.001 kilojoules

Lightning strikes are incredibly powerful releases of energy. Not only do they generate brilliant flashes of light and release intense heat (which vibrates the air to give the booming sounds of thunder), they can contain *millions* of volts of energy. It all begins in the clouds with a build-up of electrostatic energy.

1 Clouds gain charges and become polarised

Clouds are formed by water vapour rising through the air and condensing to liquid in cooler parts of the atmosphere. This rise and fall of water vapour creates a lot of particle movement within a cloud – water particles bump into each other over and over again. When this happens, friction causes them to gain or lose electrons.

The movement and the temperature differences between the upper and lower parts of a cloud can cause the cloud to **polarise**. This means that the upper part of the cloud has a positive charge while the bottom part develops a negative charge. Usually, an object can only have one overall charge, but a cloud is made up of many separate water droplets, and isn't one single 'object'.

What happens to a cloud when it polarises?

Figure 10.15 Water particles become charged as they move around in a cloud.

2 Energy leaves a cloud as lightning

The air usually has enough **resistance** to prevent electrons moving freely through it. A strong enough electrostatic field can reduce this resistance, allowing electrons and even electricity to move around.

The difference between the negative and positive regions in a cloud can become very large. This generates a strong electrostatic field. The extra electrons in the negative part of a cloud can move quickly into

Figure 10.16 Most lightning moves from one cloud to another.

positive parts of the same cloud, a nearby cloud or even the ground to balance the difference in charge. This rapid movement of electrons releases energy as light, heat and electricity – lightning.

Most lightning stays in the atmosphere, either within a cloud or from one cloud to another, and we rarely see it. Occasionally the charge is large enough to jump from a cloud to the ground, a tree or a building, and potentially cause significant damage.

What is lightning?

3 Buildings can be protected from lightning strikes

The best way to protect yourself during a lightning storm is to stay indoors. Most buildings have special protection against lightning called **grounding**. This is a way for electric current to harmlessly enter the ground. The walls and supports of most buildings are low-resistance pathways that can channel lightning away from sensitive parts of a building's structure.

Some buildings also have lightning rods – metal rods that point out of the roof. Lightning is more likely to hit the lightning rods, which are connected to low-resistance pathways, than other parts of the building, making it even safer.

Why do buildings need protection from lightning strikes?

CHECKPOINT 10.6 

1 What is required for a lightning strike to happen?

2 What has happened to an object that has a negatively charged area and another area that is positively charged?

3 What moves continuously through clouds, creating friction and charging parts of the cloud?

4 What evidence is there that lightning strike contains a huge amount of energy?

5 When you search the internet for images of lightning, the most common images show lightning with one or two thick and bright branches, and several smaller and finer branches.
   a Using your knowledge of polarisation and lightning strikes, suggest why there is usually a thick and bright branch of lightning.
   b Suggest a possible reason for the smaller, finer branches of lightning.

6 It is safer inside a building than under a tree during a lightning storm. Suggest why.

CHALLENGE

7 Lightning is a huge amount of energy flowing through the atmosphere. List some potential issues with trying to harness this energy. For each issue, try to construct a potential way to solve it. Evaluate whether lightning could be a viable source of energy for society in the future.

SKILLS CHECK

- I can explain the occurrence of lightning strikes using my understanding of electrostatic forces.

10.7 GRAVITY

At the end of this lesson I will be able to:

- **identify** that Earth's gravity pulls objects towards the centre of Earth
- **distinguish** between the terms 'mass' and 'weight'.

KEY TERMS 

gravitational force
the force that attracts physical objects with mass towards each other

mass
the amount of matter that a physical body contains

weight
the force of a gravitational field on the mass of a body

LITERACY LINK 

Write a short story about what you would do with your day if you could change the direction and size of the gravitational field around you.

NUMERACY LINK 

An object's weight on the Moon is x newtons.

If $10 + 2x = 28$, solve for x.

Every object in the universe is made of matter and has mass. This means it creates gravity, which attracts other objects. The larger the object, the stronger the **gravitational field** it makes. The closer two objects are, the greater the gravitational force acting on each one.

Gravitational forces act towards the centre of the object that creates them. That's why, no matter where you are on Earth, all objects fall towards Earth's centre.

1 Gravity is always an attractive force

All objects with mass make a gravitational field. Within this field, any other object will have a force generated on it. The size of this force depends on two things: the **mass** of the object generating it and the distance between the two objects.

Although the size of gravitational force varies, the direction does not. Any force due to gravity will *always* be an attractive force. This means that it pulls objects towards the centre of whatever mass generates the gravitational field.

Whenever you drop something, it falls to the ground, wherever you are in the world, because the attractive force is directed to the centre of Earth's mass. Why Earth? Because it's by far the most massive object that affects us every day. The Sun is much more massive, but so far away that it has little effect on people and objects compared to Earth's gravity.

Do gravitational forces push, pull or twist objects?

2 Gravitational forces change with distance

As with magnetic and electrostatic forces, you can draw field lines to represent the area where a force will be generated on an object. Remember:

- Stronger fields have field lines drawn close together; weaker fields are shown with field lines further apart.
- Gravitational field lines point in the direction in which the force is acting.

On Earth's surface, gravity is constant because the distance from the centre of Earth is about the same everywhere. The gravitational force of Earth always acts towards the centre of the planet. Further away from Earth, the gravitational force is smaller, and the field lines are further apart.

INVESTIGATION 10.7
Measuring gravity

Imagine a mattress with a bowling ball in the centre, making a hollow. If you place some marbles around the mattress, they'll eventually roll into the hollow made by the bowling ball. They might roll slowly when they're near the edge of the mattress, but they will roll faster as they get closer to the bowling ball, where the slope gets steeper. This is very similar to the way that objects are affected by Earth's gravity.

Why is gravity on Earth constant?

3 An object's weight depends on gravity

You may have heard people talk about **weight** in terms of grams and kilograms, but these are actually units of mass. What's the difference between mass and weight?

An object's mass is the amount of matter it contains. That bowling ball on the mattress contains a specific amount of matter. Putting the ball in different places – on a mattress, up a mountain, on the Moon or floating in space – won't change its mass. Mass is measured in grams and kilograms.

An object's weight is a measurement of the pull of gravity on an object. Changing the gravity acting on an object will change its weight. This is why astronauts on the Moon weigh less and can move around with huge jumps. It's also why astronauts on the International Space Station float weightlessly, even though their bodies still have mass. The metric unit used to measure weight is the newton (N).

Why don't we usually use newtons when talking about weight? It's because there are very few circumstances on Earth where the force of gravity changes. It stays constant, so we just use kilograms instead.

What is the difference between mass and weight?

Figure 10.17 Mass doesn't change wherever you are, but weight does.

Trevor's mass on Earth is 60 kg. Trevor's mass on the Moon is also 60 kg.

His weight on the Moon is about 100 N.

His weight on Earth is about 600 N.

CHECKPOINT 10.7

1 Explain what gravity is, in your own words.

2 Describe the difference between mass and weight.

3 What two things affect the amount of gravity something has?

4 If objects are attracted to one another due to gravitational forces, provide a reason why planets orbit around the Sun rather than crashing together.

5 What metric unit is weight measured?

6 How can an object's weight change if its mass remains constant?

7 On the Moon, an object weighs 60 N. Objects weigh 6 times more on Earth than they do on the Moon. Calculate its weight on Earth.

CHALLENGE

8 Choose three other planets in our solar system, then undertake some research in order to calculate your weight on each of these three planets.

SKILLS CHECK

- I can explain the difference between mass and weight.
- I can describe gravity, including what causes gravity.

10.8 UNBALANCED GRAVITATIONAL FORCES

At the end of this lesson I will be able to:

- **describe** everyday situations where gravity acts as an unbalanced force.

KEY TERMS 

acceleration
an increase in speed in the direction of movement

gravitational force
the force that attracts physical objects with mass towards each other

stationary
not moving

LITERACY LINK 

Summarise this section into a small postcard addressed to your teacher. Make sure you explain what gravity is and how it acts on objects.

NUMERACY LINK 

An object in space is affected by a +15 N gravitational force in one direction and a -18 N force in the opposite direction. What is the total force?

When an object speeds up, slows down or changes direction then the forces acting on the object are unbalanced. This means that when all the forces acting on an object are added together, they don't completely cancel each other out.

Falling objects are a perfect example of unbalanced forces, because the strongest force acting on the object is gravity pulling it down. A falling object will fall faster and faster until the forces acting on it become balanced.

1 Gravity is always acting, even on stationary objects

If an object is on the ground, not moving, you might assume that no forces are acting on it. But *every* object on Earth is within Earth's gravitational field and is affected by gravitational forces.

As you saw in section 9.1, a **stationary** object on a surface has a gravitational force pulling it downwards. The downwards force is balanced by the force of the surface pushing up on the object in response to the gravitational force. So the gravitational force acting in a downwards direction is the same size as the force acting in an upwards direction.

If the object was *not* in contact with a surface, the main force acting on it would be the force of gravity pulling it down. The forces acting on the object would be unbalanced – and the object would fall.

What are the main balanced forces acting on a stationary object?

2 Gravity causes falling objects to accelerate

Imagine you place a basketball at the edge of a table and slowly push it off the edge. The forces acting on the basketball are now unbalanced, so the ball will speed up, slow down or change direction. Which of these applies to the ball? It's not slowing down – that would mean it comes to a stop before it hits the ground. It's not changing direction either – it's always heading to the centre of Earth. That means it must be speeding up – this is **acceleration**.

In the instant before an object begins falling, it has a speed of zero – it's not moving. It moves faster and faster downwards until it reaches a surface. This change of speed is acceleration due to gravity. The gravitational force on the object is not balanced by another force, so the object accelerates in the direction of that force.

On the surface of Earth, acceleration due to gravity is constant. The speed of a falling object gets faster and faster, but it gets faster and faster at the same rate. That's because the gravitational force acting on it is always the same – the force of Earth's gravitational field.

Why do objects speed up when they fall?

3 Gravity can cause objects to slow down or change direction

Sometimes when an object is forced upwards, gravity can cause the object to slow down or change direction, rather than speed up.

Imagine that you pick up that basketball you dropped, and you take a shot at the basket. What do you expect to happen? First, the ball flies up and out, gaining height while moving closer to the basket. At some point, it stops going up and starts coming down, even though it's still moving away from you, until (hopefully) it swooshes through the basket.

The changes in the ball's flight are all due to unbalanced forces. After you release the ball, you can't put any additional force on it – it's literally out of your hands! When the ball is in flight, the largest force, gravity, will pull the ball back to the ground. It doesn't matter how much force you initially give the ball – after it has left your hands, gravity will immediately start slowing it down and eventually pull it back to Earth.

Gravity is always an attractive force, pulling objects towards the centre of Earth. But it's the overall sum of unbalanced forces that determines how objects move, and when they fall to the ground.

What is an example of gravity causing an object to slow down or change direction?

Figure 10.18 When you make a free throw, you need to account for gravity acting on the basketball.

CHECKPOINT 10.8 

1 On the surface of Earth, in which direction does a force due to gravity point?

2 How would an object with overall unbalanced forces acting on it move differently to an object with overall balanced forces acting on it?

3 Give an example of an object on Earth that has balanced forces acting on it, and describe how it would move.

4 Give an example of an object on Earth with unbalanced forces acting on it, and describe how it would move.

5 Provide an example where unbalanced forces allow an object to move away from the ground, and an example where unbalanced forces move an object towards the ground.

6 You experience unbalanced forces any time you walk upstairs or drop something. For one of these situations, describe the source of the forces at work.

CHALLENGE

7 Drones, rockets, space shuttles and aeroplanes all generate an upward thrust in order to lift off the ground. Choose one example to research and describe the mechanism it uses to move against the force of gravity. Outline any positives and any negatives of your chosen method of lift-off.

SKILLS CHECK

- I can describe how gravity acts on objects when they are moving and when they are stationary.
- I can give examples of everyday situations where gravity acts as an unbalanced force.

CHAPTER SUMMARY

Forces are the result of objects interacting with each other or with a field.

Fields
areas in which non-contact forces act on certain objects

Gravitational forces affect all objects made of matter.

▼ Magnetic forces affect any objects that contain ferrous metals.

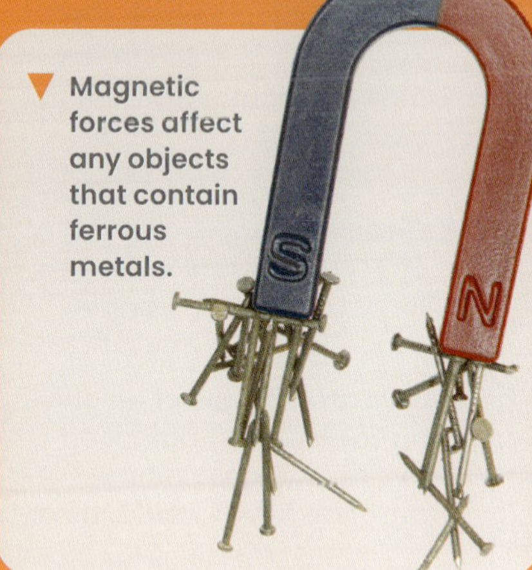

▼ Electrostatic forces affect any objects with a positive or negative charge.

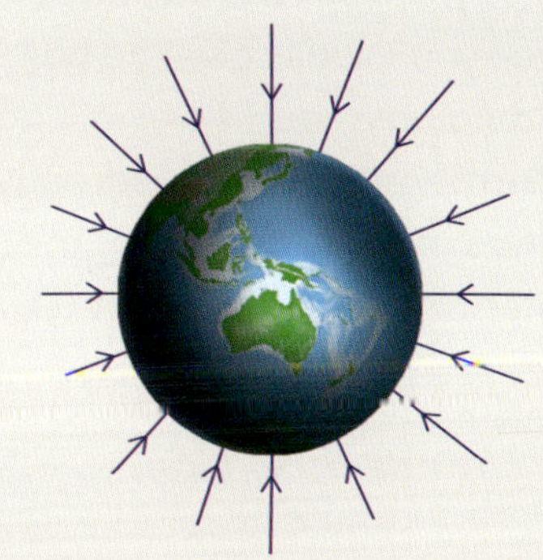

▼ Objects with high mass gravitationally attract objects with lower mass.

▼ Like magnetic poles repel and unlike poles attract.

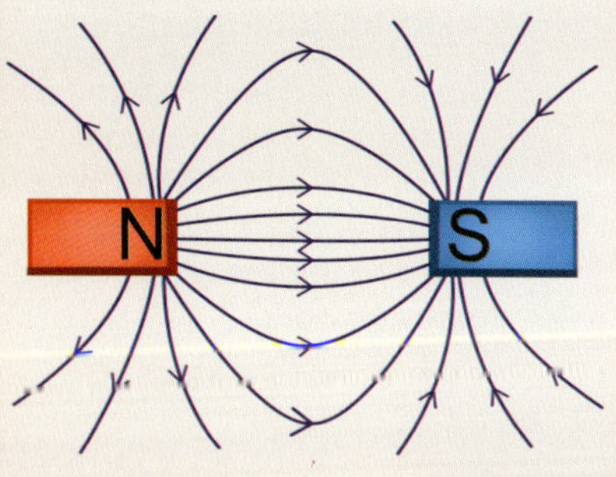

▼ Like electrostatic charges repel and unlike charges attract.

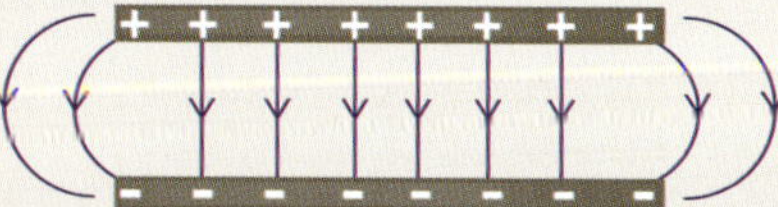

▼ **Electromagnets** are used in devices such as speakers and medical equipment.

Clouds slowly build up charge, then discharge the energy as lightning.

▲ The gravity of Earth pulls everything on its surface towards its centre.

★ FINAL CHALLENGE ★

LEVEL 1 ★☆☆☆☆☆ 50xp — LEVEL UP!

1. In your own words, provide a definition of the word 'field'.
2. Copy and complete the following sentence: Like poles __________ each other, while unlike poles __________ each other.
3. Explain what an electrostatic charge is.

LEVEL 2 ★★☆☆☆☆ 100xp — LEVEL UP!

4. Draw a diagram showing the magnetic field lines of:
   a. Two like poles close to one another
   b. Two unlike poles close to one another
5. Draw a diagram showing how gravity acts for two people standing on opposite sides of Earth.

LEVEL 3 ★★★☆☆☆ 150xp — LEVEL UP!

6. List three everyday devices or technologies that use either magnets or electromagnets.
7. You have two positively charged objects and you bring them close together. Describe what will happen.
8. An object that has been dropped will speed up as it falls. Explain why.

LEVEL 4 ★★★★☆☆ 200xp — LEVEL UP!

9. Describe the conditions that must be present in order for lightning to occur.
10. Using your understanding of gravity, explain why your weight would be different on different planets.

LEVEL 5 ★★★★★★ 300xp — LEVEL UP!

11. Rubbing a balloon on your hair causes an electrostatic charge to form. Describe what is happening to the atoms of the balloon and your hair when this occurs.
12. Design a device that could be used to protect cars from lightning strikes. Draw a diagram and show all required materials.

11 ENERGY

In physics, energy is the ability to do work, but what does this actually mean? Work might be moving something against a force, such as gravity. When you do something against gravity, even just stand up, you use energy.

Energy is closely related to mass. More mass equals more energy. Energy never disappears or runs out, it just moves or changes into a different kind of energy, over and over again, forever!

1 LEARNING LINKS

What do you already know about energy?

How many different ways do we use electricity each day?

How many different light sources do you have in your home or school?

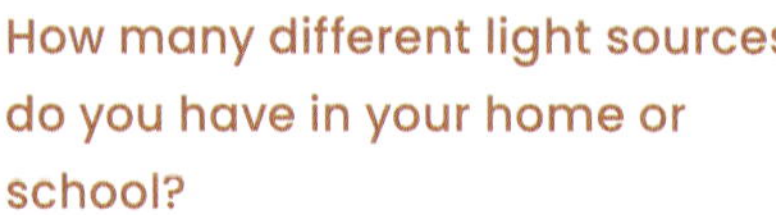

How does heat move from one object to another?

2 SEE-KNOW-WONDER

List three things you can **see**, three things you **know** and three things you **wonder** about this image.

3 CRITICAL + CREATIVE THINKING

Alphabet: Compile a list of words from A to Z that are about energy.

Commonality: List the things that a circuit and a rollercoaster have in common.

Variations: How many ways can you think of to make light?

4 THE MOST INVENTIVE!

Nikola Tesla was a very famous and fascinating scientist. Tesla invented the remote control, and worked on alternating current (which provides the electricity in your house) and the modern car motor. Tesla also had plans to provide free wireless electricity to the world – something that's still not possible today.

Tesla was a fierce rival of American inventor Thomas Edison. The two men even created rival electricity modes. Tesla was an incredible scientist but he lacked social skills, so many of his inventions were misunderstood or thought to be hoaxes.

11.1 KINETIC AND POTENTIAL ENERGY

At the end of this lesson I will be able to:

- **identify** objects that have energy because of their motion (kinetic)
- **identify** objects that have energy because of other properties (potential).

KEY TERMS 

mass
the amount of matter that a physical body contains

motion
the change in position of an object over time

stationary
not moving

transfer
move from one place or object to another

transform
change from one type to another

NUMERACY LINK 

Janine's bicycle has kinetic energy when it's moving and potential energy when it's stopped.

She rides her bike from 8am to 10am, travelling 20 km, then stops for an hour.

From 11am to 2pm she travels 30 km, then stops for an hour again.

Finally she rides 50 km back home, which takes her 3 hours.

Draw a travel graph to represent Janine's day.

The rollercoaster you're riding slowly starts to climb up the track. The clicking gears mark each metre you rise. By the time you reach the highest point, your heart is really pumping. There's only a moment when you're still, before you rush down the other side, in the grip of gravity. Down and back up again, over and over until finally the ride comes to a stop.

What just happened, and can you do it again?

1 A moving object has kinetic energy

The kinetic energy of an object such as a rollercoaster is the energy that it has because of its **motion**. The amount of energy in a moving object depends on its **mass** and its speed.

Kinetic energy can be **transferred** from one object to another and **transformed** into other kinds of energy.

What type of energy do moving objects have?

Figure 11.1 The greater the mass of a moving object, the greater its kinetic energy. The greater the speed of an object, the greater its kinetic energy.

The bicycle and the truck are travelling at the same speed, but the truck is heavier, so it has more kinetic energy.

Both people weigh the same amount, but one is moving more quickly, meaning one has more kinetic energy.

2 Potential energy is stored until release

Potential energy is stored energy that can be used to do work. This work might be hammering a nail, shooting an arrow or cooking dinner. The natural state of objects is to have the lowest energy. There are different ways that energy can be stored and released to do work.

Gravitational potential energy is the energy an object has because of its position or height. An object lifted high has the potential to do work. When the mass falls because of gravity, the object moves and gains speed as it falls. This is gravitational potential energy being transformed into kinetic energy.

Figure 11.2 The greater the height of an object, the more gravitational potential energy it has. The greater the mass, the more gravitational potential energy.

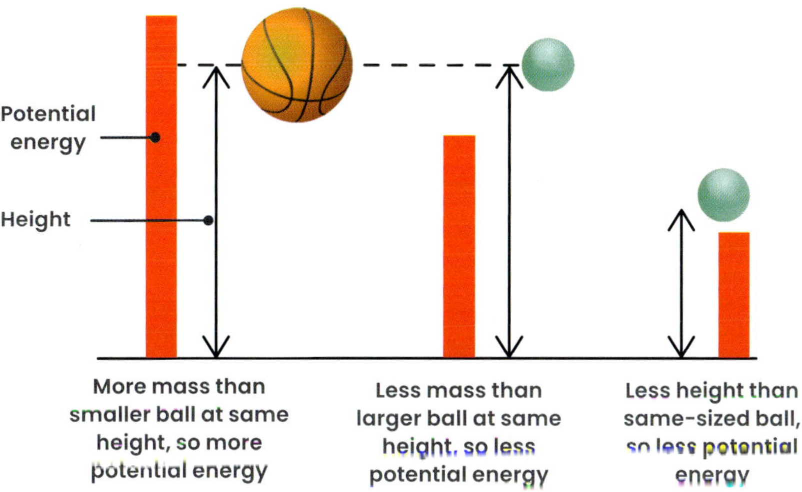

Elastic potential energy can be stored up in an object, such as a spring or an elastic band. If you pull back on an elastic band, you have stored up some energy in the elastic. When you let it go, it returns to its original shape. The stored energy is transformed into kinetic energy. The stronger the elastic band or spring, the greater the elastic potential energy. Also, the more the elastic is stretched, or the spring is compressed, the greater the potential energy.

Chemical potential energy is stored in the bonds between atoms in molecules. Chemical reactions break the bonds and release the energy holding them together. One very common chemical reaction is fire. Heat and light energy are released when the bonds between the atoms that make up wood are broken and rearranged.

What are some ways that energy can be stored?

INVESTIGATION 11.1
Rolling balls

CHECKPOINT 11.1

1 Explain the difference between kinetic and potential energy.

2 Identify which of these objects have kinetic energy.
   a cup
   b bouncing ball
   c stretched elastic band

3 When does an aeroplane have more kinetic energy – sitting on the runway, taking off or in flight?

4 A truck, a car and a motorcycle are all travelling at 60 km/h. Which has the most kinetic energy? Suggest why.

5 Explain the difference between gravitational, chemical and elastic potential energy.

6 Discuss which objects in question 2 could have potential energy.

7 A boy is launching balls of paper to a target using an elastic band slingshot. The problem is that he can't reach the target, no matter how far he pulls the elastic band back. Can you give him advice about what he can do differently?

CHALLENGE

8 Research and compare the amount of potential energy stored in stretching an elastic band (elastic potential) compared to a litre of petrol (chemical potential).

SKILLS CHECK

- I can explain what kinetic energy is.
- I can explain what potential energy is, including naming some types of potential energy.

11.2 HEAT ENERGY

At the end of this lesson I will be able to:

- **describe** the transfer of heat energy by conduction, convection and radiation
- **identify** situations where conduction, convection and radiation occur.

KEY TERMS 

conduction
transfer of heat from one place to another in solids

conductor
a material that is good at transferring heat or electricity

convection
transfer of heat from one place to another in gases and liquids

insulator
a material that is poor at transferring heat or electricity

matter
particles that make up all physical substances; they have mass and take up space

LITERACY LINK 

Using scientific terms, explain what happened in the kitchen disaster at the beginning of this section. These words will help: *conduction, convection, radiation, insulation.*

NUMERACY LINK 

The heat energy radiating from a stove is equal to $5(10 - x)$ joules, where x is the number of metres you are from the stove.

Expand the expression $5(10 - x)$ using the distributive law.

You're cooking a fantastic soup for your family when you burn your hand on a pot handle! Why did it burn you? The handle wasn't over the heat source. Next, you open the pot lid and suddenly your glasses are fogged with steam. This is not going well!

You take a break outside, only to discover that your favourite seat is too hot to sit on. What could be going on here?

1 Heat is conducted through solid objects

Heat is transferred through solids by **conduction**. In a solid, all the particles in the material are vibrating in place. The amount of energy of a substance has to do with how much its particles are vibrating.

When a solid is heated, it gains energy and the particles vibrate more. As more particles vibrate, the heat moves through the solid.

Figure 11.3 Objects will even out in temperature to reach their lowest energy state. Hotter areas will cool and cooler areas will heat until an even temperature is reached.

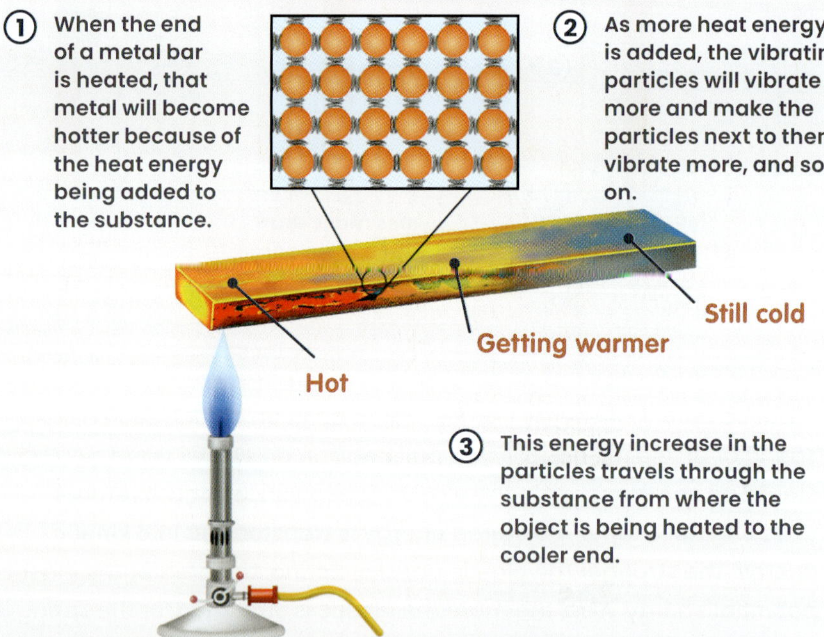

Some solids transfer heat better than others. Substances that transfer heat well are called **conductors**. Metals are usually good conductors of heat energy.

Substances that don't transfer heat well are called heat **insulators**. Substances that are good heat insulators include plastics, glass, foam, rubber, gases and water.

How does heat energy move through solids?

❷ Convection currents move heat through liquids and gases

Liquids and gases don't conduct heat. The particles in liquids and gases can flow past each other and don't just vibrate in place. Heat is transferred through liquids and gases by **convection**.

When you add heat energy to a liquid or gas, groups of heated particles move upwards in a convection current. Colder particles move down towards the heat source, and then join the convection current as they warm up. This continuing movement of warm particles through the convection current eventually warms the entire mass of gas or liquid.

How does heat energy move through liquids and gases?

Figure 11.4 You can see convection currents in a pot of soup cooking on a stove.

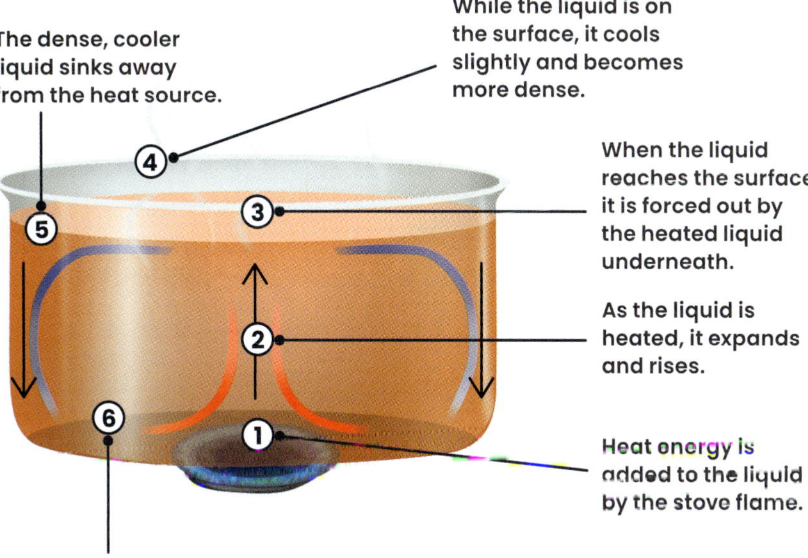

❸ Heat radiates through a vacuum

Heat energy can also transfer from one place to another without **matter**, by radiation. This is how the heat of the Sun travels through the vacuum of space to reach Earth.

Many different types of energy can be transferred by radiation. Heat energy, or infrared radiation, causes objects to heat up when placed near a fire or left outside on a hot day.

What is radiation?

INVESTIGATION 11.2A
Conduction–heat energy transfer in a solid

INVESTIGATION 11.2B
Convection–heat energy transfer in a liquid

INVESTIGATION 11.2C
Radiation and colour

CHECKPOINT 11.2 ✓

1 Copy and complete these sentences using the words *conduction*, *convection* and *radiation*.

a When hot coffee is stirred with a spoon, the spoon gets hot due to __________.

b Warm air over the beach rises while cooler dense air from the ocean rushes in due to __________.

c A metal skewer gets so hot that you drop your marshmallow in the campfire because of __________.

d You try to sit on a rock in the evening after a warm day, but you can't sit on it because of __________.

e You sit on a rock in the evening after a warm day so that you can keep warm by __________.

2 Using your knowledge of heat energy, explain why the handles on some metal pots and pans are made of plastic.

CHALLENGE

3 The water in your friend's plastic water bottle always seems to be warm before she has finished drinking. How could you fix this with items you already have?

SKILLS CHECK

- I can describe and draw a basic diagram of:
  - convection
  - conduction
  - radiation.

11.3 ELECTRICAL ENERGY

At the end of this lesson I will be able to:

- **relate** electricity with energy transfer in a simple circuit.

KEY TERMS 

current
the overall movement of energy in one direction

electric circuit
a path along which electrons flow

load
any part in a circuit that receives power

resistance
how much a substance opposes an electric current through it

LITERACY LINK 

Design an investigation to determine what materials are conductors and what materials are insulators. Write an aim, hypothesis, risk assessment and method for the investigation.

NUMERACY LINK 

An electric current moves through two materials. The first reduces it to $\frac{6}{7}$ of its speed, and the second reduces it to $\frac{4}{9}$ of its speed.

$\frac{6}{7} \times \frac{4}{9} =$ __________

Figure 11.5 Batteries are a convenient and portable way to store chemical potential energy. They come in many sizes and are made of a variety of materials.

We can see the transfer of energy in circuits every time we push a button or flick a switch. We can see the light energy and feel the heat energy from a light bulb.

What is happening inside the wires? Is it something we can see through a microscope, or is it invisible? Will we need a model to describe this science?

1 Circuits transform chemical energy into electricity

An **electric circuit** is an energy converter. The stored chemical energy in a battery is transformed into electric potential energy by the circuit.

Figure 11.6 Electrons move through the the electrical circuit via the electrical wires in a smoke detector.

In an electric circuit, electrons move through the wires and components of the circuit. The electrons receive electrical potential energy from a battery. They then carry this energy around the circuit, but give away the energy when they reach a load. The load could be a light bulb, a motor or a heater coil. The electrons transfer their energy to the component, which will convert that electrical energy into another type. For example, a light globe will convert electrical potential energy into heat and light energy. The electrons then continue through the circuit, returning to the battery, where they will receive more energy, and repeat the cycle again.

What type of energy is stored in a battery?

2 Energy and current are different

A battery gives electrical potential energy to the electrons in a circuit, but this is not the same as the **current** of the circuit. The current is how fast the energy moves through the circuit. A battery might hold a lot of potential energy, but the circuit may deliver it slowly, so the current is small.

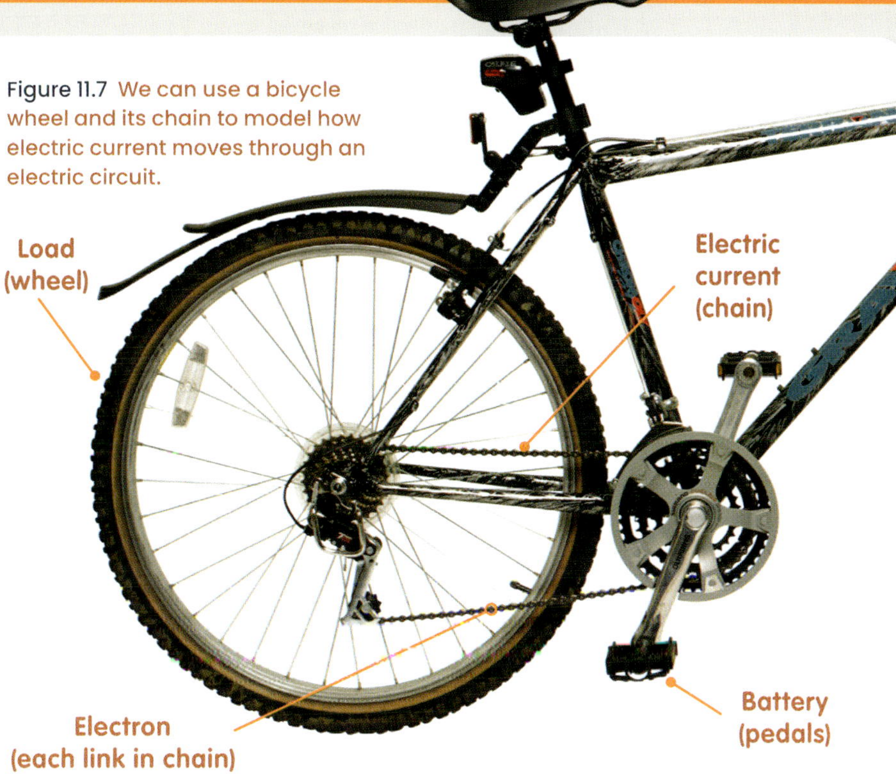

Figure 11.7 We can use a bicycle wheel and its chain to model how electric current moves through an electric circuit.

A bicycle wheel and chain can be used to model an electric circuit. The current is the entire chain, and each link of the chain represents an electron. The pedals are the battery pushing the electron around the circuit, while the wheel is the load in the circuit, such as a light bulb. As soon as you push the pedals and they start to turn, the back wheel begins to move. A link in the chain might take a long time to go all the way around, but the wheel will turn immediately and for as long as the pedals do.

Does a circuit transfer energy quickly or slowly?

3 Materials provide resistance to electric currents

Something else that affects the speed of a current is the **resistance** of the circuit. Some materials allow electrons to move easily through them, while others stop or slow the current.

A material that contains movable electrons (or other charged particles) is called an electrical conductor. These substances have the lowest resistance to current moving through them. Pure silver, gold and copper are good electric conductors. Most wiring is made of copper because it's cheaper than gold and silver, and easier to get.

A material that doesn't contain movable electrons is an insulator. These substances have the greatest resistance to current moving through them. Plastic, ceramic and rubber tend to be good insulators.

Electric wiring is covered in insulation, which acts as a barrier. The current can't pass through the insulation, and so the wires are much less dangerous to people and other electrical objects.

What is electrical resistance?

CHECKPOINT 11.3 

1 Match the electrical objects to their energy types. (There could be more than one correct answer.)

speaker	light energy
toaster	chemical energy
light	sound energy
battery	kinetic energy
fan	heat energy

2 The model of a bicycle chain has been used to describe energy transfer in an electric circuit.
- a Describe what the pedal, chain and wheel represent in a real circuit.
- b What does this model explain well and what is left out? Suggest improvements.

3 Electrical tape is made of a flexible plastic with sticky adhesive on one side.
- a What is electrical tape used for?
- b Give reasons why plastic is a suitable material for electrical tape.

CHALLENGE

4 Investigate different types of batteries and their uses. List strengths and weaknesses for each type of battery.

SKILLS CHECK

- I can describe how a circuit demonstrates energy transfer.
- I can explain the difference between a conductor and an insulator.

11.4 MAKING ELECTRIC CIRCUITS

At the end of this lesson I will be able to:

- **construct and draw** circuits containing a number of components to show a transfer of electricity.

KEY TERMS 

component
a part of an electric circuit in an electrical device

diode
an electrical component that allows electric current to flow in only one direction

electric circuit
a path along which electrons flow

LITERACY LINK 

Write a practical guide to reading circuit symbols and creating circuits, aimed at younger students.

NUMERACY LINK 

Dion and Yvette buy LEDs to build a circuit. Dion spends $2.50 less than Yvette, and together they spend $14.50.

If Yvette spent x, write an equation to represent the total cost of the LEDs.

Have you ever wondered what is going on inside your electronic devices? Every device you own, from the simplest torch to the most sophisticated computer, works because of the electrical **components**.

Electrical components are part of **electric circuits**, which can be shown using circuit diagrams.

1 Electricity will only flow through a complete circuit

Electric circuits can be open or closed. A closed circuit is also known as a complete circuit.

A closed or complete circuit allows electricity to flow. An open or incomplete circuit is like a hose that has been cut, so water cannot flow to the sprayer or sprinkler. Any gaps will cause an incomplete circuit and stop the current from flowing through the circuit.

Many circuits contain switches, which are used to open and close circuits. When you turn on a light, the switch changes the open circuit to a closed circuit and allows energy to flow through the wires to operate the light bulb.

What is another name for a closed circuit?

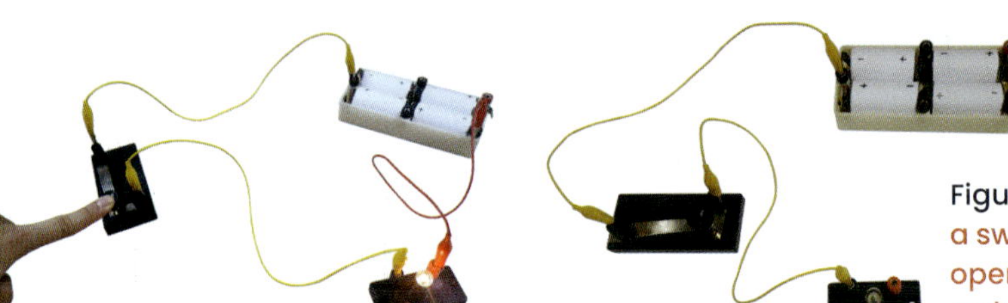

Figure 11.8 Turning on a switch changes an open circuit (right) to a closed circuit (left).

2 Electric circuits can be planned and drawn using diagrams

An electric circuit can be recorded as a picture called a circuit diagram. Circuit diagrams use symbols to show how the different electrical components are connected.

Diagrams are used to record circuits that would be too confusing to show as pictures of each component. They are more efficient to draw, easier to read, and so circuits can be built more quickly from them.

How is an electric circuit recorded?

Figure 11.9 Circuit diagrams show the components of electric circuits in a simple and standard way.

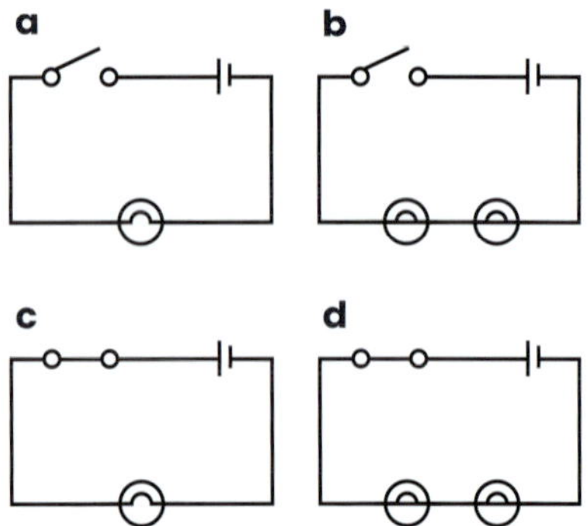

3 Circuit components are shown using standard symbols

Component symbols are nearly always shown the same way. This means that anyone who knows these symbols can read a circuit diagram. Table 11.1 shows symbols for the most common components, and what each component looks like.

Table 11.1 Common electrical components and their symbols

Component	Picture	Symbol
Open switch	ON OFF	
Closed switch	ON OFF	
Lamp (e.g. light globe)		
Light-emitting **diode** (LED)		
Cell		
Battery (multiple cells)		
Wire		

Follow these steps when drawing circuit diagrams:

1 Use a sharp pencil and a ruler.

2 Draw the symbols in the correct places. Symbols should not be drawn at corners.

3 Use a ruler to connect the symbols with wires. Wires should generally only do 90° angle changes.

4 Make sure that the wires don't cross.

5 Check the circuit by tracing around from the battery through each component and back to the battery.

How is a closed switch drawn?

INVESTIGATION 11.4
Investigating conductors and insulators

CHECKPOINT 11.4

1 a Identify each of the circuits in Figure 11.9 as open or closed.
b In which of the circuits will the bulbs light up?

2 Draw these components in a simple circuit: battery, open switch, bulb.

3 Draw the components in question 2 in a closed circuit and add arrows to show the direction of the flow of energy.

4 Electricity will only flow through a complete circuit. Explain why.

5 Give some advice about how to draw an accurate circuit diagram.

CHALLENGE

6 Design and create a card for a special occasion using LEDs, wires and a switch. Draft the circuit diagram and check with your teacher before building your circuit. Be creative!

SKILLS CHECK

- I can draw a simple closed circuit that shows wires, a switch, lamp and battery.
- I can identify the different components of a circuit from their symbols.
- I can create a circuit from a circuit diagram.

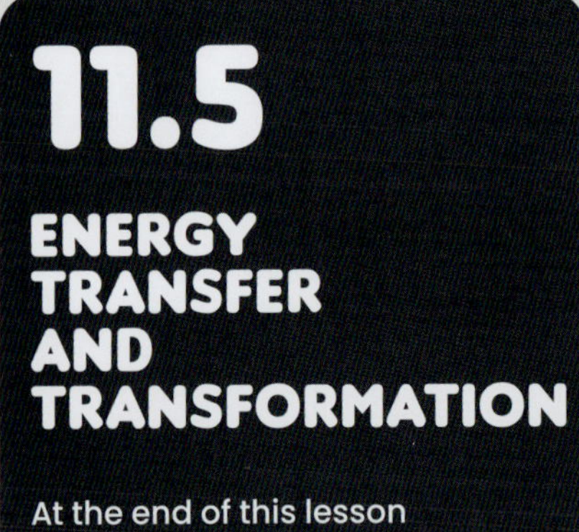

11.5 ENERGY TRANSFER AND TRANSFORMATION

At the end of this lesson I will be able to:

- **investigate** some everyday energy transformations that cause change within systems.

Figure 11.10 A Rube Goldberg machine performs a simple task in a complicated way, using many transfers and transformations of energy.

KEY TERMS 

system
a set of simple things that work together as a more complex whole

transfer
move from one place or object to another

transform
change from one type to another

LITERACY LINK 

Write a step-by-step account of the energy transformations involved in ten-pin bowling.

NUMERACY LINK 

Kenan builds a Rube Goldberg machine, and determines that it transforms energy using the formula $y = \frac{x^2}{2}$.

Use the formula to copy and complete this table of values.

x	0	2	4	6	8
y					

All energy on Earth comes originally from the Sun. Nuclear reactions in atoms power our Sun, and this energy is **transformed** into light and heat. The Sun then radiates an enormous amount of light and heat energy into space in all directions.

Only a small amount of this energy reaches Earth, but it is then stored, released, **transferred** and transformed over and over again.

1 Energy causes change in systems in many ways

Energy transfer and transformations cause change in **systems**. These changes are happening all the time in the world around us.

Energy transfers happen when one object passes energy to another object. This changes those objects. An example is when you throw a ball – the kinetic energy of your hand transfers to the ball when you let it go.

Energy transformations happen when one form of energy is changed to another form of energy in the same object. This also causes change in that system. An example is when the gravitational potential energy of a ball changes into kinetic energy as it falls and gains speed.

What causes change in systems?

2 Energy can transfer and transform in many ways

Rube Goldberg machines, such as the one in Figure 11.10 , are made up of a series of objects arranged to transfer energy in a sequence. Each small part triggers the next until eventually the end result is achieved. These machines mostly just exist in cartoons (and some video clips), but examples of energy transfer and transformation are everywhere.

Figure 11.11 How many energy transformations are happening here?

Imagine that you threw a book and it fell to the ground with a bang. How many energy transfers and transformations would happen?

There are many different forms of energy, and Table 11.2 shows a few examples. In a suitable situation, any one of these forms could be transformed into one of the other forms.

Table 11.2 Examples of different forms of energy

Energy form	Description
Gravitational potential energy	The potential of an object to do work because of its relative position
Elastic potential energy	The energy stored in an object when it is stretched or deformed
Chemical potential energy	The energy stored in the bonds between atoms, released as heat, sound or light during chemical reactions such as burning
Nuclear potential energy	The energy stored in the nucleus of atoms. It may be released quickly as a nuclear explosion or controlled slowly in a nuclear reaction
Kinetic energy	The energy that an object has because of its movement
Heat energy	The motion of atoms and molecules in a substance or object
Light energy	A form of electromagnetic radiation emitted by hot objects. It is usually visible light
Sound energy	The vibration of particles that can be detected by the ear
Electrical energy	The energy generated by moving electrons

Describe one energy transformation.

CHECKPOINT 11.5

1 Identify the following as either a transfer or transformation of energy.
- a kicking a ball
- b lighting a candle
- c hitting a drum
- d a light bulb turning on

2 Identify what situation could result in the following transformations of energy.
- a chemical potential energy → heat and light energy
- b electrical energy → sound energy
- c chemical potential energy → heat and kinetic energy
- d gravitational potential energy → kinetic and sound energy
- e electrical energy → heat and light energy

3 Consider these two columns of words. What could the titles of each column be? Explain your reasoning in sentences.

Title 1	Title 2
Coal	Movement
Natural gas	Heat
Wood	Nuclear
Food	Light
Wind	Electrical
Uranium	Chemical

CHALLENGE

4 Create your own Rube Goldberg machine using everyday objects. Record the energy transfers and transformations in your machine as flowcharts.

SKILLS CHECK

- I can describe the difference between the transfer and transformation of energy.
- I can give at least one example of an everyday energy transformation.

CHAPTER SUMMARY

Energy
A non-moving object has potential energy that transforms into kinetic energy when it moves.

The amount of energy in a moving object depends on its mass and speed.

Heat is a form of energy that can be transferred in three different ways

conduction
transfer of heat from one place to another in solids

convection
transfer of heat from one place to another in liquids and gases

radiation
transfer of heat from one place to another that does not rely on any contact between the heat source and the heated object

Circuits transform the chemical energy inside batteries into electrical energy.

Energy transfers and transformations happen every day, such as when fire transforms chemical energy into heat energy

★ FINAL CHALLENGE ★

1. Petrol contains chemical potential energy. What is the main type of energy that this is converted to when used by a car?
2. What two types of energy does Earth receive from the Sun?

LEVEL 1 ★☆☆☆☆☆ 50xp LEVEL UP!

3. Give an example of an object that you have encountered today that contains:
   a Gravitational potential energy
   b Chemical potential energy
   c Elastic potential energy
4. In your own words, explain the difference between current and energy.

LEVEL 2 ★★☆☆☆☆ 100xp LEVEL UP!

5. Copy and complete the following sentence, choosing the correct words from the underlined options.
   For an object to have maximum kinetic energy, it must have a large/small mass and a high/low speed.
6. Most wires used in electric circuits are coated in rubber or plastic. Suggest why.

LEVEL 3 ★★★☆☆☆ 150xp LEVEL UP!

7. Draw a diagram of a pot of water being heated on a stove. Label where conduction, convection and radiation are occurring.
8. Explain why electricity will not flow through Christmas lights if one of them is broken.
9. Draw a flow chart showing the energy transformations that occur when a battery-powered torch is turned on.

LEVEL 4 ★★★★☆☆ 200xp LEVEL UP!

10. Do you think solids can experience convection? Explain your answer.
11. Draw a circuit diagram that shows a lightbulb connected to a battery that can be operated by a switch.

LEVEL 5 ★★★★★★ 300xp LEVEL UP!

12 ENERGY AS A RESOURCE

We need energy for everything we do. Living things need it to survive and grow, households need energy for light, heat and appliances, and businesses need it to produce all of the things that we use every day.

We require so much energy, but we're not always good at producing it or conserving it. Most of the energy we use is wasted. An example of wasted energy is the heat produced by many light bulbs – we don't use that heat, just the light. Ways to make appliances and processes use energy more efficiently is one focus of science and technology.

LEARNING LINKS

What do you already know about energy as a resource?

What does it mean when an energy source is sustainable?

How is electricity generated?

How do some devices transform electrical energy to heat energy, light, sound or movement, e.g. hair dryers, light bulbs, doorbells and fans?

2 SEE-KNOW-WONDER

List three things you can **see**, three things you **know** and three things you **wonder** about this image.

3 CRITICAL + CREATIVE THINKING

Disadvantages: List some disadvantages of a hair dryer, then suggest improvements for each disadvantage.

Interpretations: Think of some ways to explain this situation: A small pinch of powder thrown into a bin causes a huge explosion.

B-A-R: Consider a normal household light bulb. What could you make *bigger*, what could you *add* and what could you *replace* in order to improve it?

THE MOST FAMOUS!

One of the most famous and recognisable equations in the world is Einstein's formula $E = mc^2$. It seems like a simple formula, but it tells us some amazing things about the way things work.

The formula tells us that energy (E) is equal to mass (m) times the speed of light (c) squared. That means that all things with mass have a colossal amount of energy.

12.1 ENERGY EFFICIENCY

At the end of this lesson I will be able to:

- **identify** that most energy conversions are inefficient and lead to the production of heat energy, e.g. in light bulbs.

KEY TERMS 

efficient
reduces or avoids waste

friction
a force opposing the motion of surfaces in contact

joule
a unit of energy for kinetic and potential energy

law of conservation of energy
a law stating that energy cannot be created or destroyed

transfer
move from one place or object to another

transform
change from one type to another

LITERACY LINK 

Prepare a short news article about the law of conservation of energy, including what it is, some examples and why it is an important law in physics.

NUMERACY LINK 

An energy efficient car uses 8 litres of fuel to travel 200 km. How far could it travel using 20 litres?

Figure 12.1 A bicycle is a system that transforms the energy in your muscles into energy for movement.

Get on your bike! When you do a physical activity such as cycling, your muscles change the stored chemical energy your body makes from food to kinetic energy. This energy is transferred from the pedals, through the bicycle chain to the wheels.

And you're off! But, as you ride, you hear an annoying squeak in the gears. When you stop to check your bike, you notice that the wheels feel hot and so do the brakes. Why is this?

1 Energy cannot be created or destroyed

All the energy we use has come from somewhere, and it can't be destroyed. Energy can be stored, **transformed** to other energy types, or it can spread out in the environment, and **transfer** between objects.

This is the **law of conservation of energy**: energy cannot be created or destroyed. It can also be stated like this:

total initial energy input = total energy output

When energy transforms to another type, some of it often 'escapes' into the environment. This is usually as heat, sound or vibration. This energy is generally caused by parts of a system rubbing together and causing **friction**, like the brakes on your bike. So unless the device is made to create heat or sound, then some energy is wasted.

Some systems transfer less energy to their environment than others. These systems are more **efficient**.

What is the law of conservation of energy?

INVESTIGATION 12.1
Energy efficiency

❷ No energy transfer is perfectly efficient

When looking at energy transfers and changes, we're often interested in a specific system. This could be a house, a bicycle, a human body or any other space in which a process takes place.

No system is perfectly efficient. This means that waste energy is always produced when energy is transferred or transformed. This waste energy moves out of that system and into the larger environment.

Efficiency measures the proportion of useful energy left after a transfer or transformation. It can be calculated as a percentage:

$$\text{efficiency} = \frac{\text{useful energy output (J)}}{\text{total energy input (J)}} \times 100\%$$

A good example is wasted heat energy. Old electric light bulbs are not very efficient – they produce light, but also a lot of heat. A typical light bulb might have an input of 100 **joules** (J) of electrical energy, but have an output of only 10 joules of light energy. The efficiency calculation for the light bulb is:

$$\text{efficiency} = \frac{10\text{ J}}{100\text{ J}} \times 100\% = 10\%$$

Figure 12.2 Early light bulbs were inefficient and produced a lot of heat energy as well as light.

How can energy efficiency be calculated?

❸ Sankey diagrams show energy transfers in systems

A Sankey diagram can show all the energy transfers in a system. These diagrams show all the energy flows into and out of a system as lines and arrows of various sizes. Thicker lines and arrows mean more energy.

A Sankey diagram can also show waste energy leaving a system. This energy is shown as a curved arrow pointing in a different direction to the useful energy.

Figure 12.3 The Sankey diagram for a light bulb shows that most of the electrical energy transforms to heat rather than light.

How can energy transfers be shown using a diagram?

CHECKPOINT 12.1

1 Identify the useful and non-useful types of energy produced by:
   a a smartphone
   b a washing machine
   c riding a bicycle.
2 Draw a Sankey diagram for each of the situations in question 1 to show the energy input and outputs.
3 Explain why energy transformations aren't 100% efficient.
4 Calculate the amount of waste heat energy for an energy-efficient light bulb that uses 70 J of electrical energy to produce 60 J of light.
5 Calculate the efficiency of the energy-efficient light bulb in question 4.

CHALLENGE

6 Investigate perpetual motion machines in a web search. Explain how they would break the law of conservation of energy.

SKILLS CHECK

- I can explain how and why energy is lost during energy transfer or transformation.
- I can use the example of a light bulb to describe how the production of heat energy can demonstrate energy inefficiency.

12.2 ENERGY SOLUTIONS

At the end of this lesson I will be able to:

- **research** ways in which scientific knowledge and technological developments have led to finding a solution to a contemporary issue, e.g. improvements in devices to increase the efficiency of energy transfers or conversions.

KEY TERMS 

insulation
material that acts as a barrier to heat flow

passive design
building design that considers climate and location to make the best use of natural heating and cooling

thermal
to do with heat

LITERACY LINK 

Create a social media campaign that encourages people to be more energy efficient. Create a slogan as well as three key messages for your campaign.

NUMERACY LINK 

A solar power system costs $5000 to install, but saves the homeowner $750 per year. How long will it take until the cost of the system is paid back in full?

Figure 12.4 In the future, homes and workplaces will need to be more energy efficient.

Imagine a home of the future. It's made from reused and recycled materials and exists in harmony with the surrounding environment. Every appliance communicates to keep you comfortable, while being as efficient as possible.

Would you like to live there? Let's find out how you can.

1 Passive design helps regulate a building's temperature

Passive design is building design that considers the climate and where a building will be located to keep it at a comfortable temperature without using appliances for heating or cooling.

A building made with passive design has features such as:

- windows and doors that are placed to allow good air flow
- few windows facing the hot afternoon sun
- high and low vents so that cool, fresh air replaces hot, stale air during summer and reduces heat loss during winter
- roof angles that maximise shade during summer and heat absorption during winter
- plants that give shade and cool the air, like the canopy in a rainforest.

These features can reduce or eliminate the need for heating and cooling that relies on electricity.

What does passive design aim to do?

2 Building materials can help control temperatures

Insulation acts as a barrier to heat flow. It keeps homes warm in winter and cool in summer. Insulation is rated using a system called R-values – materials with higher R-values give better insulation.

There are two different types of insulation: bulk and reflective. Bulk insulation relies on air trapped in fibres to slow the transfer of heat energy, such as when you wrap yourself in a blanket to keep warm.

Bulk insulation can be made of glass wool, cellulose fibre, polyester, wool or polystyrene. Reflective insulation uses its shiny surface to reflect heat energy away.

The colour of building materials also affects how much heat energy a building absorbs. Dark, dull surfaces absorb more heat energy than light, shiny surfaces.

What type of building material acts as a barrier to heat flow?

3 Energy sources should be renewable and used efficiently

A lot of the energy used in Australian households is electrical energy. If household electricity came mostly from renewable sources, demands on non-renewable energy sources could be reduced.

Solar energy is a common renewable energy source in Australia. Solar **thermal** systems can be used with hot water systems, and they transfer the Sun's heat energy to water. Solar power cells absorb the Sun's energy and transform it to electrical energy. This energy can be stored in batteries for use at night. Solar cell technology is becoming more efficient and affordable.

Light bulbs, washing machines and refrigerators are examples of appliances that need electrical energy to function.

Energy-efficient appliances function using minimum energy. Three ways to reduce the energy use in your house are:

- avoiding unnecessary use of appliances
- choosing energy-efficient appliances
- switching appliances off when not in use.

What is a common renewable energy source in Australia?

Figure 12.5 This Australian house uses passive design, such as the design of its roof, to help control the temperature inside.

INVESTIGATION 12.2
Insulation and heat transfer

CHECKPOINT 12.2

1 Copy and complete the following sentences.
  a Passive design is ______________.
  b Household solar cells can ______________.
  c Choosing energy-efficient appliances is important because ______________.
  d Improved building materials allow ______________.
  e Insulation is used in buildings to ______________.

2 James says he designed his home with white roofing because it will help keep the building cool. Xing says she chose black roof tiles because this colour stops the heat passing through the tiles. What do you think?

CHALLENGE

3 Look around your home and make a list of any energy-saving features. Research methods to increase the energy efficiency of your home and suggest how these measures could be put in place.

SKILLS CHECK

- I can describe some ways that homes can be more energy efficient.
- I can make suggestions about improving energy efficiency in buildings, devices or other technology.

12.3 HEAT ENERGY AND THE ENVIRONMENT

At the end of this lesson I will be able to:

- **discuss** the implications for society and the environment of some solutions to increase the efficiency of energy conversions by reducing the production of heat energy.

KEY TERMS 

circulate
move freely or continuously through an area

smog
a form of severe air pollution

temperature inversion
weather conditions that stop fresh air from circulating

NUMERACY LINK 

Coulson measures the air temperature in his neighbourhood over the course of an autumn day.

It climbs from 9°C at 9am to 15°C at 3pm, falls to 10°C by 6pm, and is 8°C when Coulson takes his final measurement at 9pm.

Draw a line graph to represent the temperature changes over the day.

The energy we use at home and in factories often generates waste heat energy. This can worsen pollution, which affects air quality.

To reduce pollution, we need to use energy wisely and generate energy that doesn't waste a lot of heat energy.

1 Warm pollutants can affect air quality

Under normal conditions, the Sun's energy heats the ground, which then heats the air just above the ground. This warmer air rises as cooler air sinks below it. In this way, fresh air can **circulate**.

During a **temperature inversion**, the layer of air close to the ground is cooler than the air above it, which prevents fresh air from circulating. The warm air is sometimes a result of pollutants such as those from car exhausts and factories, which can make **smog** in the air. This inversion can be made worse in a place such as a valley, where air circulation is prevented even more by the surrounding hills.

Figure 12.6 In normal conditions (left), air is able to circulate. In a temperature inversion (right), warm air traps cooler air, preventing circulation.

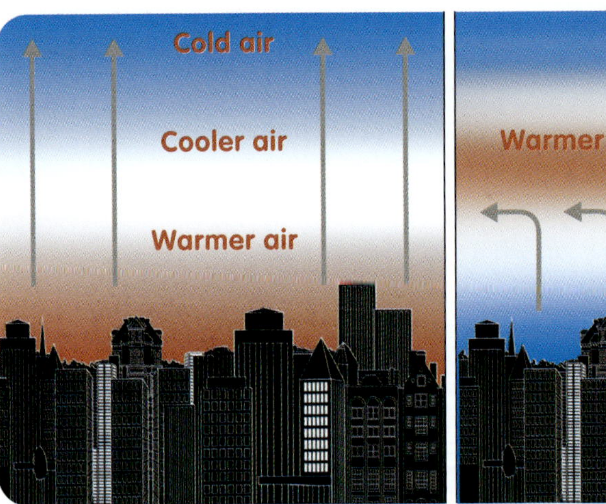

Smog contains a substance called ozone, which is a molecule made up of three oxygen atoms. Increased ozone levels can be bad for health. Breathing in smog can cause problems such as:

- difficulty breathing and lung damage
- coughing and throat irritation
- asthma symptoms.

Children, elderly people and those with respiratory conditions are at most risk of health problems from smog.

What can cause smog to be trapped around cities?

2 Reducing heat energy waste has many advantages

There are two main ways to reduce heat energy waste in industry and at home. The first is to use improved insulation materials. Better insulation reduces heat loss, so less energy is required to heat buildings. The other way is to recover 'waste' heat using heat pumps or steam recyclers. This waste energy can be used to keep us warm, or be recirculated to avoid release into the environment.

Reducing heat energy waste helps protect our health and the environment. But there are advantages for industries too, such as saving money on energy and repairs, and having lower emissions of other pollutants.

What are some advantages of reducing pollution?

3 Renewable energy sources make less heat

The electrical energy we use is generated in a variety of ways. It can be made from non-renewable and renewable energy resources.

Non-renewable energy resources such as coal, oil and natural gas are burned to release heat energy. This heat energy is transformed to electrical energy in fuel-fired power stations. A lot of this heat energy leaves the system in the process.

Renewable energy sources, such as wind power, hydro-electricity and solar power, don't make heat as burning fossil fuels does. This is because these energy sources can directly turn turbines to generate electrical energy, rather than heating water to steam to turn turbines.

Which energy sources produce minimal waste heat energy?

Figure 12.7 When coal is burnt in a power plant, the heat is used to boil water that turns a turbine.

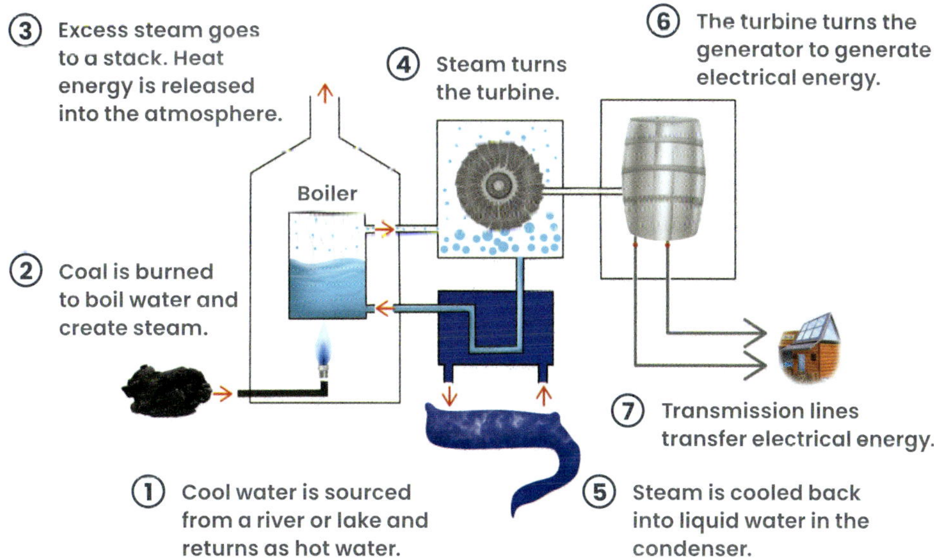

INVESTIGATION 12.3
Atmospheric convection and the inversion layer

CHECKPOINT 12.3

1 Explain what heat energy waste is and how it is generated.

2 Describe some health issues caused by smog.

3 True or false? Write the correct statement if a statement is false.
  a Wind, hydro-electricity and solar energy are renewable sources of energy.
  b Waste heat energy can't be reused.
  c An atmospheric submersion layer can trap polluted cold air around cities.

4 Andrea wants to buy an electric car because she thinks it would help reduce the use of fossil fuels. Gabriel disagrees. Who do you think is right? Justify your choice.

CHALLENGE

5 Research the impacts of thermal pollution on a waterway. Which animals are most affected? How could these impacts be minimised? Brainstorm solutions and select one option to develop further. Justify why you have selected this option.

SKILLS CHECK

- I can explain how energy inefficiency can affect society and the environment.
- I can suggest some solutions that could improve energy efficiency.

CHAPTER SUMMARY

Energy as a resource
Energy cannot be created or destroyed

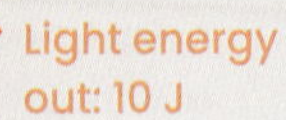

Reducing heat energy waste has many advantages
- lower energy costs
- lower maintenance costs
- lower emissions of other types of pollutants
- better workplace efficiency.

Electrical energy in:
100 J
Light energy out: 10 J
Heat energy out: 90 J

When energy transforms to another type, some of it escapes into the environment. This is usually as heat, sound or vibration.

Kinetic energy out: motion of bicycle

Kinetic energy in: muscles turn pedals

Waste energy out: heat (friction) and sound

Three ways to reduce the energy use in your household

avoid unnecessary use of appliances

choose energy efficient appliances

switch appliances off when not in use

Waste energy in the environment can cause pollution, such as smog

★ FINAL CHALLENGE ★

1. Explain what is meant by passive design, in your own words.
2. What is a Sankey diagram used to show?

LEVEL 1

50xp

LEVEL UP!

3. Outline the difference between the transfer and transformation of energy.
4. Describe how solar thermal systems operate.

LEVEL 2

100xp

LEVEL UP!

5. Think of three pieces of advice that you could you give someone who wants to have a more energy efficient home.
6. Renewable energy sources create less wasted heat energy than non-renewable sources; suggest why.
7. Identify some health conditions caused by the inhalation of smog.

LEVEL 3

8. Draw a diagram of a race car, then annotate where it is both using and losing energy.
9. Why do you think energy efficiency is so important? Justify your response.

LEVEL 4

10. Design a café that operates as energy efficiently as possible. Annotate your design to detail its features.
11. State the law of conservation of energy and explain how it relates to energy efficiency.

LEVEL 5

CLASSIFICATION

13

Scientists use classification systems to organise and identify all living things. This allows them to compare and investigate species, including how closely related they are.

When the first platypus was sent from Australia to scientists in England in the late 18th century, they thought it was a prank, and that someone had sewed a duck bill onto a large water rat. Given a platypus has a bill, is furry but lives in water, lays eggs and produces venom, it does seem a bit strange! After their initial surprise, scientists were later able to classify the platypus as a monotreme – a special kind of mammal that lays eggs.

1 LEARNING LINKS

What do you already know about classification?

Living things can be organised into groups based on their observable features. How many examples of this can you think of?

Living things are different from non-living things. How do we know this?

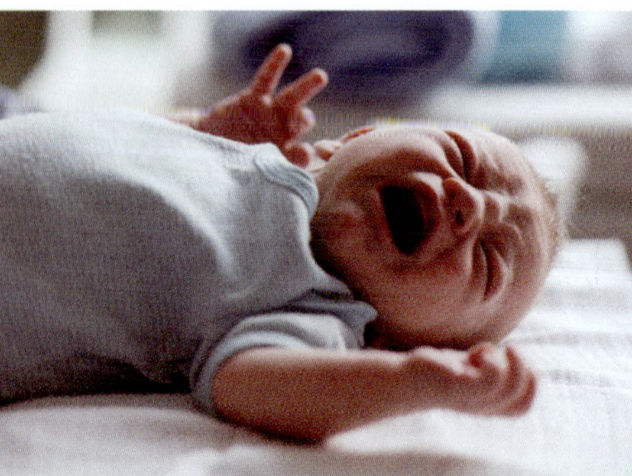

2 SEE-KNOW-WONDER

List three things you can **see**, three things you **know** and three things you **wonder** about this image.

3 CRITICAL + CREATIVE THINKING

Prediction: What could happen if supermarkets decided not to organise products into groups?

Variations: How many different ways can you classify the entire contents of your pencil case?

Interpretations: Imagine that scientists discover a horse that has a horn on its forehead. Think of some different explanations for its existence.

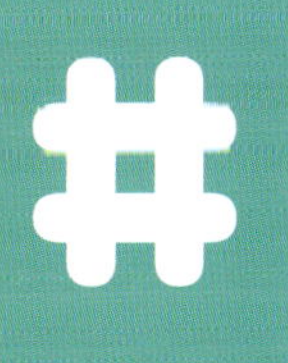

4 THE CLOSEST!

The closest known relative to humans is a species of great ape called the bonobo. The bonobo is an endangered species – these apes are at risk of becoming extinct. They are found in the Congo in Africa and are omnivores, which means they eat both plants and animals, like humans do.

Bonobos are social animals and show empathy, kindness, compassion and patience. Bonobos can recognise themselves in a mirror, which means they are self-aware, like all the great apes.

13.1 CLASSIFYING LIVING THINGS

At the end of this lesson I will be able to:

- **identify** reasons for classifying living things.

KEY TERMS 

classification
the process of sorting things into groups or classes

organism
an individual animal, plant or other living thing

taxonomy
the area of science to do with classifying organisms

LITERACY LINK 

Write a speech that explains why classification is important, giving reasons why scientists classify organisms.

NUMERACY LINK 

Helena classifies all of her music into six different categories and numbers them from 1 to 6.

If she rolls two 6-sided dice to choose a song, what is the probability that both dice will show the number 6?

Figure 13.1 There are many different types of bear on Earth. Classification helps us identify them by organising them into groups.

Malayan sun bear

Classification is the process of sorting of things into groups.

Different classification systems exist in many areas in your daily life. For example, your school library classifies books using the Dewey decimal system, which is useful for finding both subjects and authors. You might use a classification system to store your music or video files.

❶ Classification means sorting things into groups

Things can be classified for many different reasons, depending on who is doing the arranging, sorting or classifying.

Many scientists use classification. They do so by looking at how the objects they study are similar:

- Chemists classify substances by their physical properties and their chemical reactions with other substances.
- Geologists classify rocks according to how they were formed and their mineral content, such as sedimentary, igneous and metamorphic.
- Astronomers classify objects in the universe by observing properties such as size, density and whether they give off light.

What kinds of things can be classified?

❷ Living things are classified by their characteristics

Biologists classify living things. The science of grouping living things is called **taxonomy**, and a biologist who specialises in classification is a taxonomist.

Millions of different types of **organism** live on Earth. All living things:

- are made of cells
- are made of molecules containing carbon
- have biological characteristics in common, such as being able to grow, move, reproduce and respond to stimuli.

Many other characteristics are shared by only some organisms. Plants and animals have very different characteristics, of course, but the differences between types of animal are also very significant. Consider the differences between dogs and octopuses!

What is the name of the science of grouping living things?

❸ The importance of classification

One of the most important reasons to classify living organisms is to work out what relationships exist between different groups. Is a cat closely related to a tiger? They have similarities, but some very significant differences. Classification systems help biologists learn how these two animals, or any other two living things, are related.

A classification system helps biologists identify newly discovered organisms. These discoveries happen more often than you might think – the more we explore the world, the more new organisms we find. Many of these, such as bacteria, are tiny – but new plants, insects, fish and other animals are also discovered every year.

A classification system helps biologists from different countries to communicate. An Australian wildlife scientist may use different everyday names for trees and plants than a Chinese, French or Indonesian scientist, but if they use the same classification system, they can identify the organism more easily.

What are the benefits of classifying living things?

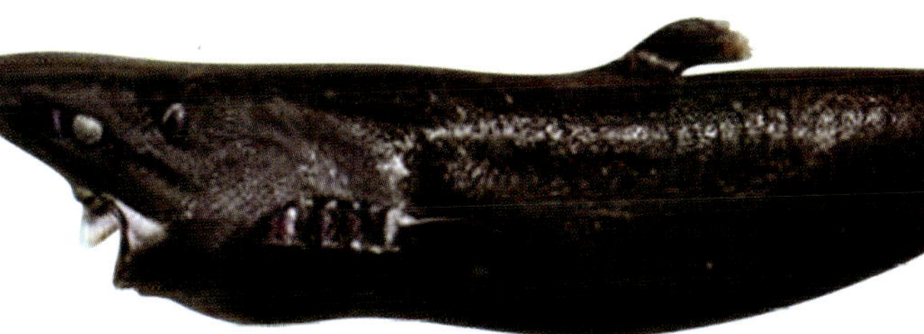

Figure 13.2 The ninja lanternshark is a small shark with body parts that glow in the dark. It lives deep in the ocean and was not discovered by scientists until 2015.

INVESTIGATION 13.1
Observing and classifying

CHECKPOINT 13.1

1 Describe classification in your own words.
2 Give at least three reasons for classifying living things.
3 Explain what is meant by the term 'taxonomy'.
4 Identify four situations where things are classified in everyday life.
5 Classify the following objects into three groups: an apple, a car, a fish, a basketball, a mobile phone and a drop of blood. Justify your choice of classification.
6 Give some characteristics of living things.
7 Classification can help identify which species are closely related. Suggest how.

CHALLENGE

8 When we classify living things, the first two categories that usually come to mind are plants and animals. Undertake further research to identify the other categories of living things.

SKILLS CHECK

- I can describe what classification is.
- I can give three reasons why classification is useful.

13.2 THE LINNAEAN CLASSIFICATION SYSTEM

At the end of this lesson I will be able to:

- **identify** reasons for classifying living things

KEY TERMS 

naturalist
someone who studies nature and its history

species
a single, specific type of living organism

LITERACY LINK 

Create a mnemonic to remember the order of classification: domain, kingdom, phylum, class, order, family, genus, species. An example is 'Dear Katy Perry Can Often Find Green Shoes'.

NUMERACY LINK 

There are 8 levels in the Linnean classification system.

What are the first six multiples of 8?

Figure 13.3 All plants and animals are in the same domain, but brush-tailed rock wallabies are the only type of animal of their species.

Scientists can classify living organisms in many different ways. What's most important is that every scientist uses the same system, so that their findings can be compared and they can share information.

A system for grouping living things was first proposed by Swedish **naturalist** Carolus Linnaeus in 1753. It's known as the Linnaean classification system and has been used by scientists ever since.

❶ Eight Linnaean levels, from domain to species

Carolus Linnaeus' system classifies organisms by levels called taxa, and each of these levels is divided into groups. These groups are based mainly on cell and structural characteristics, and also on how the organisms behave and reproduce.

At the highest level, these groups are huge – a single group might contain all plant life in the entire world. As you go down the levels, the groups become more specific.

At the top of the system, the most general classifications are domains and kingdoms. The characteristics used to separate these groups are based on cell structure. At the bottom of the system are genuses and **species**. This is where you can identify individual types of organisms.

Here is how the brush-tailed rock wallaby is classified:

Classification level	Classification for brush-tailed rock wallaby
Domain	Eukaryota (living organisms with specialised cells)
↓	↓
Kingdom	Animalia (animals)
↓	↓
Phylum	Chordata (animals with a backbone)
↓	↓
Class	Mammals
↓	↓
Order	Marsupialia (marsupials)
↓	↓
Family	Macropodidae (wallabies, kangaroos, pademelons)
↓	↓
Genus	*Petrogale* (wallabies)
↓	↓
Species	*penicillata* (brush-tailed rock wallaby)

How many levels are in the Linnaean classification system?

2 Every species has a two-word name

All rock wallabies are part of the *Petrogale* genus, and the brush-tailed rock wallaby's species name is *penicillata*. This means that its full scientific name is *Petrogale penicillata*.

This naming is another important part of the Linnaean system: every species is given a two-word scientific name, in addition to its everyday name in English or other languages. These names are developed according to some strict rules:

- The first word is the genus name and begins with a capital letter.
- The second word is the species name and begins with a lowercase letter.
- The name must be shown in italics (or underlined in handwriting).

The words in these names are usually Latin or Ancient Greek. Because those languages are no longer actively used, they don't change over time like modern languages. This is useful because the meanings stay stable and consistent. Even if scientists don't speak Latin, they can look up the words and find the same meaning each time.

What two things make up a scientific name?

Figure 13.4 Every organism has a two-part scientific name, which describes its genus and species.

Homo sapiens
Humans

Canis familiaris
Domestic dogs (all breeds)

Felis domestica
Domestic cats (all breeds)

Staphylococcus aureus
Golden staph bacteria

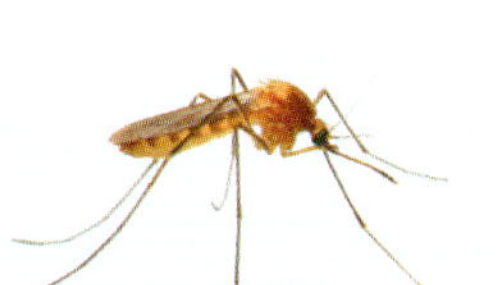

Anopheles gambiae
African malaria mosquitoes

Pisum sativum
Peas

Triticum aestivum
Common wheat

CHECKPOINT 13.2 

1. Explain the difference between a species and an organism.
2. Would you expect there to be more organisms in a species or a genus? Explain your answer.
3. Describe the Linnaean classification system in your own words.
4. Copy and complete these sentences.
   In a scientific name, the first word is the __________ and begins with a __________. The __________ word represents the __________ and begins with a __________. The name must be written in __________ or __________.
5. Explain why it's important for all scientific names to be in the same language.
6. Would you expect the red kangaroo to be in the same genus as the brush-tailed rock wallaby? Why or why not?

CHALLENGE

7. Undertake research to identify the scientific names of the following animals: polar bear, black rat and great white shark.

SKILLS CHECK

- I can give a description of the Linnaean system of classification.
- I can describe how a species' scientific name is formed.

13.3 CLASSIFICATION KEYS

At the end of this lesson I will be able to:

- **design** and **construct** simple keys to identify a range of living things.

KEY TERMS 

dichotomous
divided into two parts

key
a system for identifying characteristics

LITERACY LINK 

Create a branching key or dichotomous key for five of your favourite types of chocolate. When you have finished, ask a classmate to check it for you and provide feedback.

NUMERACY LINK 

Joachim designs a classification key using 18 different branching options. Express 18 as a product of its prime factors.

Classification systems are based on **keys**. A key is a system for identifying important characteristics.

Keys for identifying living things may use pictures, words, numbers and instructions. They are used mainly to find the names of organisms.

1 Some keys divide groups using branches

A common type of key is a branching key, which has two or more branches at each level to separate groups. As you move down the branches, you identify characteristics until you determine the right organism.

Consider a small group of animals: a bird, an earthworm, a lizard and a cat. You can use this branching key to identify each organism:

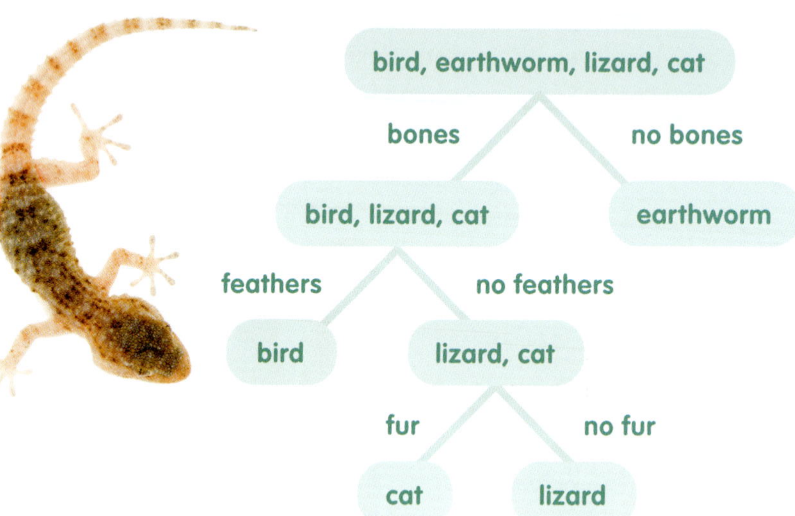

How does a branching key work?

2 Dichotomous keys use questions

Another type of key is a **dichotomous** key. *Dichotomous* means 'divided into two parts', so this key uses statements or questions with only two choices at each step. When a statement is correct, you follow directions to reach the next question, until you determine the organism.

Let's look back to the earlier group of animals – bird, earthworm, lizard and cat. A dichotomous key for identifying this group may look like this:

1. a Skeleton of bone .... go to statement 2
   b Does not contain bones .... earthworm
2. a Covered in feathers .... bird
   b Not covered in feathers .... go to statement 3
3. a Covered with dry scales .... lizard
   b Covered with fur .... cat

What are two types of classification keys?

❸ Keys identify a set of important characteristics

The most important part of designing a key is deciding on which characteristics it uses to identify organisms.

The characteristics chosen for a key depend on the organisms to be classified. At the kingdom level, the steps may be very general, such as whether an organism has cell walls or not. At the species level, the steps may be very specific, such as whether an insect has spots on its wings.

Consider the largest group of animals on Earth, the phylum Arthropoda, or arthropods. *Arthropoda* means 'jointed feet', and all arthropods have legs with joints as well as an exoskeleton, which is outside their bodies.

Arthropoda is a phylum, so the next level down in the classification structure is 'class'. Table 13.1 shows the five classes within Arthropoda, and that they are grouped by their structural characteristics. Can you develop a key that would allow you to identify the class of any arthropod?

Table 13.1 The five classes of arthropod (phylum Arthropoda)

Class	Characteristics	Examples
Insects	3 body parts (head, thorax, abdomen) 6 legs (3 pairs) 0, 1 or 2 pairs of wings 1 pair of antennae	Fly, mosquito, dragonfly
Crustaceans	3 body parts (head, thorax, abdomen) 10 legs (5 pairs) No wings 2 pairs of antennae	Crab, lobster, prawn
Arachnids	2 body parts (cephalothorax, abdomen) 8 legs (4 pairs) No wings No antennae	Spider, tick, scorpion
Centipedes	Many body segments 1 pair of legs on each segment Flat body cross-section No wings 1 pair of antennae	Centipede
Millipedes	Many body segments 2 pairs of legs on each segment Rounded body cross-section No wings 1 pair of antennae	Millipede

Why do different keys include different characteristics?

INVESTIGATION 13.3
Supermarket classification key

CHECKPOINT 13.3

1. Explain the role of a classification key.
2. Describe some structural features of humans that might be used to classify them.
3. Explain the difference between a branching key and a dichotomous key.
4. While exploring the Amazon rainforest, you find a previously undiscovered species of arthropod. It's bright pink, has 10 spiky legs, a head, thorax and abdomen, and it cannot fly. What class would you put it into?
5. Create a branching key that can be used to classify all the different shoes worn by students in your class.

CHALLENGE

6. Do some research to compare the classification of the domestic cat and the domestic dog, and comment on their shared characteristics.

SKILLS CHECK

- I can name at least two types of classification key.
- I can create a simple key of my own to classify objects or organisms.

13.4 CLASSIFYING ANIMALS

At the end of this lesson I will be able to:

- **classify** a variety of animals based on similarities and differences in structural features
- **outline** the structural features used to group living things, including animals.

KEY TERMS 

invertebrate
organism without a backbone or spinal cord

vertebrate
organism with a backbone or spinal cord

LITERACY LINK 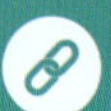

Use this sentence starter to write a half-page opinion piece: 'Invertebrates are far superior to vertebrates because ...'

From the largest whale to the smallest insect, there are more than 800 000 different animal species in the kingdom Animalia. All animals have many cells and must eat other organisms for energy.

The first question to consider when classifying animals is whether they have backbones.

Figure 13.5 Blue whales are the largest vertebrates in the world. Their skeletons couldn't support their weight on land – they weigh up to 140 tonnes.

1 Vertebrates are animals with backbones

Animals with backbones are called **vertebrates**, named after the small, oddly shaped bones in the spine called vertebrae.

In the classification system, vertebrates are members of the phylum Chordata. This name is used because all vertebrates have a nerve cord inside their backbone, which sends and receives information between body parts.

Table 13.2 shows the five classes of phylum Chordata. The identifying characteristics are their body coverings, how they control their temperature, how they reproduce and how they move.

What is the defining feature of vertebrates?

Table 13.2 The five classes of vertebrate (phylum Chordata)

Class	Body covering	Temperature control	Reproduction	Structure for movement	Examples
Fish	Slimy scales	Gain heat from surroundings	Lay eggs	Fins to move through water	Salmon
Amphibians	Naked skin	Gain heat from surroundings	Lay eggs in water	Young (tadpoles) have fins; adults have legs	Frog
Reptiles	Dry scales	Gain heat from surroundings	Lay leathery eggs	Legs for walking on land (except snakes)	Goanna
Birds	Feathers and scales (on feet)	Produce their own heat	Lay hard-shelled eggs	Legs for walking on land and wings for flying	Kookaburra
Mammals	Hair or fur	Produce their own heat	Lay eggs	Legs with different shapes for walking, swimming or flying	Echidna
			Give birth to underdeveloped young		Human
			Give birth to fully developed young		Dog

INVESTIGATION 13.4
Investigating features of marine mammals

2 Invertebrates are animals without backbones

The prefix *in-* means 'not', so **invertebrate** means 'not a vertebrate'. In other words, these animals do not have backbones – or any bones at all.

There are many different phyla of invertebrates. They can be grouped by structural features. Most invertebrates have exoskeletons, outside their bodies. These support them as they move, but are not made of bone. Invertebrates such as worms don't even have exoskeletons – they have inside structures that support them in other ways.

Table 13.3 Some of the many phyla of invertebrates

Phylum	Characteristics	Examples
Cnidarians	Soft body with one opening	Anemone, coral, jellyfish
Annelids	Segmented worm, round cross-section, lives on land	Earthworm
Platyhelminths	Segmented worm, flat cross-section, freshwater or in another organism	Tapeworm, planarian worm
Nematodes	Unsegmented worm, round cross-section, often in another organism	Round worm, heartworm
Poriferans	Collection of cells, some with hair-like structures, single opening	Sponge
Molluscs	Shelled animal	Snail, clam, oyster
Echinoderms	Body parts arranged around a central point, many 'feet'	Sea star, sea urchin, brittle star
Arthropods	Jointed legs, hard exoskeleton, segmented body	Insect, spider, prawn

What kinds of animal don't have a backbone?

Figure 13.6 All insects are invertebrates – they don't have backbones or spinal cords. Most have exoskeletons instead.

CHECKPOINT 13.4

1 Describe the difference between an invertebrate and a vertebrate.
2 Identify the phylum that vertebrates are members of.
3 Give three examples each of vertebrates and invertebrates.
4 Most invertebrates have exoskeletons - suggest why.
5 Which phyla do the following invertebrates belong to? Use Table 13.3 to help you.
   a earthworm
   b snail
   c a mussel (a shelled animal)
   d lobster
6 What class of vertebrate am I?
   a I have scales on my feet but not on my back. My body is covered with feathers not fur. I lay eggs and have legs.
   b I have no feathers or scales but I am a hairy organism. Wings and fins are not my thing, but I do have useful legs.

CHALLENGE

7 Why aren't animals classified according to the environment they live in?

SKILLS CHECK

- I can explain the difference between a vertebrate and an invertebrate.
- I can give at least two examples each of vertebrates and invertebrates.

13.5 CLASSIFYING PLANTS

At the end of this lesson I will be able to:

- **classify** a variety of plants based on similarities and differences in structural features
- **outline** the structural features used to group living things, including plants.

KEY TERMS 

angiosperm
a plant that produces seeds in fruit or flowers

bryophyte
a plant that doesn't contain vascular tissue

gymnosperm
a plant that produces seeds in cones

spore
a tiny part of some plants that is used to reproduce

tissue
a group of cells with a similar structure and function

tracheophyte
a plant that contains vascular tissue

vascular tissue
plant tissue that transports fluid and nutrients

LITERACY LINK 

Identify an angiosperm near your home or school, and write a description of it that is exactly 77 words long.

NUMERACY LINK 

A Norfolk Island pine is 59 metres tall. Convert 59 m to cm.

Figure 13.7 Mosses and liverworts are bryophytes. These plants don't have inside support and grow out rather than up.

Kingdom Plantae includes almost 400 000 species of plant. There is an incredible variety of plants in the world, including ferns, mosses and flowering plants. Plants have different types of **tissue**, and they have a variety of seeds, fruits and flowers.

1 Vascular tissues support plants and transport nutrients

Like other living organisms, plants are usually classified by their structure. The first identifying feature is whether a plant has **vascular tissue**. This transports sugars, water and minerals through tiny tubes in the plant, and also supports the plant. It works in a similar way to the bones and blood vessels of humans and other mammals.

Most plants have vascular tissues, and are called **tracheophytes**. They are supported from inside and can grow very tall. Water moves from the soil into the roots, stem and then the leaves. Glucose is made in the leaves and travels to the rest of the plant, to be used for energy or stored as starch.

Plants that do not contain vascular tissue, such as mosses and liverworts, are called **bryophytes**. They do not have support inside them, so these plants can't grow very high and instead appear as soft coverings on surfaces such as rocks and soil. They don't have true roots or leaves. They reproduce using **spores** and grow in areas where water is freely available, such as next to rivers or waterfalls.

What is vascular tissue?

❷ Seeds, fruits and flowers can be used to classify plants

Tracheophytes can be further divided into plants that produce seeds and those that do not.

Ferns don't have seeds, and instead reproduce using spores. These are small round spheres on the underside of leaves. Spores fall to the ground and grow into heart-shaped plants, which produce male and female cells. Ferns need water for fertilisation and only grow in areas where water is plentiful, such as rainforests.

Gymnosperms produce naked seeds that develop in cones. They have needle-like or thin leaves, and often grow into tall trees with woody trunks. Because they have large root systems, they can collect water from underground supplies and can grow in many different types of environments. Pines, firs, cycads and ginkgoes are all gymnosperms.

Angiosperms produce seeds from flowers and fruits. There are many different types of flowering plants, and this is the largest group of plants found on Earth. Wattles, roses, grass, wheat and eucalypts are all angiosperms.

Figure 13.8 Norfolk Island pines, are gymnosperms, which produce cones.

Table 13.4 Characteristics used to identify some classes of plant

Class	Contains vascular tissue?	Produces seeds?	Method of producing seeds	Examples
Bryophytes	No	No	–	Moss, liverwort
Ferns	Yes	No	–	Fern
Gymnosperms	Yes	Yes	Cones	Pine tree
Angiosperms	Yes	Yes	Flowers and fruits	Orange tree

What kinds of plant do not produce seeds?

CHECKPOINT 13.5 

1 Identify some common characteristics used to classify plants.

2 Suggest reasons why scientists need to classify plants.

3 Explain the role of vascular tissue in plants.

4 Bryophytes cannot grow very high, preferring to cover areas low to the ground; suggest why.

5 Is a cherry tree an example of an angiosperm or a gymnosperm? Explain your answer.

6 Why do ferns have spores?

7 a Who am I? I produce seeds from flowers and fruit.

b Who am I? I don't have seeds and instead reproduce using spores.

c Who am I? I produce seeds that develop in cones.

CHALLENGE

8 Undertake some further research into the vascular tissue of plants. Draw a diagram to show what vascular tissue is made of.

SKILLS CHECK

- I can describe why it's important to be able to classify plants.
- I can state some of the ways that plants are classified.

13.6 MUSHROOMS, YEASTS AND OTHER FUNGI

At the end of this lesson I will be able to:

- **outline** the structural features used to group living things, including mushrooms, yeasts and other fungi.

KEY TERMS 

antibiotic
a substance that kills or slows the growth of bacteria

decomposer
an organism that breaks down and recycles decaying matter

parasite
an organism that gets nutrients from the body of another organism, which harms the other organism

spore
a tiny part of some plants that is used to reproduce

symbiont
an organism that lives with a host organism, and both organisms benefit

NUMERACY LINK 

Chelsea mixes yeast in two different bowls. The temperature of Bowl A is 25°, while the temperature of Bowl B is 20 less than double the temperature of Bowl A.

Write a number sentence to represent and then calculate the temperature of Bowl B.

When he constructed his classification system, Carolus Linnaeus only recognised two kingdoms: animals and plants. He thought that organisms such as mushrooms and yeast were plants.

Since that time, scientists have classified these organisms in their own category: kingdom Fungi.

1 Fungi are not the same as plants

Fungi are similar to plants in a number of ways. They can't move around, and they reproduce using **spores**. However, they are different from plants in some important ways:

- Fungi can live in many different places, such as air, water, soil and on plants and animals.
- Instead of roots, fungi have hair-like filaments that grow into whatever they live upon.
- Fungi do not create their own food using photosynthesis.

One of the most important characteristics of fungi is how they receive nourishment. They are **decomposers**, breaking down dead organisms into simple compounds. Fungi do this using enzymes that break down dead plant and animal matter from their surroundings. They then absorb those nutrients through their filaments.

Some fungi are **parasites** that feed on living things. They harm the organisms on which they live so that they can digest the damaged parts. Other fungi are **symbionts**, and actually help the organisms on which they live. A symbiont living on a plant's roots will take sugar and oils from the plant, but the plant gains useful minerals from the fungi when it decays.

What are the main characteristics of fungi?

2 Many fungi are part of our diets

There are over 70 000 different species of fungi, and many of them can be eaten. Mushrooms, truffles and morels can all be eaten. Mushrooms are common in Australia and around the world, and they grow in many different environments. Truffles and morels are rarer, and are often considered an expensive delicacy. These fungi grow underground in close relationships with the roots of trees such as oaks.

The fungus that has the biggest part in our diets is yeast. Yeasts are single-celled microscopic fungi that feed on sugars and produce carbon dioxide. Unlike most fungi, they are single-celled organisms. They are used to bake bread, because the carbon dioxide gas they produce when they break down the bread flour makes the dough rise. Yeasts are also used to break down sugars in plants to produce some drinks.

How do yeasts cause bread to rise?

Figure 13.9 For many breads, yeast is used to break down flour into sugars. This produces bubbles of carbon dioxide gas , making the dough rise.

INVESTIGATION 13.6
Growing mould

❸ Fungi can cause and cure diseases

Some fungi can cause diseases such as athlete's foot and ringworm. Tinea pedis, or athlete's foot, is a common skin infection, mostly growing between the toes. A number of different fungi are responsible and you can become infected by walking on surfaces that have microscopic spores on them. There are other forms of tinea, on different parts of the body, caused by these fungi.

Fungi have a great variety of medical uses. British scientist Alexander Fleming made the first **antibiotic** medicine from penicillin fungus in 1928. Since then, antibiotics have become a vital part of treating diseases caused by bacteria.

Yeasts are another fungus important in medicine. Scientists can insert useful genes into them. When the yeast reproduces, it can be used to make medicines. For example, if the yeast cells are combined with the human gene for producing insulin, they can be used to make human insulin, which is used to treat diabetes.

What are some medical uses for fungi?

Figure 13.10 Tinea pedis is a painful fungal infection that can grow on your toes.

CHECKPOINT 13.6 ✓

1 Suggest why fungi were given their own category separate to plants.

2 Describe the role of fungi in medicine.

3 Create a Venn diagram to compare fungi to plants. In the centre of the overlapping circle, identify characteristics they have in common. On each side, identify ways in which they are unique.

4 Provide some uses of the fungus known as yeast.

5 Explain how fungi get their energy (hint: it's not through photosynthesis).

6 There are good and bad things about fungi. Summarise these and give your opinion on whether fungi are helpful or harmful organisms.

CHALLENGE

7 Fungi are decomposers; they break down dead organisms. Research other types of decomposers, summarise your research and suggest the importance of decomposers to life on Earth.

SKILLS CHECK

- I can describe the difference between plants and fungi.
- I can describe at least two characteristics and examples of fungi.

13.7 THE MICROSCOPIC KINGDOMS

At the end of this lesson I will be able to:

- **identify** some examples of groups of microorganisms
- **outline** the structural features used to group living things, including bacteria.

KEY TERMS 

algae
organisms living in water that make food by photosynthesis

bacteria
microscopic, unicellular organisms that can live in a range of environments

eukaryote
an organism with a nucleus and structures inside its cell(s)

microorganism
an organism only visible under a microscope

plankton
microscopic, animal-like organisms that float or drift in water

prokaryote
an organism without a nucleus or structures inside its cell(s)

protist
a microorganism that is not a fungus or bacteria

LITERACY LINK 

Write a creative story that features a protist, bacteria and archaea. it can be about anything you like and can be serious or humorous.

The Linnaean classification system has been changed many times since it was constructed, as we learn more about living things. For example, when scientists invented microscopes they saw that there are many different types of **microorganism**.

We now know that microorganisms do not all belong to the same kingdom. Their simplest classification includes two kingdoms: Protista and Monera.

1 Protists are very simple organisms

One of the microscopic kingdoms is Protista. Organisms in this kingdom are sometimes called protozoa or **protists**, and they are made of a simple, single cell. All protists are **eukaryotes**: organisms with a nucleus and special structures with their cells.

Most protists are single-celled organisms that live in water. A few of them have more than one cell, but they do not have specialised cells. Instead, they consist of large masses of identical cells.

Some protists, known as **algae**, look like plants and make their own food by photosynthesis. Others, known as **plankton**, are more like animals because they absorb their food from the water around them. There's even a third type of protist, moulds, that acts like fungi.

What are some examples of protists?

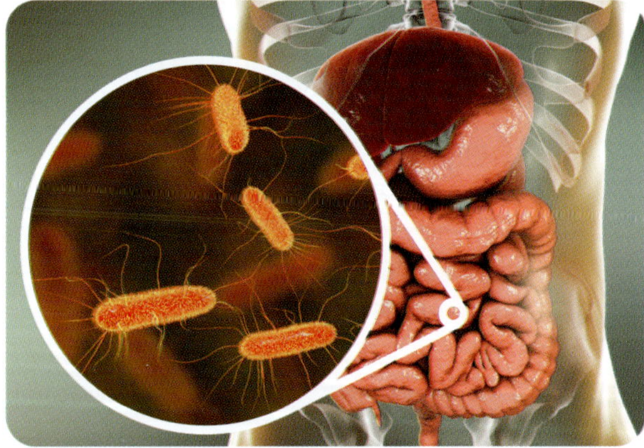

Figure 13.11 *E. coli* is a type of bacteria found inside the human digestive system.

2 Bacteria are the most common organisms on Earth

The microscopic kingdom Monera includes **bacteria**. They are **prokaryotes**, organisms that don't have a nucleus or special structures in their cells.

Bacteria are the most common organisms on Earth. A single millilitre of water contains a million bacteria. Most bacteria benefit life because they are decomposers, like fungi, which are important in every ecosystem. For example, the human digestive system relies on bacteria in the gut for digesting food. Other bacteria are harmful and cause infections and diseases.

Bacteria reproduce by splitting into two identical cells. When conditions suit the bacteria, some can reproduce every 20 minutes. If this happens, then one cell can become more than 16 million in just eight hours! Bacteria can quickly infect another organism in this way.

There are several different ways of classifying bacteria. The two that are most commonly used are their shape and how they gain nutrition:

- Bacteria can be classified into five different basic shapes. The three most common shapes are spheres, cylinders and spirals. Some less common bacteria are shaped like commas and corkscrews.
- Some bacteria make their own food, either by photosynthesis (like a plant), or by changing chemicals in their environment into food. Others act like animals and eat other bacteria to gain nutrients.

What are the main ways to classify bacteria?

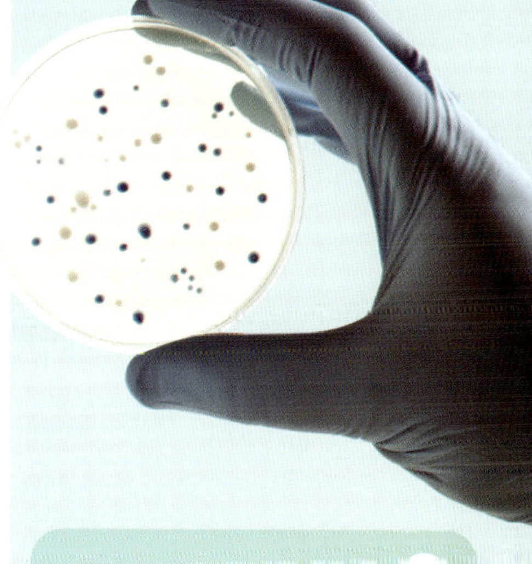

3 Archaeans can survive in extreme environments

Archaeans are single-celled microorganisms that also belong to kingdom Monera. These prokaryotes have a similar size and structure to bacteria, but they have different molecules in their cell walls. Archaeans live in very different environments to bacteria. They are sometimes called extremophiles because they can live in very hot, salty and acidic environments.

Some archaeans live in deep-sea hydrothermal vents and volcanic mineral springs. The temperatures in these areas are greater than 90°C! Others live in very salty environments, such as the Dead Sea in Israel, or in very acidic soils.

Some archaeans live in less extreme environments, although humans would not find these places comfortable. Archaeans can live in hot springs, swamps and marshes. A few archaeans live in the digestive systems of humans and other animals.

Scientists have found it difficult to study archaeans, because they live in places that are hard to reach. However, they have managed to find out some things about these organisms. Archaeans reproduce by splitting into cells, they have distinct shapes, and most can move in the water by using their whip-like tails. They can have rectangular or square shapes and, unlike bacteria, do not seem to cause diseases or harmful infections in other organisms.

Why are archaeans also known as extremophiles?

Figure 13.12 The Morning Glory hot spring in Yellowstone National Park in the USA is so colourful because of a rare kind of archaean that lives there.

CHECKPOINT 13.7

1 In your own words, explain what a microorganism is.

2 Summarise some of the different ways that microorganisms can be categorised.

3 Bacteria are the most common organism on Earth. Suggest why this is the case.

4 List some positives and negatives of bacteria.

5 Suggest how bacteria are able to reproduce so quickly.

6 Explain what is meant by an 'extremophile' and give an example of one.

7 Compare how algae and plankton obtain their energy.

CHALLENGE

8 Research the discovery of penicillin. Summarise why penicillin is so important in medicine, and suggest what would be different in modern medicine if it had never been discovered.

SKILLS CHECK

- I can describe the difference between a prokaryote and eukaryote.
- I can describe key features of protists, bacteria and archaea.

CHAPTER SUMMARY

All of the millions of types of organisms on Earth have biological characteristics in common, such as being able to grow, move, reproduce and respond to stimuli. This allows us to classify them.

Classification level	Classification for brush-tailed rock wallaby
Domain ↓	Eukaryota (living organisms with specialised cells) ↓
Kingdom ↓	Animalia (animals) ↓
Phylum ↓	Chordata (animals with a backbone) ↓
Class ↓	Mammals ↓
Order ↓	Marsupialia (marsupials) ↓
Family ↓	Macropodidae (wallabies, kangaroos, pademelons) ↓
Genus ↓	*Petrogale* (wallabies) ↓
Species	*penicillata* (brush-tailed rock wallaby)

All species of living things can be classified under the Linnaean system using a two-word name.

▲ *Homo sapiens* (Humans)

▲ *Canis familiaris* (Domestic dogs)

A classification key can be used to identify the distinguishing features of organisms.

- bird, earthworm, lizard, cat
  - bones → bird, lizard, cat
    - feathers → bird
    - no feathers → lizard, cat
      - fur → cat
      - no fur → lizard
  - no bones → earthworm

Protists, bacteria and archaea are all types of microscopic organisms.

◄ Animals that have a backbone or spinal column are classified as vertebrates; animals that don't are invertebrates.

▼ Plants are classified depending on whether they have vascular tissues and whether they produce seeds, fruits or flowers.

▼ Fungi play an important role in the environment by decomposing dead matter.

★ FINAL CHALLENGE ★

LEVEL 1 ★☆☆ ☆☆☆ 50xp — LEVEL UP!

1. State some reasons why scientists use classification in all areas of study, particularly in biology.
2. If you come across an organism and want to identify it, what steps could you take to name it?

LEVEL 2 ★★☆ ☆☆☆ 100xp — LEVEL UP!

3. Give two reasons why vascular plants grow much taller than non-vascular plants.
4. Describe the main differences between prokaryotic cells and eukaryotic cells.
5. Fungi were once classified with plants. Explain why they are now classified in their own kingdom.

LEVEL 3 ★★★ ☆☆☆ 150xp — LEVEL UP!

6. Insects all make noises using different parts of their bodies.
   a Suggest some reasons why insects make noises.
   b Using three different examples, describe how insects make noises.
7. Using the basic description of all insects on page 205, draw a diagram of an insect.

LEVEL 4 ★★★ ★☆☆ 200xp — LEVEL UP!

8. Describe the characteristics that all animals in phylum Chordata possess.
9. State the one characteristic shared by all vertebrates.
10. Name the five main groups or classes of vertebrates. For each, describe three characteristics and one example.

LEVEL 5 ★★★ ★★★ 300xp — LEVEL UP!

11. The three groups of mammals are monotremes, marsupials and placentals. Explain how these groups are different and what they have in common.
12. The majority of marsupials are only found on the Australian continent. Research reasons for this and then evaluate the evidence.

14 CELLS

All living things are made of cells, which is why cells are commonly known as the 'building blocks of life'. However, a cell in an animal looks and acts very differently to a cell in a plant.

There are many different types of cells – even within the human body. They all look really different because they have very different jobs to do. Nerve cells send messages around your body from and to your brain, so they have special features. Blood cells carry oxygen around your body, so they look different too.

1 LEARNING LINKS

What do you already know about cells?

What kind of materials make up living things?

How do plants and animals interact for survival?

2 SEE-KNOW-WONDER

List three things you can **see**, three things you **know** and three things you **wonder** about this image.

3 CRITICAL + CREATIVE THINKING

What if ... you found a way to stop cells from ever dying?

Prediction: What could happen if the nerve cells in your body stopped working?

Alternatives: List ways that you could see a cell without using a light microscope.

4 THE BIGGEST AND THE SMALLEST

The biggest cell in the human body is the female egg cell (ovum). These cells are about as wide as the thickness of a human hair. The body's smallest cell is the male sperm cell.

Females are born with all the egg cells they will ever have. After they reach puberty, males create millions of new sperm every day. In females that may have babies, an egg cell will be released once each month. If sperm are around, they will race against each other to try to beat them to the egg. The winning sperm will dissolve the outside of the egg and wiggle in to possibly start a new life.

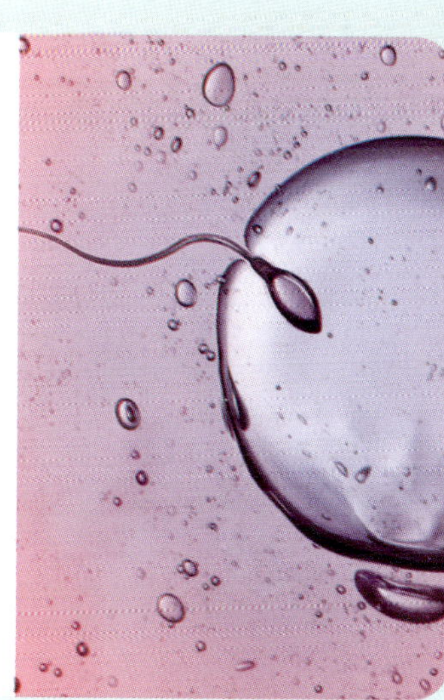

14.1 WHAT ARE CELLS?

At the end of this lesson I will be able to:

- **identify** that living things are made of cells.

KEY TERMS 

cell
the smallest functional unit of an organism

microscope
an instrument used to look at objects too small to see with the naked eye

theory
an explanation that can be supported or disproved using evidence

LITERACY LINK 

Identify three adjectives (describing words) from this section. Suggest replacement words for all three.

NUMERACY LINK 

Your body contains trillions of cells.

A trillion is 10^{12}; write this as a numeral.

Cells are the smallest structural and functional units of living things. This means that they are the smallest part of a living thing that can carry out the functions needed for life, all on its own.

All living things, from the largest and most complex to the smallest and most complex organism, are made of cells.

1 Living things have one or more cells, which come from other cells

Cells were first identified in 1665 by Robert Hooke, an English scientist. He was using a simple microscope to study the bark of a cork tree. The tiny structures he could see through the microscope reminded him of the tiny rooms that monks slept in, also called cells.

Even after cells had been discovered, scientists weren't sure whether all organisms were made of cells. It wasn't until 1839 that scientists developed the 'cell theory', which was based on three ideas:

- All living things are made of one or more cells.
- Cells are the basic building blocks of life.
- Cells arise from cells that already exist.

This **theory** helped people to start thinking about living things, and where life comes from, in a scientific way. All of the evidence and discoveries made since then have confirmed the cell theory.

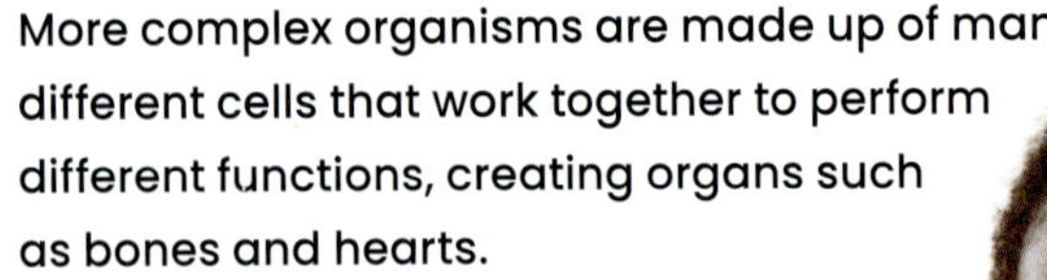
What does the cell theory tell us about living things?

2 Cells are the building blocks of life

Cells are the basic building blocks of all living things. Sometimes a single cell is a complete – but very simple – organism itself. More complex organisms are made up of many different cells that work together to perform different functions, creating organs such as bones and hearts.

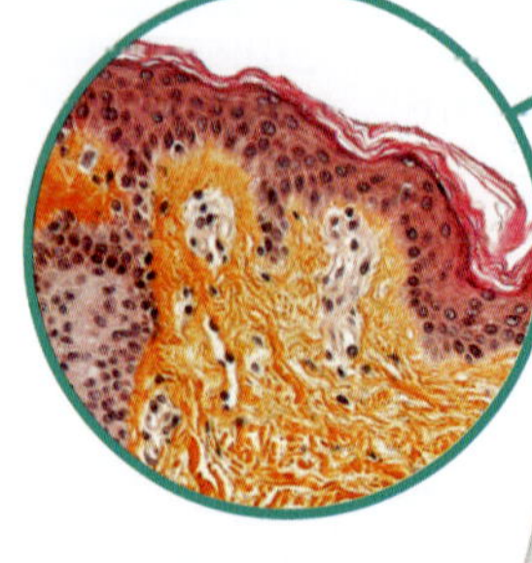

Figure 14.1 All living things are made of cells and there are different types of cells, each with different purposes and jobs. This skin cell is different to a nerve cell or a red blood cell.

Your body is made up of trillions of cells. They provide structure, take in nutrients from food, change those nutrients into energy, as well as other functions. Your cells contain your genetic material and can make copies of themselves. This allows you to grow new cells, such as for hair and fingernails.

Plants are also made up of cells. Their cells have some different features to animal cells, because plants don't have organs or the ability to move around. But the cells of plants still do many similar things to those of animals, and many of them have the same functions, such as transporting nutrients.

How many cells are in the human body?

INVESTIGATION 14.1
Examining cells under a microscope

Figure 14.2 An onion is made of many cells. The cells in a thin layer of onion membrane can be seen using a microscope.

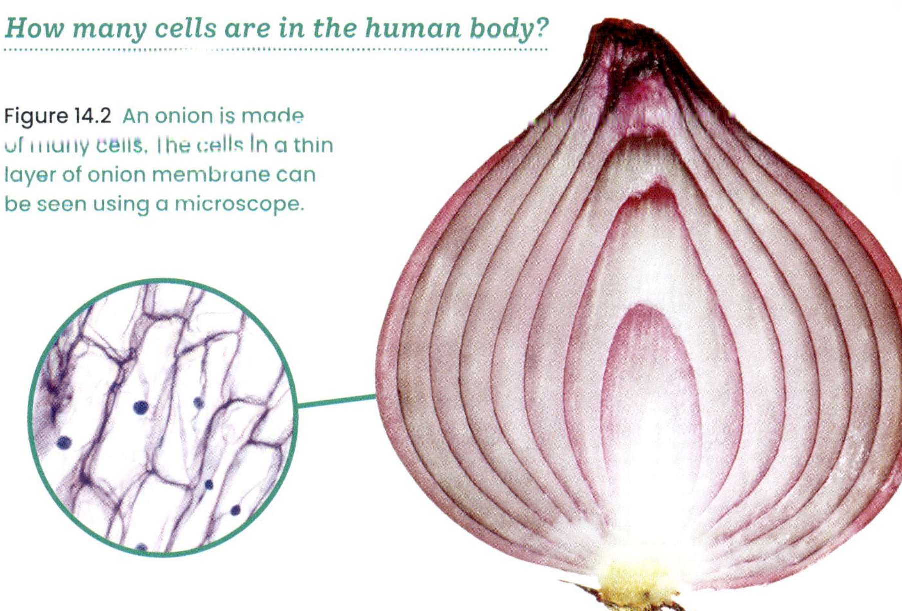

3 Microscopes allow us to see and identify cells

Cells can vary a lot in size. The largest cells in the world are ostrich eggs – a single unfertilised egg cell is about 15 cm wide and weighs more than a kilogram! Red blood cells are only 0.008 mm wide – a line of 125 red blood cells is only 1 mm long. Most cells are about this size – far too small for you to see with just your naked eye.

A **microscope** can be used to see cells. The word comes from two Ancient Greek terms: *micro*, which means 'small', and *scope*, which means 'to see'. So a microscope is a tool for looking at small things.

The most common type of microscope in the science laboratory is a light microscope. It allows scientists to study cells by shining a bright light through an extremely thin slice of tissue taken from an organism. The image is magnified by the microscope's lenses, which scientists look through.

What is the name of the scientific instrument used to study cells?

CHECKPOINT 14.1

1 Cells are described as the building blocks of life. Suggest why.

2 Copy and complete the following sentences.
Cell theory is made of ________ ideas: all ________ things are made of ________; cells are the ________ and cells come from ________ that already ________.

3 Identify three different types of cells from the text.

4 Name the scientist who first identified cells.

5 Explain why a microscope is necessary to view cells.

6 Rewrite the main ideas of cell theory in your own words.

7 Approximately how many cells are there in the human body?

CHALLENGE

8 Research and draw a diagram of a red blood cell. Explain how the structure of the red blood cell helps in its function (job).

SKILLS CHECK

- I can explain what cells are.
- I can identify what kinds of things are made of cells.

14.2 THE STRUCTURE OF CELLS

At the end of this lesson I will be able to:

- **identify** structures within cells, including the nucleus, cytoplasm, cell membrane, cell wall and chloroplast, and describe their functions.

KEY TERMS 

cell membrane
a thin layer around a cell that controls the substances coming in and out

cell wall
a stiff layer around a plant cell that supports it

chloroplast
a small organelle that allows plants to make food

cytoplasm
a jelly-like fluid in which the other parts of a cell sit

nucleus
the control centre of a a cell; DNA is found inside the nucleus

organelle
a cell structure that has a membrane around it (usually)

LITERACY LINK 

Investigate a cell structure of your choice. Describe its function in the cell and include a picture.

NUMERACY LINK 

Michaela mixes 150 mL of water with 15 mL of purple dye to create a stain. Express this as a ratio in its simplest form.

Think about a town you know – what keeps it running? Many parts of a town have different jobs. At the town hall, people make decisions about the running of the town. Factories make items and farms grow food sold in shops, the sewage works treat waste and so on.

In the same way, cells are made up of different parts that each have a particular role.

❶ Animal and plant cells have a nucleus, cytoplasm and cell membrane

Each part of a cell has a particular function that helps keep the cell healthy.

There are many different cell structures, just as there are different organs in the body. There are some structures that every cell has, while there are others that only some organisms have.

All animal and plant cells have these three structures:

- The **nucleus** is the central part of the cell, and it controls all the cell's activity. DNA, the genetic material, is found inside the nucleus. The nucleus is an example of an **organelle**, a structure surrounded by a membrane.
- The **cytoplasm** is a jelly-like fluid in which other cell structures sit. Many of the chemical reactions in a cell happen in the cytoplasm.
- The **cell membrane** is a flexible envelope that surrounds the cell. It controls substances coming in and out of a cell.

What are the main cell parts and why do cells need them?

Figure 14.3 A typical animal cell under a microscope, with some parts labelled.

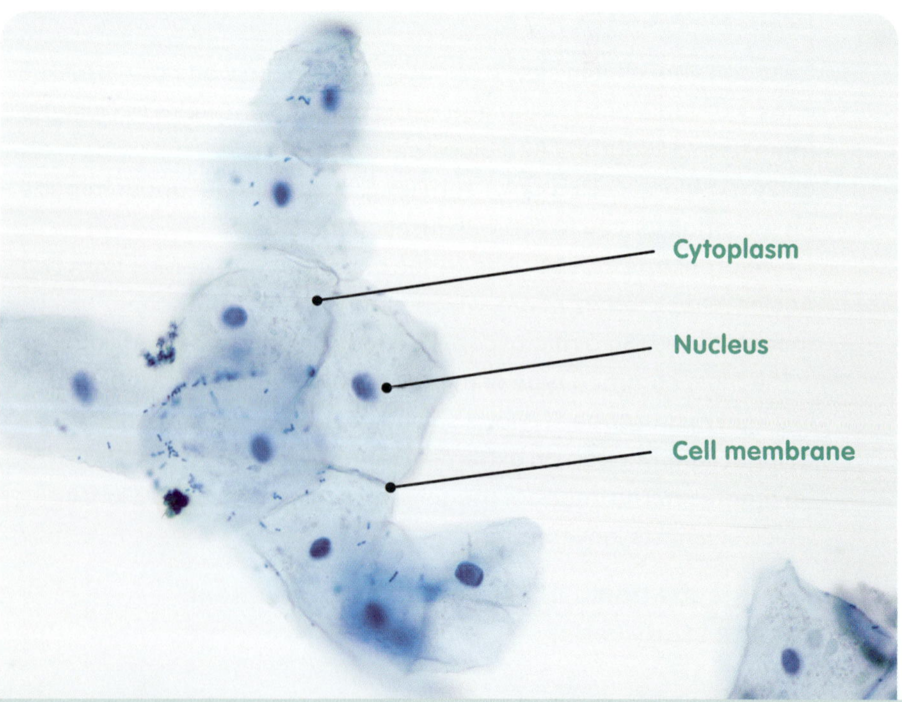

2 Plant cells have a cell wall and chloroplasts

There are many obvious differences between plants and animals. No-one is likely to confuse a cat with a cactus, or a shark with seaweed!

Some differences are not obvious, such as the structures inside plant cells. Plant cells have a nucleus, cytoplasm and cell membrane, just like animal cells, but they have two parts that animal cells do not:

- The **cell wall** is a stiff layer around the cell, outside the cell membrane. It helps to protect and support the cell.
- **Chloroplasts** are organelles that act as energy producers. They allow plants to make sugar using the Sun's energy.

Animal cells don't have these structures because they don't need them. Animals have a skeleton or structure to support themselves, and they can move around to get food.

How are plant cells different to animal cells?

Figure 14.4 A typical plant cell under a microscope, with some parts labelled.

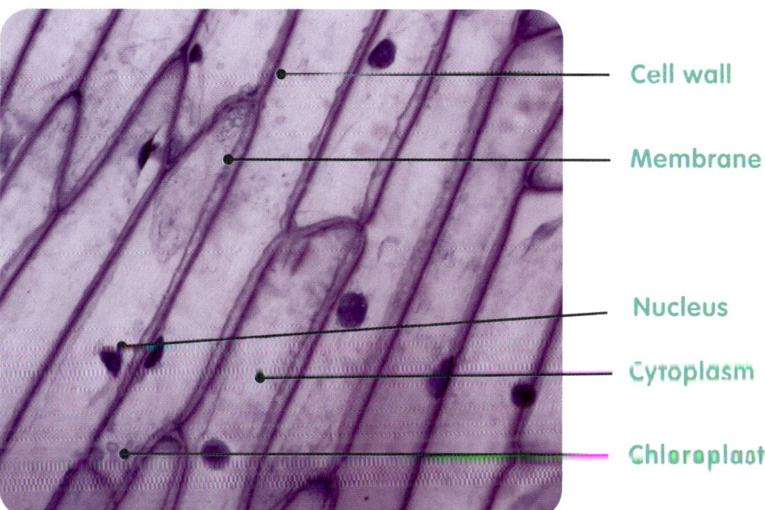

3 Staining cells shows their parts

Scientists use chemicals called stains to artificially colour some of the parts of cells. This makes them easier to see under the microscope. In Figure 14.4, a purple stain has been used to colour some of the cell parts.

Most cell parts can't be seen under a light microscope. They are too small. Parts of a cell that can be seen under a light microscope include the cell wall, cell membrane, nucleus and chloroplast. The cytoplasm itself can't really be seen in Figure 14.4, but you can see substances floating in the cytoplasm – they look like a grainy background inside the cell membrane.

Why do scientists use stains?

CHECKPOINT 14.2

1 Describe what organelles are.

2 Copy and complete the following table:

Structure	Function
nucleus	
cytoplasm	
cell membrane	
cell wall	
chloroplast	

3 Identify three structures that are common to both plant and animal cells.

4 Why do cells need separate structures?

5 Which organelles do plant cells have that animal cells don't?

6 What is the function of a plant cell wall?

7 Why are stains sometimes used when viewing cells under a microscope?

8 What is the function of a chloroplast? Suggest why animals do not have one.

CHALLENGE

9 Research and draw a diagram of a typical plant cell and a typical animal cell. List the differences between the two cells.

SKILLS CHECK

- I can state the function of a:
  - nucleus
  - cell membrane
  - cell wall
  - chloroplast
  - cytoplasm.

14.3 HOW CELLS MAKE ENERGY

At the end of this lesson I will be able to:

- **outline** the role of respiration in providing energy for cell activities.

KEY TERMS 

glucose
a type of sugar that is the energy source for cells

metabolism
the chemical reactions that happen in the body

mitochondria
the organelles where respiration occurs

organelle
a cell structure that has a membrane around it (usually)

respiration
a chemical reaction that converts glucose to energy

LITERACY LINK 

Provide a summary tweet of what cellular respiration is, in 280 characters or less.

NUMERACY LINK 

Glucose contains 6 parts carbon, 12 parts hydrogen and 6 parts oxygen. Write the ratio of carbon to hydrogen to oxygen in simplest form.

Respiration is a chemical reaction that happens in all living cells of plants and animals. The energy that cells need to live is stored in **glucose**, which is a type of sugar.

Through respiration, energy is released from glucose so that all the chemical processes needed for life can happen.

1 Respiration happens in mitochondria

Respiration happens in organelles called **mitochondria**, which are found in both plant and animal cells. Mitochondria are called the 'powerhouse' of cells because they take in nutrients from cells, break them down and change them into energy that cells can use. Animals and plants don't respire in the same way, and they gain energy from different nutrients, but the mitochondria have the same purpose in all species.

Mitochondria are extremely tiny. Some cells have several thousand mitochondria because they need a lot of energy, such as muscle cells in animals. Other cells have lower energy needs, so they have few mitochondria or none at all.

In which part of a cell does respiration happen?

Figure 14.5 Glucose and oxygen are needed for cellular respiration. Carbon dioxide, water and energy are produced.

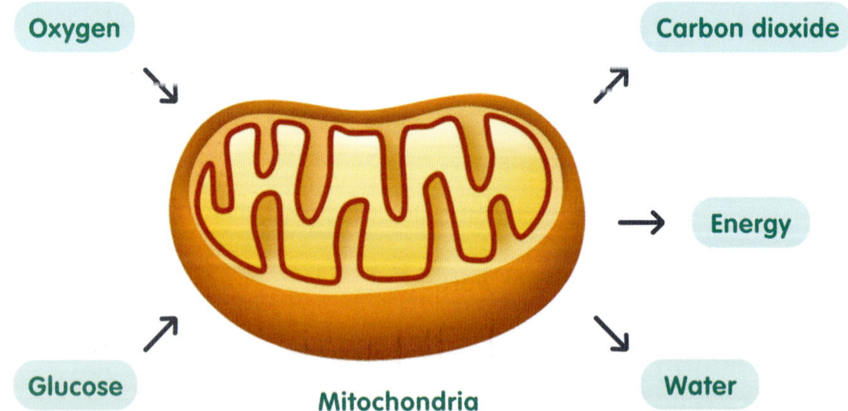

2 Respiration uses glucose and oxygen

Oxygen and glucose are needed for respiration. During respiration, glucose combines with oxygen to form carbon dioxide, water and energy. Carbon dioxide and water are the waste products.

During the day, in plants, oxygen and glucose are produced by photosynthesis. Extra glucose is stored in plants as oils, fats and starch. At night, plants do not photosynthesise because sunlight isn't available.

During the night, plants use oxygen taken in through their leaves or made during photosynthesis and they use glucose from their stored starch.

The process of respiration is different for animals. They breathe in oxygen as part of the air, and it enters their cells through their blood. The glucose needed for respiration is always in the blood. It comes from carbohydrates that animals eat and digest. Extra glucose in the blood is stored in the liver.

Figure 14.6 Glow worms are found throughout Australia and New Zealand. The energy of their respiration activates glowing chemicals in their bodies.

The energy released during respiration is used in many different ways. Animals, including humans, need it for their brains and hearts to work, and to move around. Some animals use it in other ways, such as when glow worms make light. Plants don't move around but they still use energy for cellular action, such as absorbing salts from the ground or moving nutrients through their systems.

What are the two substances that mitochondria need for respiration?

❸ Respiration is a chemical reaction

Respiration is a chemical reaction, and so it can be represented using a chemical equation. In any chemical equation, the reactants (the substances that change) are shown on the left-hand side of the arrow. The products (the substances formed) are shown on the right-hand side:

glucose + oxygen → carbon dioxide + water + energy

$$C_6H_{12}O_6 + O_2 \rightarrow CO_2 + H_2O + \text{energy}$$

Oxygen and glucose are transported into the mitochondria, where the chemical reaction happens. Carbon dioxide, water and energy leave the mitochondria. The energy is carried in a substance called adenosine triphosphate, or ATP.

Cells need to get rid of carbon dioxide because it reacts with water to make an acid. Too much acid is toxic to cells. To prevent this, animals get rid of carbon dioxide when they breathe out, while plants get rid of it through the surface of their leaves when they respire.

Why do cells need to get rid of carbon dioxide?

INVESTIGATION 14.3A
Cellular respiration in yeast

INVESTIGATION 14.3B
Energy from food

CHECKPOINT 14.3 

1 Define *respiration*.
2 What two reactants are needed for respiration to happen?
3 What are the three things generated during respiration?
4 What is the main purpose of respiration?
5 Write both a word and chemical equation for respiration.
6 What form of energy can be used by cells?
7 True or false? 'Respiration is the burning of food to release energy.' Explain your choice.

CHALLENGE

8 Research what is formed when carbohydrates break into simpler molecules.

SKILLS CHECK

- I can explain what respiration is and why it is important.
- I can identify the equation for respiration.
- I can state where respiration happens.

14.4 FORMING OF NEW CELLS

At the end of this lesson I will be able to:

- **identify** that new cells are produced by cell division.

KEY TERMS 

binary fission
a process where very simple cells divide into identical halves

chromosome
a thread-like molecule of genetic information in the nucleus of a cell

meiosis
complex cell division, where new cells are not identical to the original cell

mitosis
simple cell division, where new cells are identical to the original cell

LITERACY LINK 

Create a summary cartoon strip that shows binary fission, mitosis or meiosis.

NUMERACY LINK 

A single-celled bacteria splits half every hour to create new cells, which also split and multiply.

If you start with one cell, how many bacteria cells will you have after seven hours?

Before the cell theory was developed, scientists weren't sure how organisms were made. Some thought that living things were made out of non-living matter; for example, fleas being created from dust.

Now we understand that the cells of living organisms divide or split into new cells. This can happen in three different ways.

1 Single-celled organisms reproduce by binary fission

Some very simple organisms consist of just one cell. This cell carries out all of the functions that the organism needs to survive. These organisms don't need partners to reproduce or multiply – they just split in half, in a process called **binary fission**.

Before a cell undergoes binary fission, the genetic material inside it builds up until the amount of it doubles. The cell then splits in two, and each new cell includes half of the genetic information.

Only very simple cells, such as bacteria, can reproduce by binary fission. As well as being simple, binary fission is fast – some bacteria can split in half every 20 minutes.

How do simple cells reproduce?

Figure 14.7 Single-celled organisms such as bacteria reproduce by splitting into two identical halves.

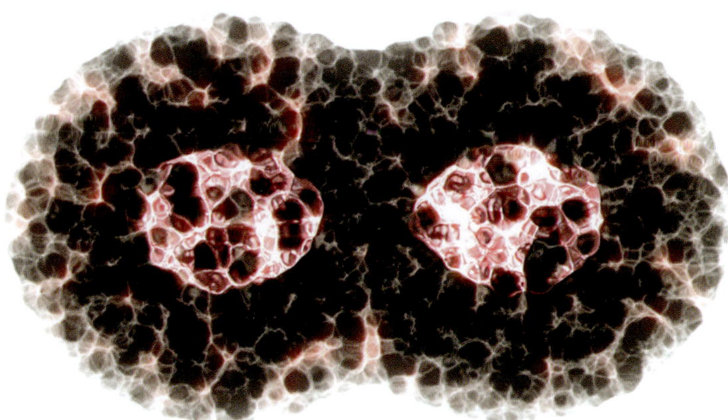

2 Mitosis is simple cell division

The cells of plants and animals are complex, and they contain genetic information. This information is in the nucleus of each cell, in threads called **chromosomes**. When new cells are formed, the chromosomes of the original or 'parent' cell are shared with the new 'daughter' cells.

The most common type of cell in your body is a somatic cell. *Somatic* comes from a Greek word meaning 'body' – somatic cells make up most of your body. Your body makes new somatic cells during **mitosis**, or simple cell division. During this process, a parent cell divides in half, creating two new daughter cells. The chromosome threads in the parent cell split in half, and each daughter cell receives an identical set of chromosomes. This means that each of these new cells is identical to the parent cell.

The major purpose of mitosis is for the organism to grow and replace worn out or dead cells. It happens faster in the growing regions of the body such as nails, hair and skin in animals, and in the shoots and roots of plants.

In what type of cell does mitosis happen?

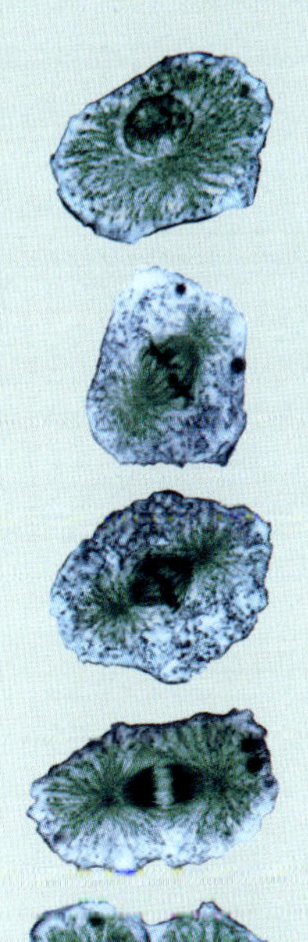

3 Meiosis is complex cell division

As well as somatic cells, your body contains reproductive cells. These cells pass half of the genetic information that makes you unique to the next generation. They include the gametes (sperm or egg) and germ cells (cells that go on to become gametes).

Sperm and egg cells are produced by a complex process called **meiosis**. First, the parent cell divides in half to create two new daughter cells, as happens during mitosis. Unlike in mitosis, each of these cells divides again so there are four daughter cells. Each is different – to both its parent *and* its sister.

When a parent cell divides the first time during meiosis, the chromosome threads don't split in half. Instead, the nucleus splits in half, and each half holds half of the threads. When the rest of the cell divides, each half carries one of the nucleus halves, which then becomes the nucleus of the new cell. This means that each daughter cell only carries half of the chromosomes of the parent cell. During reproduction, when the sperm and egg unite to form a single cell, the number of chromosomes is restored in the offspring. The fertilized egg has all the chromosomes needed to be a functioning cell.

In what type of cell does meiosis happen?

Figure 14.8 The main difference between mitosis (left) and meiosis (right) is how much of the parent cell's genetic information is transferred to the daughter cells.

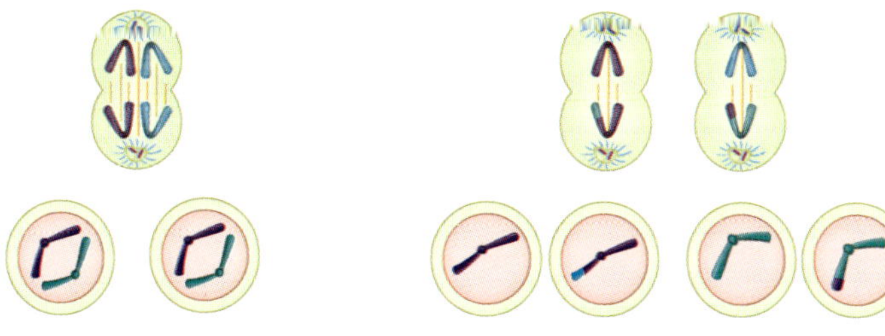

CHECKPOINT 14.4

1 Identify three ways in which cells reproduce.
2 Create a simple flow chart that outlines the steps involved in binary fission.
3 Explain the difference between mitosis and meiosis.
4 Bacteria reproduce very quickly – suggest why.
5 Where in plant and animal cells does mitosis happen?
6 What are the reproductive cells in humans called?
7 Give some reasons that cells may need to be replaced through mitosis.

CHALLENGE

8 In meiosis, why is the genetic material of daughter cells different from that of their parents? Why is this an advantage?

SKILLS CHECK

- I can explain how new cells are formed.
- I can describe the difference between binary fission, mitosis and meiosis.

14.5 UNICELLULAR AND MULTICELLULAR ORGANISMS

At the end of this lesson I will be able to:

- **identify** the differences between unicellular and multicellular organisms.

KEY TERMS 

differentiated cell
a cell that has specialised functions

eukaryote
an organism with a nucleus and special structures inside its cell(s)

prokaryote
an organism without a nucleus and special structures inside its cell(s)

protozoan
a unicellular eukaryote that moves and feeds on organic matter

unicellular
made of one cell

LITERACY LINK 

Create an A4 or A3 poster that includes key information about unicellular and multicellular organisms, including examples (and images) of each.

NUMERACY LINK 

One millilitre of water can contain a million bacterial cells. How many millilitres are there in 4.75 litres?

Organisms can have one or many cells. Single-celled organisms are **unicellular** – *uni* means 'one'. **Multicellular** organisms have more than one (and often many) cells.

Bacteria and some types of plant and fungi are examples of unicellular organisms. Multicellular organisms include humans, tigers, trees and mushrooms. Multicellular organisms can be much larger and more complex than unicellular organisms.

1 Unicellular organisms consist of one cell

Unicellular organisms are microscopic and cannot be seen with the naked eye. In these organisms, all life processes, such as digestion, feeding and reproduction, happen in one tiny cell.

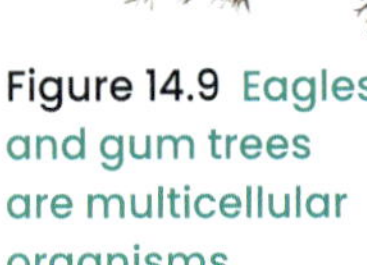

Figure 14.9 Eagles and gum trees are multicellular organisms.

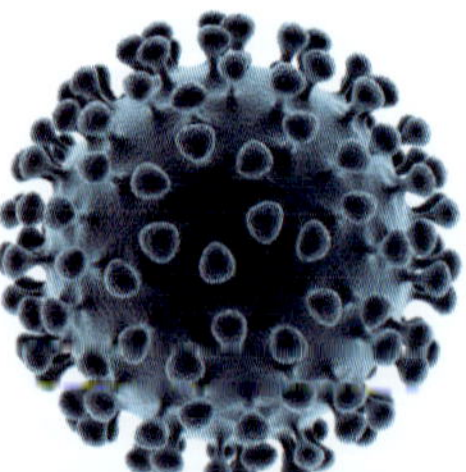

Figure 14.10 Viruses are even smaller than unicellular organisms and are difficult to see, even under a powerful microscope. They might look something like this.

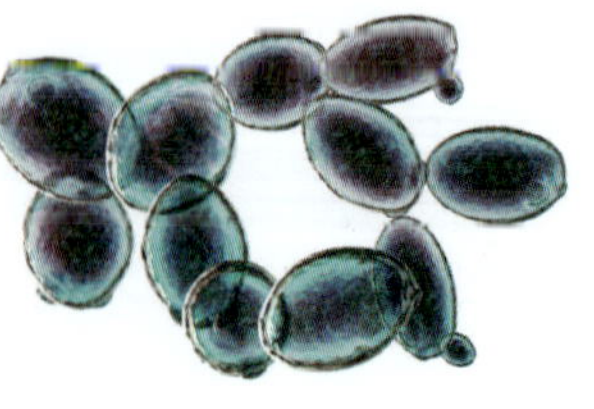

Figure 14.11 Yeast cells are unicellular organisms.

Most unicellular organisms are prokaryotes. This means they don't have the same structures as plant and animal cells. They don't have a nucleus, mitochondria or organelles contained in plant and animal cells. There are two kinds of prokaryotes: bacteria and archaeans. Bacteria are found in every environment – a single millilitre of water contains a million bacterial cells. Archaeans aren't as common, and many exist in environments we would find hostile, such as hot springs – but they also exist in our own bodies.

Unicellular organisms can also be **eukaryotes**. Eukaryotes have a nucleus and mitochondria, and some have chloroplasts, so they're similar to plant cells. In fact, some eukaryotes are simple plants, such as green algae. The patches of algae we might see in the wild are actually huge colonies of single-celled plants. Similarly, yeast is a single-celled fungus.

Prokaryotes and eukaryotes have cell parts that help them to survive and reproduce after they reach a certain size. Prokaryotes reproduce through a simple process called binary fission, eukaryotes divide and reproduce through mitosis and meiosis.

Unicellular organisms don't have senses, but they respond to various conditions such as changes in temperature and light, and to touch. Certain eukaryotes called **protozoans** can even move, propelling themselves through water or liquid using tiny hair-like or oar-like structures. They use this movement to chase down and digest other microscopic organisms.

What are the different types of unicellular organism?

2 Multicellular organisms are made up of many different cells

Most animals, plants and fungi are multicellular organisms. Some are large and some are microscopic, but they all have bodies made up of more than one cell.

Multicellular organisms have hundreds of **differentiated cells** with specific functions. These cells make up organs such as the liver, heart and kidneys, which do different things for an organism's survival.

Multicellular organisms have many advantages over unicellular organisms. The main one is that no single cell in a multicellular organism's body has to perform every function needed to survive. This leads to less work and stress for the cells, which means the organism can grow larger and live for longer.

What is the characteristic shared by all multicellular organisms?

INVESTIGATION 14.5A
Observing unicellular organisms

INVESTIGATION 14.5B
Observing specialised cells in multicellular organisms

CHECKPOINT 14.5

1 How do unicellular organisms reproduce?
2 What characteristics do unicellular organisms have in common?
3 List four differences between unicellular and multicellular organisms.
4 Discuss two benefits of multicellular organisms over unicellular organisms.
5 Explain the difference between prokaryotic cells and eukaryotic cells.
6 Do humans have prokaryotic cells or eukaryotic cells? Justify your answer.
7 Name some differentiated cells in humans.

CHALLENGE

8 Copy and complete this table by drawing an example of each type of organism in the first blank row. Undertake further research into unicellular viruses, provide a summary of your research and give at least two illustrations of their structure.

Unicellular organism	Multicellular organism

SKILLS CHECK

- I can describe at least three differences between unicellular and multicellular organisms.

14.6 TISSUES AND ORGANS IN ANIMALS

At the end of this lesson I will be able to:

- **identify** that different types of cells make up the tissues, organs and organ systems of multicellular organisms.

KEY TERMS 

differentiate
change to have a particular function

hormone
a chemical substance produced by the body that controls the activity of certain cells or organs

organ
a group of tissues with a specific function

tissue
a group of cells with a similar structure and function

LITERACY LINK 

Identify three terms from this section that you are unfamiliar with. Write definitions for each in your own words.

NUMERACY LINK 

There are around 70 different organs in the human body.

Express 70 as a product of its prime factors.

In multicellular organisms, cells work together so that organisms can survive.

In humans and animals, cells with similar functions group together to form tissues. Different tissues have different functions, and two or more tissues work together to form **organs**. Sets of organs work together as organ systems.

1 Tissues are groups of specialised cells

Before an unborn animal starts to develop, the beginning cells of its body are not specialised. These cells are called stem cells. As the animal grows, the cells **differentiate** and become specialised to carry out different functions. A group of like cells make up a **tissue**. There are four main types of tissue:

- *Epithelial tissue* forms the skin, as well as the body's inner linings.
- *Connective tissue* transports substances (such as nutrients) to where they're needed.
- *Muscle tissue* contracts and relaxes to carry out different functions.
- *Nerve tissue* transmits information between the brain and other organs.

What are the different types of tissues?

2 Organs perform the body's main functions

An organ is a mass made up of two or more tissues. The brain, heart, kidneys, liver and lungs are the main organs of the human body. Each organ has a specific function related to an organism's survival.

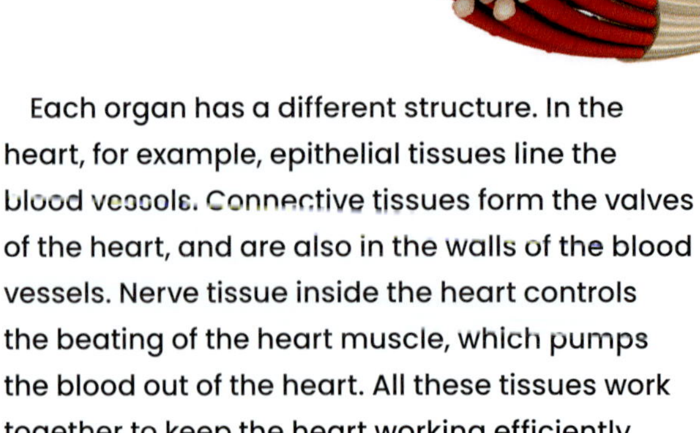

Each organ has a different structure. In the heart, for example, epithelial tissues line the blood vessels. Connective tissues form the valves of the heart, and are also in the walls of the blood vessels. Nerve tissue inside the heart controls the beating of the heart muscle, which pumps the blood out of the heart. All these tissues work together to keep the heart working efficiently.

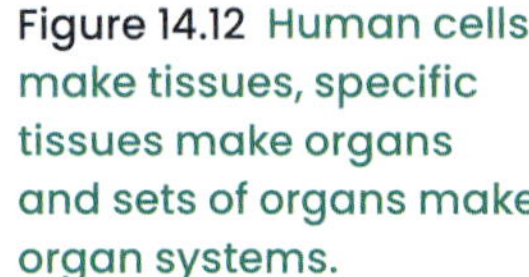

Figure 14.12 Human cells make tissues, specific tissues make organs and sets of organs make organ systems.

What are organs made up of?

❸ Sets of organs work together in systems

The function of an organ isn't useful on its own – your heart can pump blood, but that blood then has to reach the other parts of your body. Each organ is part of an organ system in an organism, along with other necessary tissues (such as blood). All of an organism's different organ systems must work with other systems.

The human body has several different organ systems, each with its own function. The following table lists major systems and their functions.

Organ system	Function	Major organs and tissues
Cardiovascular	Transports oxygen and nutrients to cells	Heart, blood
Nervous	Transmits nerve impulses between parts of the body	Nerves, brain, spinal cord
Digestive	Breaks down food so that nutrients can be absorbed	Oesophagus, stomach, liver, large and small intestines
Respiratory	Allows exchange of gases	Lungs, trachea, larynx, nasal passages
Excretory/urinary	Removes waste formed from digestion	Kidneys, ureters, urethra, bladder
Reproductive	Produces gametes (sex cells) and sex hormones	Female: ovaries, vagina, uterus, fallopian tubes Male: penis, testes, seminal glands

How many organ systems do humans have?

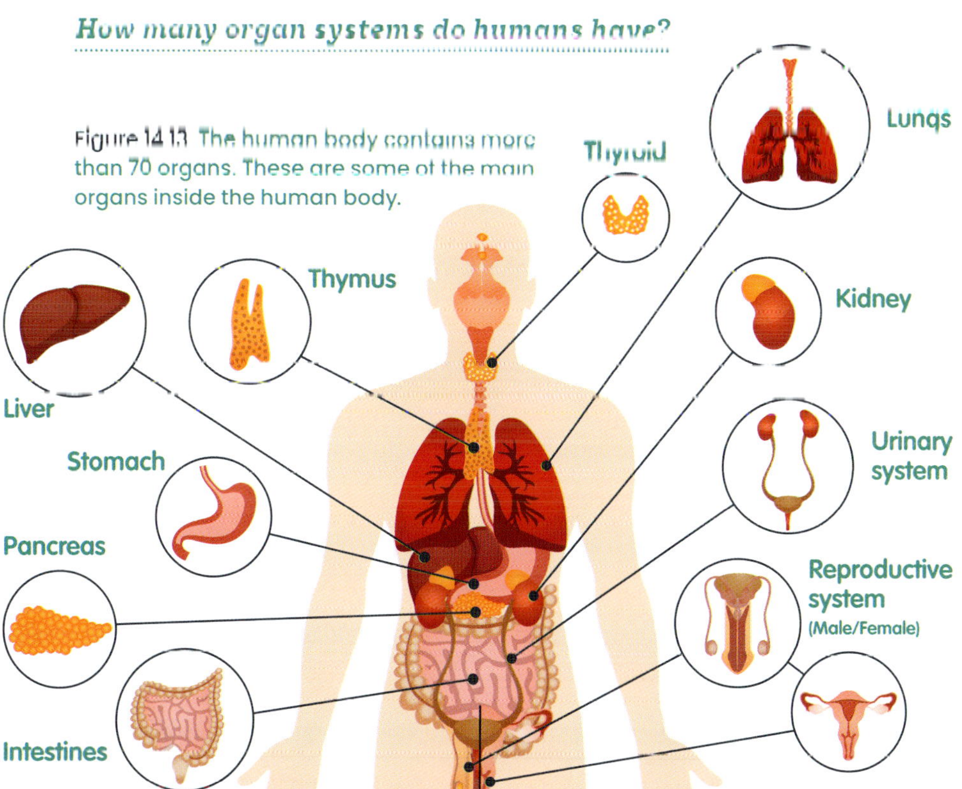

Figure 14.13 The human body contains more than 70 organs. These are some of the main organs inside the human body.

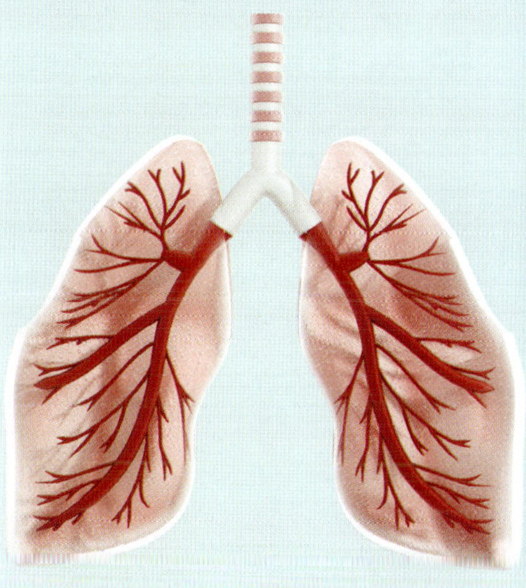

CHECKPOINT 14.6 

1. Give one example of a type of cell, an organ and an organ system.
2. Describe the role of the nervous system.
3. Which organ system does the heart belong to?
4. List the tissues that make up the heart. How do these tissues help it to function?
5. Draw a flowchart to show the levels of organisation in a multicellular organism.
6. List the functions of four tissues of your choice.

CHALLENGE

7. Draw an outline of the human body and label where each of these organs/glands are located: heart, lungs, stomach, liver, ovaries, kidneys, adrenal glands, pituitary gland, brain, thyroid gland, small and large intestine.

SKILLS CHECK

- I can explain the relationship between cells, tissues and organs.
- I can give examples of at least one type of cell, tissue and organ.

14.7 TISSUES AND ORGANS IN PLANTS

At the end of this lesson I will be able to:

- **identify** that different types of cells make up the tissues, organs and organ systems of multicellular organisms.

KEY TERMS 

epidermis
the outer layer of cells

organ
a group of tissues with a specific function

photosynthesis
the chemical reaction, powered by sunlight, that plants use to change carbon dioxide and water into sugars and oxygen

respiration
a chemical reaction that converts glucose to energy

tissue
a group of cells with a similar structure and function

LITERACY LINK 

Create an analogy for the following parts of a plant: the roots and the leaves.

NUMERACY LINK 

The stem of a plant is 12 cm high when first planted. It then grows 2 cm each month for 7 months.

Plot this information on a Cartesian plane.

Plants also have organs and tissues, but theirs are much simpler than those of animals. A group of cells with similar structure and functions make plant **tissues**, and two or more tissues work together as **organs**.

The organs and tissues carry out the vital functions of a plant such as **photosynthesis**, **respiration**, reproduction and the transport of water and nutrients.

1 Plants have four types of organs

Animals, including humans, have many different organs, all of which carry out very specialised functions. Plants only have four different types of organs. These organs are simple, but some of them carry out multiple functions:

- *Roots* absorb water and minerals from the ground. They also anchor plants in place so that they can continue to absorb nutrients.
- The *stem* is the main body of the plant – the trunk of a tree is the same as the stem of a rose. The stem supports and lifts the leaves, and it carries nutrients and water from the roots to the other parts of the plant.
- *Leaves* are where photosynthesis happens. Sunlight is absorbed through the surface of the leaves, and products such as oxygen are released, or re-used during respiration.
- Plants also have *reproductive organs*, such as flowers, fruits and seeds. These have many forms, but they all have the same function.

What are the four types of organs in plants?

Figure 14.14 The four types of plant organs are the roots, stem, leaves and reproductive organs, such as flowers.

2 Plants have a small number of tissue and organ systems

Humans have 11 different organ systems, but plants only have two organ systems.

The first organ system is the *root system*, which consists of all the organs found underground. This system absorbs nutrients and water from the ground. The roots are part of this system, and so are any underground reproductive organs of the stem. The part of a potato plant that you eat is actually part of the stem; it grows underground as part of the root system.

The second organ system is the *shoot system*, which mostly consists of the organs that grow above the ground. These parts of the plant absorb sunlight, and are where photosynthesis happens. The stem, fruit, flowers and leaves generally form the shoot system.

Plants also have four tissue systems that perform important functions, but don't combine to form organs:

- The **epidermis** is like the skin of the plant. These tissue cells form the outer surface of the leaves and the plant body.
- *Vascular tissue* transports fluids and nutrients through the plant, much like blood vessels do in your body.
- *Ground tissue* is the cells that make nutrients during photosynthesis and store nutrients for later use.
- *Meristematic cells* change to form various organs of a plant and are responsible for growth.

What are the two organ systems found in plants?

Figure 14.15 You can use a microscope to see the epidermis and vascular tissue of a plant.

INVESTIGATION 14.7
Water transport in plants

CHECKPOINT 14.7

1 How many types of organs and organ systems do plants have?
2 Identify the tissues that make up the transport system in plants.
3 Where do plants obtain water from?
4 Name two important plant organs and state their functions.
5 Name the two organ systems in plants.
6 What happens to the glucose that is made during photosynthesis?
7 Describe the role of the flowers, fruits and seeds in plants.
8 Outline the role of the vascular tissue in plants.

CHALLENGE

9 The vascular tissue has two types of specialised cells. Find out what they are and list their functions.

SKILLS CHECK

- I can explain the relationship between plant cells, tissues and organs.
- I can give examples of at least one type of plant cell, tissue and organ.

CHAPTER SUMMARY

Cells are the smallest structural and functional units of living things.

The cell includes structures, some of which are called organelles. Animal cells have structures including the nucleus, cell membrane and cytoplasm.

The cells of plants have the same structures as animal cells, as well as a cell wall and chloroplasts, which produce food.

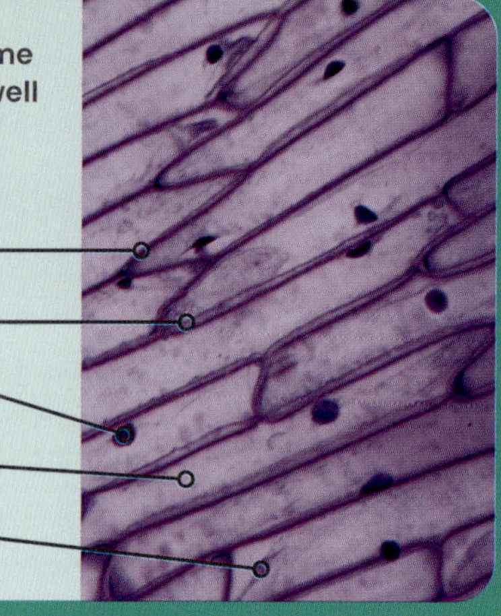

▼ Organisms can have one or many cells. Single-celled organisms, such as prokaryotes and some eukaryotes, are called unicellular – uni means one.

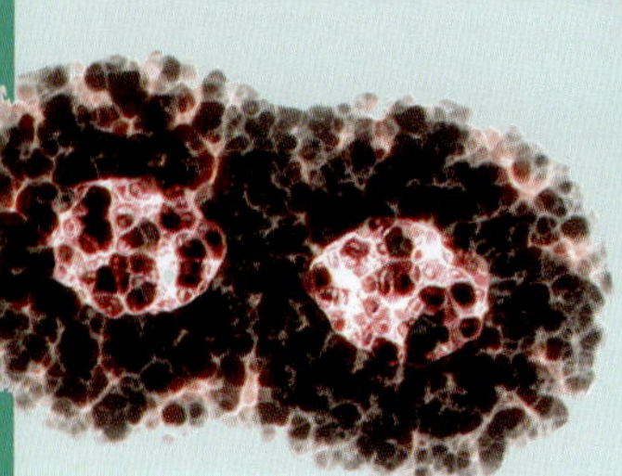

▲ Multicellular organisms, such as animals and most plants, are made up of many cells.

Organisms make energy through cellular respiration, a chemical reaction that occurs in the mitochondria of cells.

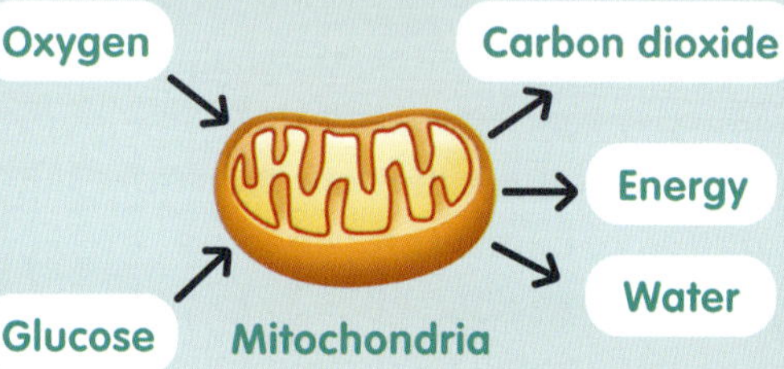

Energy → $6CO_2$ + $6H_2O$ + ATP

New cells are created through three processes – simple binary division, mitosis and meiosis.

Mitosis: simple cell division

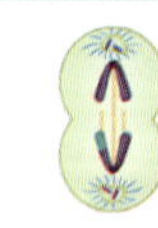

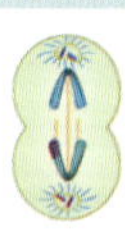

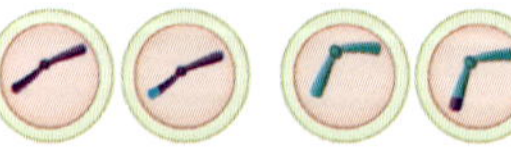

The main difference between mitosis and meiosis is how much of the parent cell's genetic information is transferred to the daughter cells.

Meiosis: complex cell division

Animal cells make up tissues, specific tissues make up organs and these organs make up organ systems. The human body contains more than 70 organs.

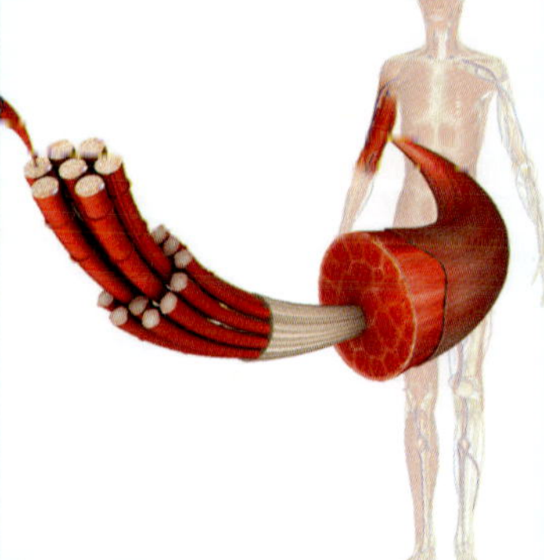

Plant cells also make up tissues and organs. There are four types of plant organ – the roots, stem, leaves and reproductive organs, such as flowers.

Flower
Leaf
Stem
Roots

★ FINAL CHALLENGE ★

1. Describe the main ideas of cell theory, in your own words.
2. Explain how the use of microscopes and stains can help you to view cells better.
3. Are all cells the same? Use two examples to explain your answer.

4. Draw a table to list the differences between plant and animal cells.
5. Choose three cell structures and explain their function. Provide a diagram of a cell showing these structures.
6. List the four levels of organisation in ascending order that are found in plants and animals. (Hint – an organ is one level of organisation.)

7. Describe what respiration is, in your own words.
8. Write the equation for respiration. Circle the reactants in blue and the products in red.
9. Explain the purpose of respiration and why respiration is an important process in all living things – what would happen if our cells did not respire?

10. Identify the three processes that lead to the formation of new cells in living things.
11. Draw a table to summarise the differences between the processes you mentioned in question 10.
12. Explain the difference between a prokaryote and eukaryote and give an example of each.

13. Name the four organs found in plants and list their functions.
14. Draw a table to summarise in detail the functions of the circulatory, digestive, respiratory and excretory systems. The headings should mention the system, the organs that work within that system, and all the functions of that system.

LEVEL 5

300xp

LEVEL UP!

15 BODY SYSTEMS

Living things are made up of systems that work together for the best chance of survival. Systems in plants support photosynthesis, using sunlight to produce energy and oxygen. Systems in animals take in food and process it to produce energy, and to support gas exchange, waste removal, sexual reproduction and more.

Organisms have evolved over billions of years, becoming extremely complex over this long time. When we investigate and learn about these systems, we discover the many intricate parts and functions of a living cell, and how they work together in multicellular organisms.

1 LEARNING LINKS

What do you already know about body systems?

How do the features and behaviours of plants and animals help them to survive in their environment?

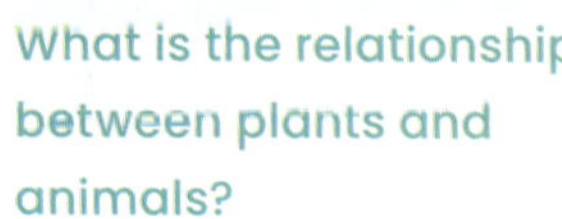

What is the relationship between plants and animals?

2 SEE-KNOW-WONDER

List three things you can **see**, three things you **know** and three things you **wonder** about this image.

3 CRITICAL + CREATIVE THINKING

Variations: In how many ways can living things excrete?

Commonality: Find as many points of commonality as you can between your heart and your lungs.

Predictions: Write a series of predictions for a situation where half the plants on Earth could no longer photosynthesise.

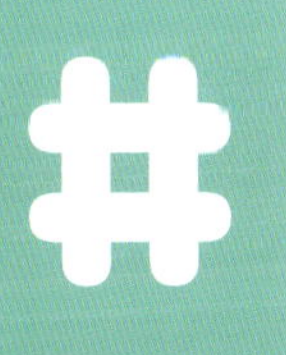

4 THE MOST BABIES!

The woman who holds the record for having the most children is Mrs Vassilyev, a peasant woman from Russia who allegedly had 69 children. 69! This was said to include 16 pairs of twins, seven sets of triplets and four sets of quadruplets. The births apparently happened in a period between 1725 and 1765. (Historical records aren't always reliable, so it's important to be sceptical about this claim.) Still, Mr Vassilyev obviously did not think that 69 children was enough – he is said to have had 18 more children with a second wife.

15.1 RESPIRATION AND PHOTOSYNTHESIS

At the end of this lesson I will be able to:

- **identify** the materials required by multicellular organisms for the processes of respiration and photosynthesis.

KEY TERMS 

aerobic respiration
how living organisms produce energy using oxygen

chlorophyll
the green pigment in chloroplasts that enables photosynthesis

chloroplasts
organelles in a plant cell that carry out photosynthesis

mitochondria
the organelles where respiration happens

photosynthesis
the chemical reaction, powered by sunlight, that plants use to change carbon dioxide and water into sugars and oxygen

stomata
pores in the surface of a leaf; the site of gas exchange in plants

LITERACY LINK 

In exactly 20 words, explain how photosynthesis is different from respiration.

NUMERACY LINK 

A garden bed is 20 m long and *x* metres wide. Each square metre of earth contains one plant. If the garden contains 3200 plants, what is the value of *x*?

Multicellular organisms are made up of many different types of cell, carrying out specialised functions for survival.

To perform these functions, cells need a constant supply of energy. Animal cells get their energy from food, and plant cells get their energy from sunlight. The energy from these sources is gained by cells using chemical reactions.

1 All cells need energy to survive

Without a continuous supply of energy, cells can't perform important functions, and they die. The only type of energy that cells can use is chemical energy. Chemical energy is stored in the bonds of glucose molecules and released when the bonds are broken.

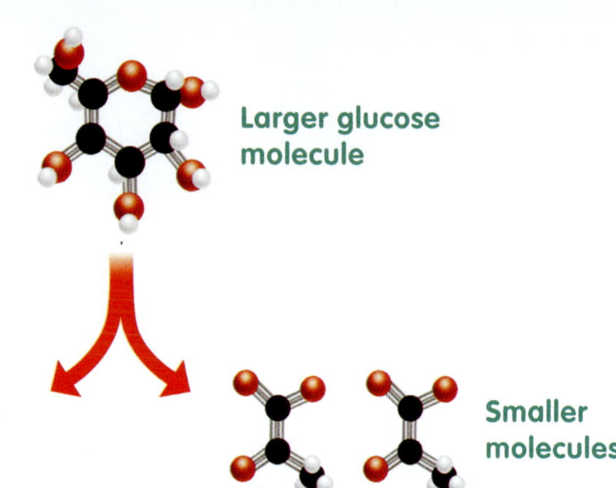

Figure 15.1 Chemical energy is released when the bonds of a molecule are broken.

A major source of chemical energy for multicellular organisms is obtained by breaking the bonds of glucose, a type of sugar molecule. Animals get most of their glucose from their food. Other organisms, such as plants, convert energy from the Sun into glucose.

What form of energy can cells use?

2 Respiration is how cells make energy

To release the energy from glucose and other energy-rich molecules, all living cells use the chemical process of respiration. Cells can carry out two main forms of respiration: **aerobic respiration**, which uses oxygen, and anaerobic respiration, which happens without oxygen.

Aerobic respiration provides multicellular organisms with most of their energy. It takes place mostly in cell organelles called **mitochondria**. Because energy is being released during respiration, the mitochondria are often called the powerhouses of a cell. The number of mitochondria in a cell is usually related to the functions of the cell.

The overall process of aerobic respiration can be summarised as:

glucose + oxygen → carbon dioxide + water + energy

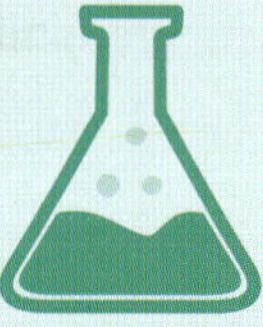

In animals, glucose is obtained from food, and oxygen is taken in from the environment across special surfaces such as lungs or gills. Both substances move into the blood and then to the cells. Plants absorb most of the oxygen they use through their leaves.

In all living cells, the chemical energy produced during respiration is transported to the parts of the cell that need it, while the carbon dioxide and water are removed from the cell.

What is cellular respiration?

3 Photosynthesis is how plants make food

Plants rely on **photosynthesis** to make their own food, using light energy, water and carbon dioxide.

One of the main reasons they are able to do this is because their cells have special structures called **chloroplasts**. Chloroplasts contain **chlorophyll**, a green pigment that absorbs light energy.

The overall process of photosynthesis can be summarised as:

$$\text{carbon dioxide} + \text{water} \xrightarrow[\text{chlorophyll}]{\text{light energy}} \text{glucose} + \text{oxygen}$$

To fuel photosynthesis, light energy from the Sun is absorbed by chlorophyll within a plant's cells. This energy is then used to change water (absorbed through the roots) and carbon dioxide (absorbed from the surrounding air or water) into oxygen and glucose (or other sugars).

The glucose produced from this reaction is either used in respiration or stored for later use. The oxygen moves into the environment through special pores (openings) in the leaves called **stomata** or it is used during respiration.

What are the substances needed for photosynthesis?

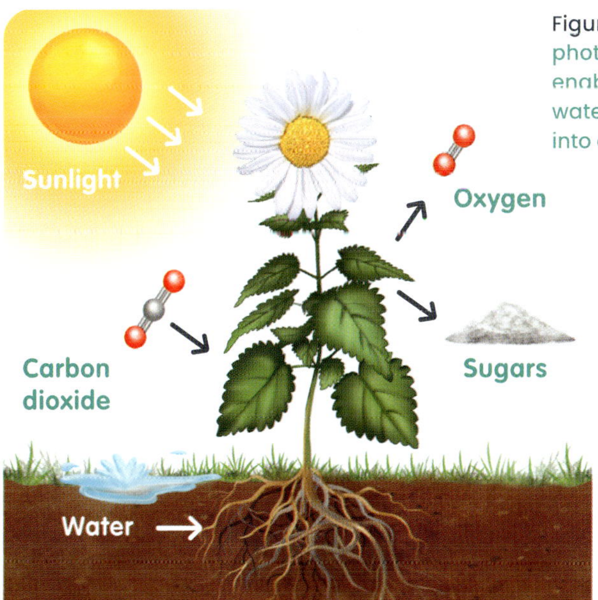

Figure 15.2 During photosynthesis, sunlight enables a plant to transform water and carbon dioxide into oxygen and glucose.

INVESTIGATION 15.1
Photosynthesis and respiration

CHECKPOINT 15.1

1 What form of energy is usable by cells?

2 Where is energy stored in molecules and how is it released?

3 Where do plants get the carbon dioxide and water for photosynthesis?

4 Where do animals get the glucose and oxygen for respiration?

5 Explain why oxygen is important in aerobic respiration.

6 Explain the role of chloroplasts in photosynthesis.

7 Identify these statements as true or false.
 a The special pores on the surface of a leaf are called stomata.
 b Carbon dioxide gas exits plant leaves into the atmosphere.
 c During photosynthesis, water is split into hydrogen and oxygen gas.
 d During photosynthesis, light energy is used to generate chemical energy.
 e Photosynthesis is an energy-producing reaction.

CHALLENGE

8 Explain this statement: 'When you eat an apple, you are also eating a little bit of sunshine'.

SKILLS CHECK

- I can state the word equations for respiration and photosynthesis.
- I can explain where multicellular organisms obtain the materials required for respiration and photosynthesis.

15.2 BODY SYSTEMS IN ACTION

At the end of this lesson I will be able to:

- **explain** that the systems in multicellular organisms work together to provide cell requirements, including gases, nutrients and water, and to remove cell wastes.

KEY TERMS 

body system
a group of organs working together

cell
the smallest functional unit of an organism

organ
a group of tissues with a specific function

tissue
a group of cells with a similar structure and function

LITERACY LINK 

Identify three terms from this section that you are unsure of. Use the internet to write a definition of each in your own words.

NUMERACY LINK 

Jermaine, a biologist, weighs 10 sheep hearts and records the following (all in grams):

239, 244, 244, 249, 250, 252, 255, 261, 265, 270

Calculate the mean, median and mode of this data set.

Body systems exist in almost every multicellular organism. These specialised organs and tissues have a common purpose.

For an organism to survive, each system must work with the others, often passing materials from one system to the next.

1 Cells work together as tissues, organs and systems

Unicellular organisms work alone – a single cell must carry out all of the major life functions. Multicellular organisms, such as humans, are made up of many different types of cell. More cells are available to share the workload, so different cells have specific functions.

Cells with similar functions form **tissues**, such as blood or muscles. Different tissues working together are **organs**, such as the heart, brain and lungs. Two or more organs connected and working together form a **body system**. The organs and tissues in each system are specialised to perform specific roles that serve a common purpose.

What is the link between a tissue and an organ?

2 Animals have several organ systems

Multicellular organisms are incredibly diverse, so the number and type of body systems can differ between organisms. Most animals, such as humans, have 10 major body systems. These are the:

- *circulatory system* – moves nutrients, gases and waste products around the body
- *nervous system* – detects, processes and sends electrical signals
- *respiratory system* – exchanges gases with the environment
- *digestive system* – breaks down and absorbs food
- *musculoskeletal system* – allows movement and provides the body with shape and support
- *endocrine system* – produces the hormones that control growth and development
- *excretory system* – removes body wastes
- *reproductive system* – produces sex cells and supports pregnancy and birth
- *immune system* – makes the white blood cells that fight diseases and infections
- *integumentary system* – protects the body from damage.

Some of these systems can be identified in other ways. For example, the body's skeleton and muscles can be considered separately as the skeletal and muscular systems, or together as the musculoskeletal system.

What are three examples of body systems in humans?

Figure 15.3 The human body consists of different systems that work together to provide cells with what they need to survive.

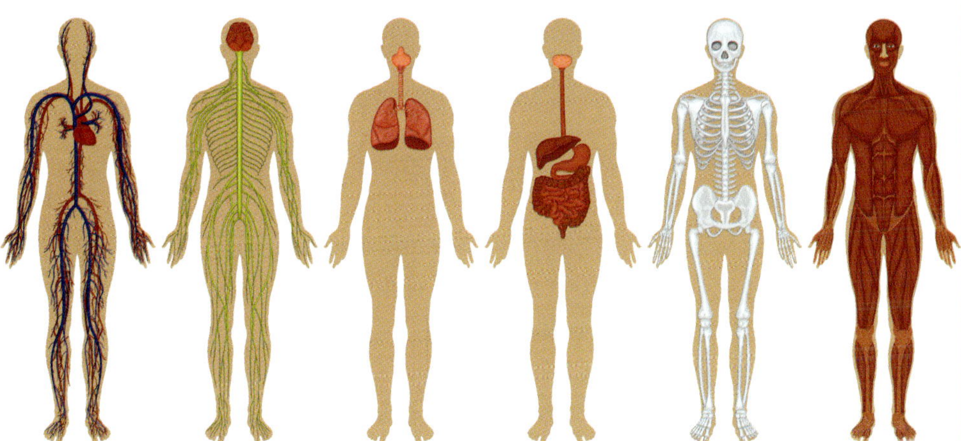

❸ Body systems work together to meet cell needs

Body systems must work together to provide cells with everything they need to function and survive, such as gases, nutrients and water. One of the best examples of this is in the human body.

The human circulatory system is connected to every other system in the body. It transports nutrients, dissolved gases and waste products between cells. Without the circulatory system, other systems would not be able to function.

How do body systems work together to give cells what they need?

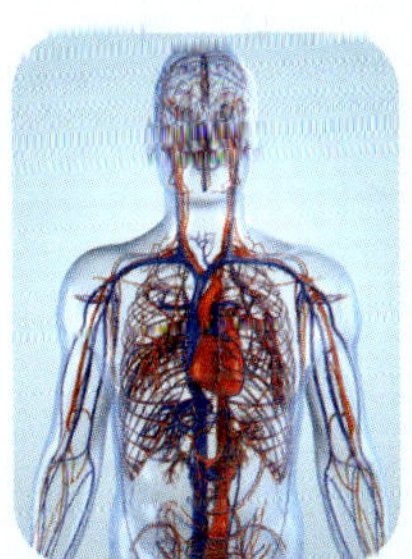

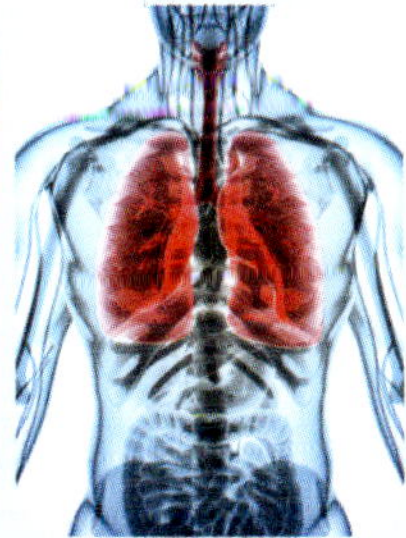

Oxygen and carbon dioxide are exchanged between the lungs and blood. The oxygen in the lungs is moved into the blood and delivered to the cells for respiration. The carbon dioxide produced by cells is taken to the lungs to be removed.

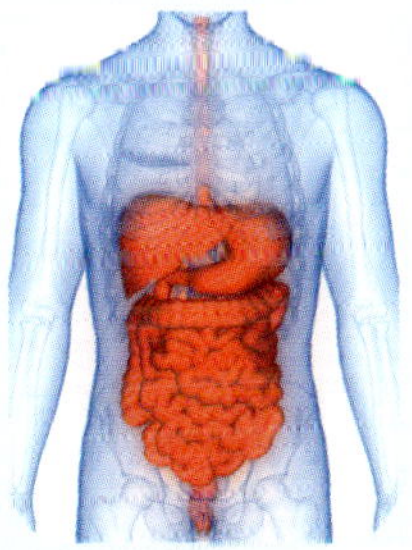

Food is broken down into nutrients by the digestive system. These nutrients are absorbed into the blood, where they are transported to the cells that need them.

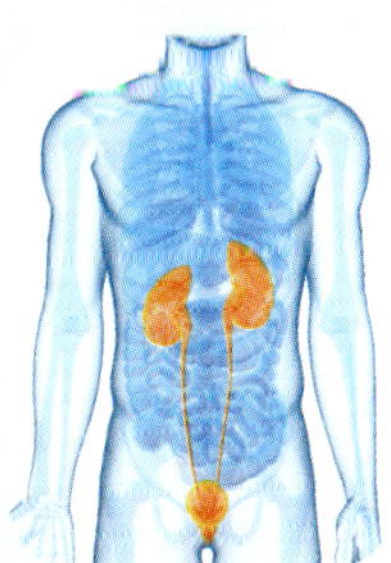

Waste products produced from cellular processes move from the cells into the surrounding blood, to be removed by the organs of the excretory system.

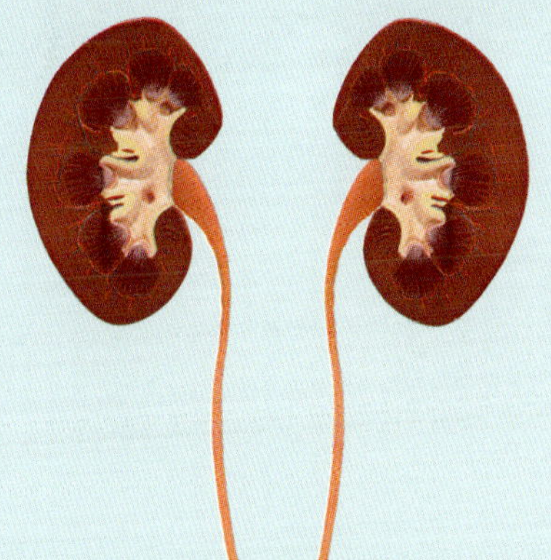

CHECKPOINT 15.2 

1 Order these terms (starting with the smallest structure) to match their organisation within multicellular organisms.

tissue, cell, organ, system

2 Describe the difference between an organ and a tissue.

3 Which system is responsible for the removal of wastes?

4 Suggest what could happen if something went wrong with the digestive system.

5 Cells require oxygen and water to survive. Suggest which body systems assist cells to obtain these materials.

6 Both the circulatory and respiratory systems are responsible for ensuring oxygen gets to our cells. Explain why.

CHALLENGE

7 Research common diseases and disorders that can affect the human heart. Prepare a short report summarising three of your choice.

SKILLS CHECK

- I can list the order of organisation from cells to systems.
- I can explain how systems work together to provide cell requirements, and provide a specific example.

15.3 CELL DIVISION

At the end of this lesson I will be able to:

- **outline** the role of cell division in growth, repair and reproduction in multicellular organisms.

KEY TERMS 

cancer
the uncontrolled growth of cells in some part of the body, which then spread to other body parts

chromosome
a thread-like molecule of genetic information in the nucleus of a cell

meiosis
complex cell division, where new cells are not identical to the original cell

mitosis
simple cell division, where new cells are identical to the original cell

LITERACY LINK 

Use the four key terms above to create a mind map. Link the terms with a line, writing along the line what the link is. You may add as many additional terms as you like.

NUMERACY LINK 

An egg contains 23 chromosomes.

23 is a prime number; list all of the other prime numbers between 1 and 23.

You started your life as an egg cell fertilised by a sperm cell. This divided into two, then four, then eight, and so on, until you eventually became the roughly 32.7 *trillion* cells that you are today.

None of this would have been possible without cell division. Cells can divide by **mitosis** or **meiosis**. These processes are important for organisms to grow, repair body tissues and reproduce.

1 Organisms grow and mature as their cells divide

Growth happens when body tissues increase in size. This isn't due to cells getting larger, but to cells dividing to create more cells. As the number of cells in a tissue increases, the size of the tissue also increases, causing the organism to grow.

Growth is influenced by many factors, but it's the genetic material inside cells that usually determines the overall height and structure of an organism. This is the reason that humans don't grow to the size of elephants!

Sometimes the genetic material inside cells is damaged by factors such as ultraviolet light or chemicals. This damage can cause them to divide uncontrollably, producing lots of unwanted cells. These cells eventually form lumps or growths called tumours, which can affect the function of surrounding tissues or organs. Some cells may spread to other parts of the body - this is called **cancer**.

What is uncontrolled cell growth called?

Figure 15.4 Gigantism happens when the pituitary gland in the brain releases too much growth hormone, causing children to grow abnormally fast and tall.

❷ New cells replace old or damaged cells

If you fall over and scrape an elbow or knee, it doesn't usually take long for the tissue to scab over and heal. This is because the cells you damaged were quickly replaced with new ones by mitosis.

Cells are constantly dividing to replace cells that are damaged or old. This important process makes sure cells in the body are healthy and can perform the functions needed for survival. Some cells, such as those cells of the stomach and intestines, only last a few days because they are exposed to really difficult conditions that wear them down. Others, such as liver cells, live much longer because they are less likely to be damaged.

Not all body cells can be replaced. Some cells, such as nerve and heart muscle cells, are unable to divide, meaning that any damage to these cells can be permanent.

Why do cells sometimes need to be replaced?

❸ Meiosis makes cells with half the full number of chromosomes

Much like other body cells, sex cells (called gametes) need to divide in order to reproduce. But instead of dividing once, they divide twice in a process called meiosis. This extra division produces sex cells that have half the usual number of **chromosomes** found in other cells of the body.

When an egg cell is fertilised by a sperm cell, the number of chromosomes is restored – two halves make a whole. Therefore, most organisms begin life with a full set of chromosomes. Meiosis ensures that organisms don't have too many chromosomes.

What process is used by sex cells to reproduce?

Figure 15.5 A normal human body cell contains 46 chromosomes. Meiosis produces either egg (ovum) or sperm cells which have 23 chromosomes each. When a sperm fertilises an egg, the full number of chromosomes is restored.

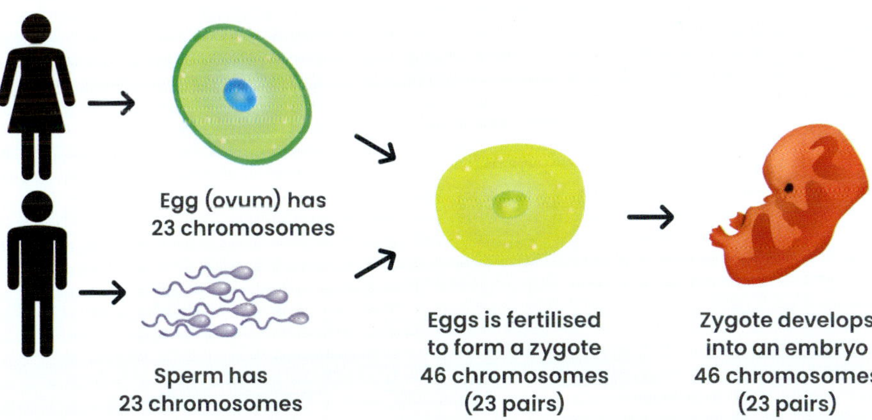

CHECKPOINT 15.3 

1 Explain why cells need to divide.

2 What is one of the main factors that influences the growth of an organism?

3 Why do tumours form?

4 Describe the difference between mitosis and meiosis.

5 Identify these statements as true or false.
   - **a** Organisms grow because their cells increase in size.
   - **b** When cells are damaged, meiosis creates new ones.
   - **c** All body cells can be replaced in cell division.
   - **d** Sex cells have a full set of chromosomes after they divide.
   - **e** All cells have the same life expectancy.

6 Cells need to be replaced when they are damaged. Explain what could happen to an organism if this didn't happen.

CHALLENGE

7 It is thought that one day scientists will be able to use stem cells to regrow human body limbs. Conduct research to find out the answers to these questions.
   - What are stem cells?
   - How are stem cells different to other cells in the body?
   - Name one place in the human body where stem cells could be found.
   - What are some of the medical benefits of stem cells?
   - Why is stem cell research controversial?

SKILLS CHECK

- I can outline the role of cell division in growth, repair and reproduction in multicellular organisms.

15.4 FLOWERING PLANTS

At the end of this lesson I will be able to:

- **describe** the role of the flower, root, stem and leaf in maintaining flowering plants as functioning organisms.

KEY TERMS 

pistil
the female reproductive organs of a flower (anther and filament)

pollen
the fine, powdery substance in the flowers of plants, which contains male sex cells

pollination
the movement of pollen from the male part of the plant (anther) to the female part (stigma)

stamen
the male reproductive organs of a flower (anther and filament)

stomata
pores in the surface of a leaf; the site of gas exchange in plants

LITERACY LINK 

If plants photosynthesise, do they need to respire? Write your answer as a short report.

NUMERACY LINK 

The seeds of a desert cactus take up to six years to grow into new plants.

If a year is 365 days, how many days go by in 6 years?

Most flowering plants have a shoot system and a root system. Together, these systems help plants to grow and reproduce.

A shoot system contains organs that you would usually find growing above the ground, such as leaves, stems and flowers. The root system contains organs that usually grow underground, such as roots.

1 Flowers contain the reproductive organs of plants

The main purpose of flowers is reproduction. Flowers contain the sexual reproductive organs of plants. Many flowers contain both male and female reproductive organs.

The male organs of a flower are the **stamen**. It is made up of the:

- *filament* – the stalk that supports the anther
- *anther* – the organ that generates **pollen**, a fine powdery substance that contains the male sex cells of the plant.

The female organs of the flower are the **pistil**. It is made up of the:

- *stigma* – the organ where pollen germinates
- *style* – the stalk that connects the stigma and ovary
- *ovary* – the organ that stores the female sex cells (ova or eggs).

For plants to reproduce, the pollen coating the anthers needs to be moved to the female parts of the flower to fertilise the ovum. This is called **pollination**.

Sometimes this movement is assisted by the wind or rain, but it usually requires the help of pollinating birds and insects, such as bees. Plants attract these organisms by producing flowers that are bright and colourful, smell nice and contain sugary nectar that the birds and insects eat.

Once fertilised, the ova (eggs) become seeds and the ovary swells and enlarges to become a fruit. The seeds in fruit can grow into new plants when conditions are suitable. For some plants, this can take years to happen.

What is the main purpose of flowers?

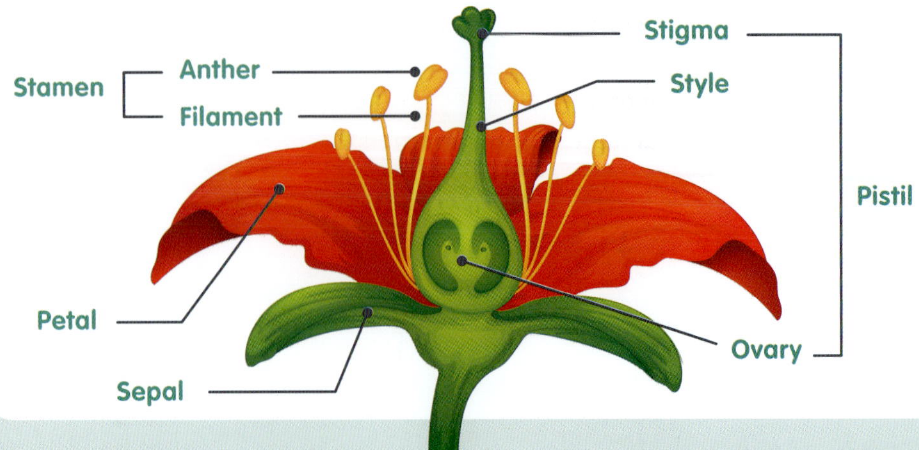

Figure 15.6 Flowers contain the reproductive organs of the plant. The pistil contains the female parts and the stamen contains the male parts.

❷ Stems and roots are for support, transport and growth

People tend not to think about the stems and roots of flowering plants – it's the pretty, sweet-smelling flowers that get all the attention. But the stem and roots perform many vital functions, and plants would collapse and die without them.

The stem is the main body of the plant. It does similar things for the plant that the skeletal and circulatory systems do for humans. These include:

- *support* – helping a plant to stand up and hold the weight of leaves, flowers and fruit
- *transport* – connecting the root and shoot systems of the plant. Water, sugar and other substances move through the stem to provide each system with its requirements
- *growth* – allowing buds to grow from the stem and form into new branches, leaves or flowers.

Roots absorb water and other nutrients from the soil. Water is necessary for photosynthesis and provides the plant with the fluid that dissolves and moves substances around its structure.

Another function of roots is to anchor a plant to the ground. Without roots, trees would fall over in strong winds, and marine plants would wash away in ocean currents.

What are the main functions of the stem and roots of a plant?

❸ Photosynthesis happens in leaves

Leaves could be called the solar panels of plants, because their main role is to perform photosynthesis. Leaves have many features that make them perfect for carrying out this process. They are often flat, which increases their surface area, allowing them to absorb more sunlight. They are thin, so carbon dioxide can travel easily into the cells from the environment. Plants contain green pigments called chlorophyll which absorb light energy from the Sun. Veins in plants allow water and other substances needed for photosynthesis to travel to the leaf cells.

If you look at a leaf under a microscope, you will probably notice round pores called **stomata**. These open and close to allow plants to exchange gases, such as oxygen and carbon dioxide, with their environment. Water can also pass through these pores and water loss is sometimes an unwanted consequence of gas exchange.

What is the main function of leaves?

Figure 15.7 Stomata are tiny pores on the surface of leaves that allow plants to exchange gases with their environment.

INVESTIGATION 15.4
Dissecting a flower

CHECKPOINT 15.4

1 Which body system of a plant contains organs that are usually found above the ground?

2 What are the male organs of a flower called?

3 What are the female organs of a flower called?

4 Describe how a fruit is formed.

5 What would happen to most plants if they didn't have a stem?

6 Photosynthesis needs carbon dioxide, water and sunlight energy. Explain how leaves help a plant to obtain these resources.

7 Roots are usually found underground, so they don't receive the light necessary to carry out photosynthesis. Explain why they are still really important to the process of photosynthesis.

CHALLENGE

8 Use the internet to describe how insects and birds assist plants to reproduce.

SKILLS CHECK

- I can describe the main roles of the flower, stem, leaf and root in the body system of a flowering plant.

15.5 MUSCLES, BLOOD AND BONES

At the end of this lesson I will be able to:

- **describe** the role of the circulatory and musculoskeletal systems in maintaining a human as a functioning multicellular organism.

KEY TERMS 

blood vessel
tube such as a vein or artery that carries blood in the body

cartilage
connective tissue that holds bones together

tendon
connective tissue that connects muscle to bone

LITERACY LINK 

Interview someone in your class who has broken a bone or damaged a muscle. Find out what happened, how long it took to mend and what the experience was like.

NUMERACY LINK 

A man's heart beats 115 417 times one day.

Write this number in words.

We all rely on body systems to move. These systems are made up of specialised organs and tissues, such as bones and muscles, working together.

Like other processes in the body, movement requires a constant supply of energy. The circulatory system is vital to this energy production because it supplies all cells with oxygen for respiration.

1 The circulatory system moves materials around the body

Your heart, **blood vessels** and blood make up your circulatory system. This system delivers oxygen, nutrients and other substances to every tissue in your body. It also helps your body to remove waste products, such as carbon dioxide.

Oxygen passes from the lungs into the blood, and is then transported through veins to the heart. An adult human's heart can beat more than 115 000 times a day, making it the hardest working muscle in the body.

The heart has two 'filling' chambers, called atria, and two 'pumping' chambers, called ventricles. Entry to these chambers is controlled by special valves. Oxygen-rich blood comes in through the atria, is pumped out through the ventricles, and is then distributed through arteries to the rest of the body.

Figure 15.8 The heart pumps blood throughout the body.

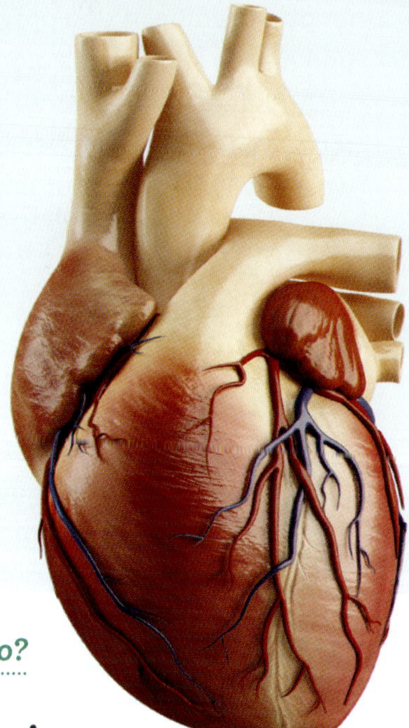

What does the circulatory system do?

2 The skeletal system supports and protects the body

The human skeletal system consists mostly of bones and **cartilage** - a connective tissue that holds the bones together. Bones are made up of living cells and so they need oxygen and nutrients to survive, grow and repair.

Human bones come in all shapes and sizes. More than half of them are in the hands and feet. The largest bone is called the femur, and it is in the top part of the leg. The smallest bones - the stapes- are in the ears.

Functions of the human skeleton include:

- *support* – Without a skeleton, the human body wouldn't be held upright and would collapse into a big blob of tissue and water.
- *protection* – The brain is protected by the skull, the vertebrae protect the spinal cord and the ribs protect the heart and lungs.
- *muscle attachment* – Muscles are attached to the bones by special fibres called **tendons**.
- *blood cell production* – Red blood cells and other blood components are made in the marrow at the centre of bones.
- *mineral storage* – Vital elements such as calcium and phosphorus are stored in the bones and released when they are needed.

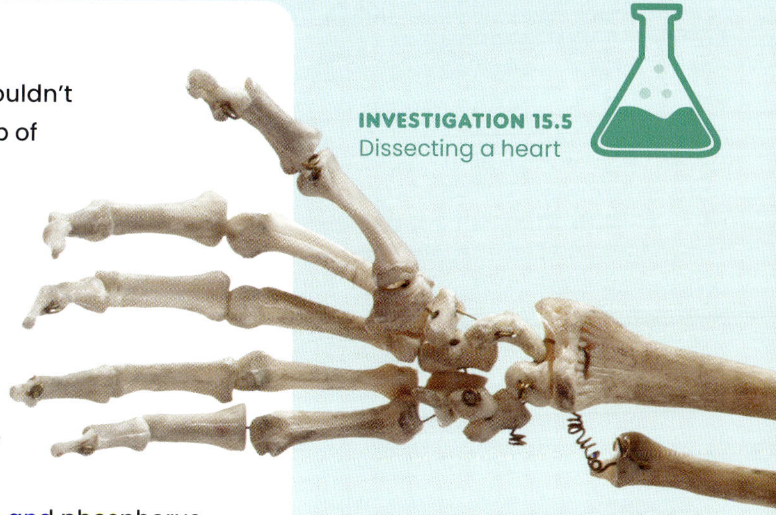

INVESTIGATION 15.5
Dissecting a heart

What does the skeletal system do?

3 The muscular system allows the body to move

The main role of the muscular system is movement. Without working muscles, you wouldn't be able to walk around, your heart wouldn't beat, and food would take much longer to break down and move through your digestive system. Muscles let you smile when you're happy and frown when you're not.

Much like the bones of the skeletal system, muscles can vary in size and shape. The largest muscle in the human body is the gluteus maximus - this is the scientific name for a muscle in the buttocks. The smallest muscles are in the ear.

Muscles work by contracting and relaxing. When muscles contract, the fibres in them shorten and thicken. When they relax, the fibres become longer and thinner. Many muscles work in pairs to coordinate body movements - when one contracts, the other relaxes. When the biceps of your arm contracts, the triceps on the opposite side of your arm relaxes, pulling the forearm up. When the biceps relaxes, the triceps contracts, pulling the forearm down.

What does the muscular system do?

Figure 15.9 Muscles, bones and blood work together in systems that maintain essential functions.

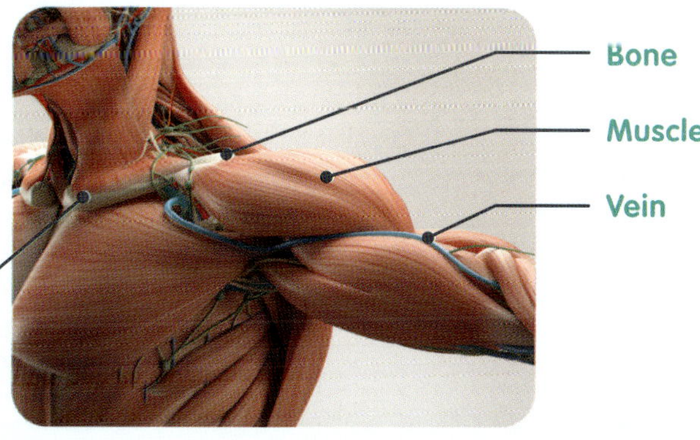

CHECKPOINT 15.5

1 What organs make up the circulatory system?
2 Explain how the muscular and skeletal systems work together in the body.
3 What does the circulatory system transport around the body?
4 One of the roles of the skeletal system is organ protection. Suggest why.
5 What is the difference between a vein and an artery?
6 Where are the biggest and smallest muscles in the body?
7 Give three examples of muscles in the human body and where they are located.

CHALLENGE

8 Use the internet to research voluntary and involuntary muscle movements. Make a summary of each.

SKILLS CHECK

- I can describe the role of the circulatory system.
- I can describe the role of the skeletal system.
- I can describe the role of the muscular system.

15.6 ENERGY IN, WASTE OUT

At the end of this lesson I will be able to:

- **describe** the role of the digestive, excretory and respiratory systems in maintaining a human as a functioning multicellular organism.

KEY TERMS 

digestion
the physical and chemical processes that break down food in the body

enzyme
a chemical that speeds up a reaction (e.g. digestion of food)

excretion
the elimination of cellular waste from the body

LITERACY LINK 

Summarise the roles of the respiratory, digestive and excretory systems into one sentence each.

NUMERACY LINK 

Farmer Farrah feeds each of her cows 20 kg of feed each day. If she has 13 cows, how much feed will she need each day to keep her herd healthy?

Did you know that cows have four stomachs? They need them to break down the tough plant material they eat.

Humans have just one stomach, and as part of the digestive system it works closely with the excretory system to take food in, get all the things the body needs and get rid of the rest. The respiratory system does something similar, taking in oxygen and then getting rid of carbon dioxide.

1 The digestive system releases the nutrients in food

Think of your most recent meal and why you ate it. Were you hungry? Did it just look yummy? Did you want to obtain the nutrients for important cellular processes? If you answered 'yes' to the third question then your mind and your body are on the same page!

Humans and other animals need to consume food to obtain the nutrients inside it. These are essential for energy production, growth, tissue repair and basically all other cellular processes. However, food can't just move directly into the cells. It first needs to be broken down into smaller molecules during **digestion**.

The digestive system in humans is made up of specialised organs that break down food using mechanical (physical) and chemical processes. Digestion begins in the mouth, as the teeth mechanically break food into smaller pieces while **enzymes** work on reactions that dissolve them. The food breaks into smaller and smaller pieces as it travels through the stomach, and then the nutrients move into the small intestine, where they are absorbed through the walls into the bloodstream. Any undigested matter is expelled from the body.

What does the digestive system do?

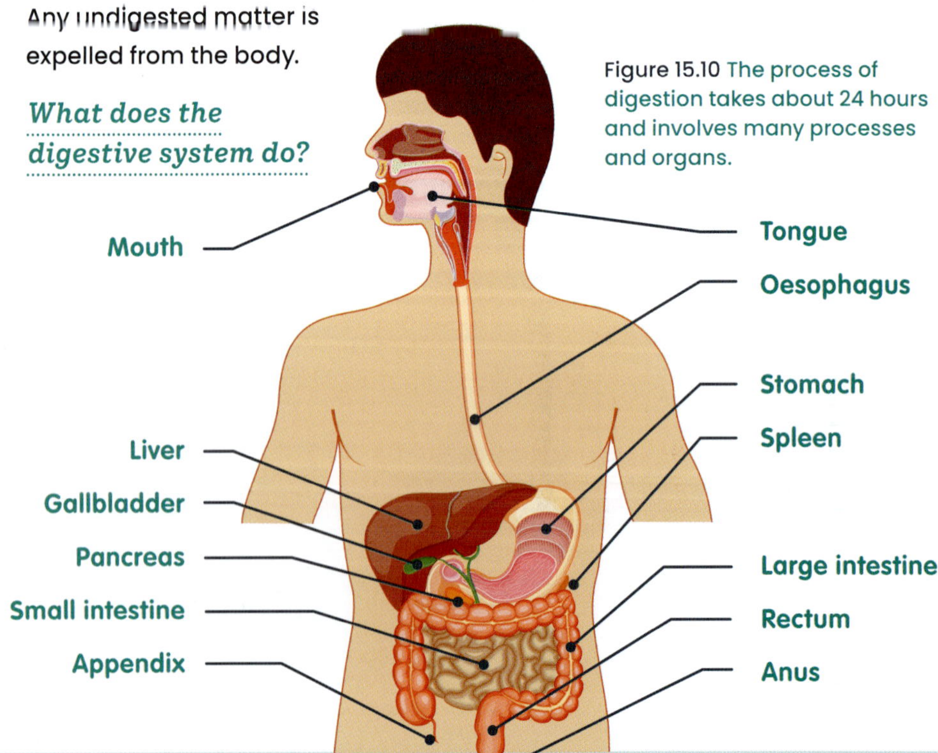

Figure 15.10 The process of digestion takes about 24 hours and involves many processes and organs.

2 The excretory system removes waste from the body

Cells are constantly undergoing chemical reactions. These reactions produce cell wastes that need to be removed from the body before they build up and cause harm. The elimination of cell waste from the body is called **excretion**. Most wastes leave the body in faeces or urine, although some leave in sweat or the breath.

Specialised organs remove different types of cellular waste, and these organs make up the excretory system. Many parts of the human body have some role in excretion, including the skin, lungs and liver. However, there are some specific organs that make up the excretory system:

- The kidneys are two bean-shaped organs behind the lower part of your abdomen. They filter all blood to remove harmful wastes. These wastes are then excreted in urine.
- The urinary bladder is the organ that collects the urine excreted by the kidneys.
- The large intestine is where undigested food particles collect. Any remaining usable water is absorbed through the wall of the intestine, and the remaining solid waste is excreted through the rectum and anus as faeces.

What does the excretory system do?

3 The respiratory system processes oxygen

Your respiratory system gathers and processes oxygen, a molecule essential to life. You breathe in (inhale) oxygen with air, and release carbon dioxide and water vapour when you breathe out (exhale). This gas exchange is only possible because of the special structures of the respiratory system.

During inhalation, air enters the nostrils or mouth and moves into the *trachea* (windpipe). It then travels into two branching *bronchi* and into smaller passageways called *bronchioles*. From here, the air is passed into clusters of tiny air sacs called *alveoli*. Each alveolus is moist, thin and contains many tiny blood vessels called capillaries.

Oxygen from the air moves through the walls of the alveoli and into the capillaries, where it is then transported in the blood to the cells for aerobic respiration. At the same time, carbon dioxide and water move from the blood into the airways, where they then take the opposite route to leave the system during exhalation.

What does the respiratory system do?

CHECKPOINT 15.6 

1 What is the main purpose of the digestive system?

2 The lungs are more like sponges than balloons. Suggest why.

3 What is the role of the kidneys in excretion?

4 Which body system contains the bronchi, and where in the body are they located?

5 Explain how the excretory system ensures the human body is able to function effectively.

6 Give some examples of how wastes can leave the body.

CHALLENGE

7 The trachea is surrounded by C-shaped rings of cartilage. Use the internet to research the purpose of these rings.

SKILLS CHECK

- I can describe the role of the digestive system.
- I can describe the role of the excretory system.
- I can describe the role of the respiratory system.

Figure 15.11 The human respiratory system is made up of airways that connect to the lungs.

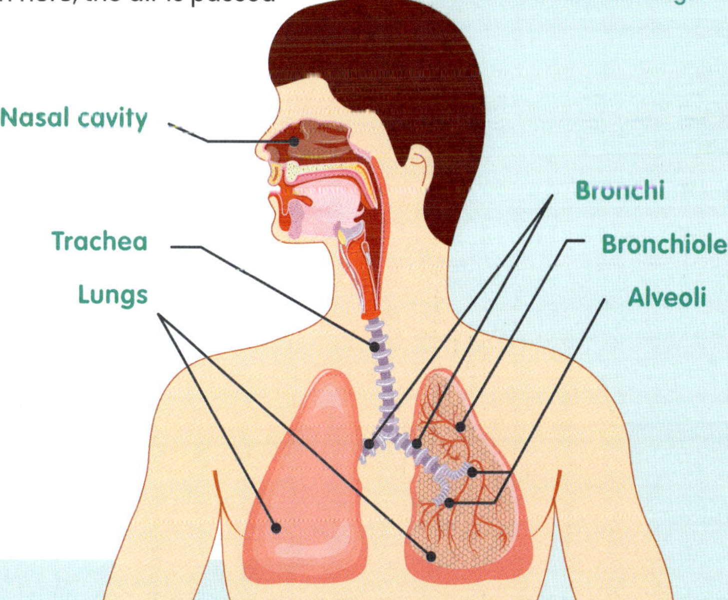

15.7 HUMAN REPRODUCTION

At the end of this lesson I will be able to:

- **outline** the role of the reproductive system in humans.

KEY TERMS 

embryo
an early stage of development of a baby

fertilise
join a sperm with an ovum

foetus
a later stage of development of an unborn baby

zygote
the first single cell of new life

LITERACY LINK 

Identify three adjectives (describing words) in this section. Suggest an alternative word for each adjective you identify.

NUMERACY LINK 

The average development period for a human baby is 40 weeks. How many days are there in 40 weeks?

Most humans have either a male or female reproductive system.

To produce a new human, a sperm from a male must find its way to a female ovum (egg) and wiggle inside.

❶ The human reproductive system produces new life

The function of the human reproductive system is the production of new life. Before technology such as IVF, new life always began with sexual intercourse between a male and a female. If sexual intercourse happens around the time of ovulation, then the male sex cell (sperm) has a chance to **fertilise** the female sex cell (ovum). The new single cell, or **zygote**, quickly divides and becomes a ball of cells called an **embryo**.

About five days after fertilisation, the embryo moves out of the fallopian tube into the uterus. It then embeds itself into the lining of the uterus, where it receives nourishment from the blood vessels and can develop into a **foetus**. The foetus then grows and develops, receiving nutrients and removing wastes from the mother via the placenta and the umbilical cord.

A human baby develops in the uterus for about 40 weeks. After this, different hormones make the muscles of the uterus contract to push the baby out through the vagina. The amniotic sac bursts, fluid is expelled and then the baby is born. About a third of births in Australia are by caesarian section, where a baby is removed surgically through its mother's abdomen.

What is the role of the human reproductive system?

❷ The male reproductive system

Humans mostly have the same organs – we all have hearts, lungs and brains. The only system that differs between humans is the reproductive system, because male humans have different reproductive organs than female humans.

The testes produce sperm (male sex cells) and the male hormone testosterone. They are in the scrotum, which provides them with some protection and allows them to stay at the right temperature while outside of the body.

When ejaculation happens, sperm move from the testes through the vas deferens (sperm duct) to the seminal gland, where seminal fluid is added. This fluid contains mainly water and glucose to moisten and feed the sperm. The prostate and Cowper's glands both add male hormones. The fluid is now known as semen. Semen travels through the urethra, a tube within the penis, to the outside.

What are the key parts of the male reproductive system?

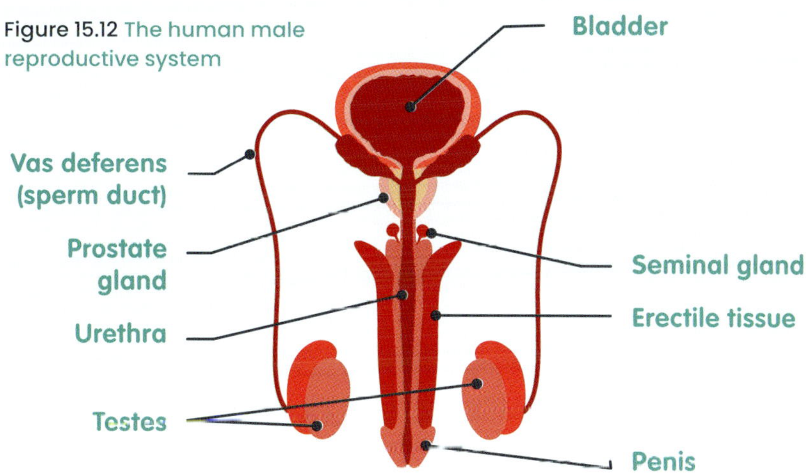

Figure 15.12 The human male reproductive system

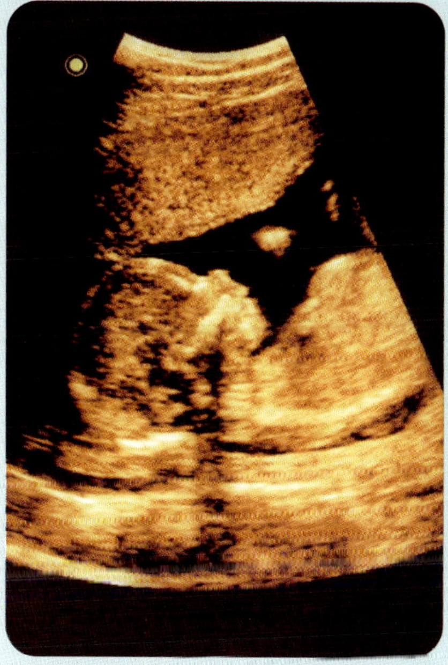

3 The female reproductive system

The ovaries are the main organs of the female reproductive system. They produce ova (eggs) and the female hormone oestrogen. In a fertile female, one ovum is usually released from one ovary each month, during ovulation.

The ovum moves down the fallopian tube. If the egg does not meet a sperm, it continues down the fallopian tube and moves through the uterus. Hormones have made the lining of the uterus thicken with blood and tissue.

If fertilisation does not happen, the ovum moves out through the vagina. The uterus lining breaks down and also moves out through the vagina. This is known as menstruation or a period. On average, this happens once every 28 days after the beginning of puberty and lasts until menopause, which usually happens between the ages of 45 and 55.

What are the key parts of the female reproductive system?

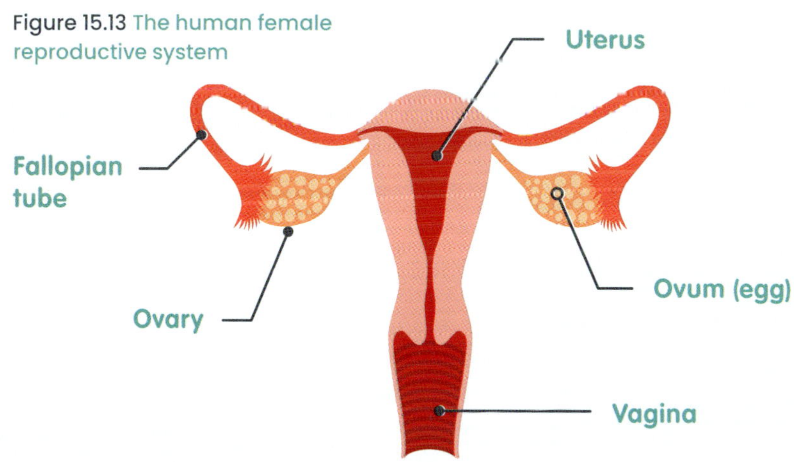

Figure 15.13 The human female reproductive system

CHECKPOINT 15.7

1 Describe the role of the human reproductive system in one sentence.

2 What are the male and female sex cells called?

3 How long does the time from conception to birth usually take?

4 Sperm have tails called flagella – suggest why.

5 What substances make up semen?

6 What is menstruation and why does it happen?

7 In which part of the female reproductive system does the embryo embed and become a foetus?

CHALLENGE

8 Find out how long pregnancy is for different mammals such as a gorilla, a chimpanzee, a cat, a dog and an elephant.

SKILLS CHECK

- I can describe the role of the human reproductive system.
- I can compare the female and male reproductive systems.

CHAPTER SUMMARY

Multicellular organisms make energy through the process of cellular respiration.

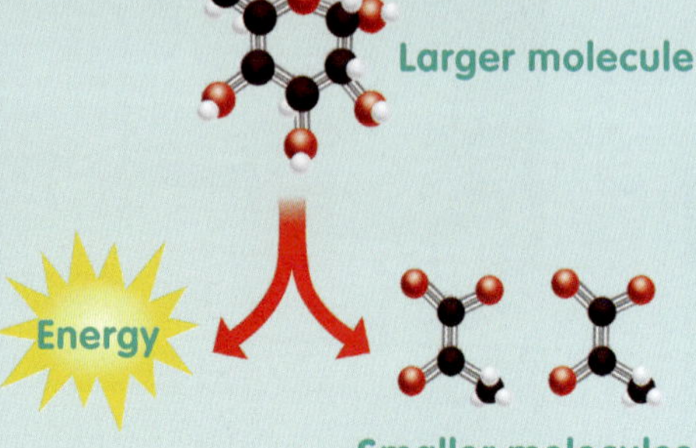

Animals gain their energy from glucose and oxygen.

glucose + oxygen → carbon dioxide + water + energy

Plants use photosynthesis to capture energy from light and produce food from carbon dioxide and water.

$$\text{carbon dioxide} + \text{water} \xrightarrow[\text{light energy}]{\text{chlorophyll}} \text{glucose} + \text{oxygen}$$

Flowers contain the reproductive organs of plants. They reproduce with the help of wind, water or animals, which move pollen from the male to the female organs during pollination.

Cells combine to make tissues, organs and organ systems. Animals can have more than ten major body systems that work together to support life.

The respiratory system, which includes the lungs and trachea, extracts oxygen from air and transports it into the blood.

The muscles of the muscular system contract and relax, allowing the body to move.

The digestive system breaks down food using multiple organs to release nutrients.

The circulatory system consists of the heart, blood and blood vessels. It moves materials around the body.

The skeletal system consists of bones and cartilage. It supports and protects the body.

Organisms grow and mature as their cells divide. New cells replace old or damaged cells.

Unlike other cells, sex cells (gametes) divide twice in meiosis, producing cells with half the usual number of chromosomes.

The excretory system removes waste from the body through faeces, urine, sweat and breathing.

The male reproductive system generates sex cells, which fertilise the eggs created by the female reproductive system to create new life.

★ FINAL CHALLENGE ★

1. Bones must be both strong and rigid as well as somewhat flexible. Suggest why.
2. What do the equations for photosynthesis and cellular respiration have in common?
3. Why do the muscular and skeletal systems need to work so closely together?

LEVEL UP!

LEVEL 1

★☆☆
☆☆☆

50xp

4. What is the role of the flower in a flowering plant?
5. Draw a labelled diagram of the various parts of a flowering plant.
6. Explain how the respiratory and circulatory systems work together.

7. Describe the physical and chemical differences between the food that goes into your mouth and the waste that is eliminated.
8. What is respiration? When does it happen? Where does it happen?

9. Describe the role of the excretory system and suggest what could happen to someone with kidney failure.
10. Give an example of two body systems that work together and discuss how they do so.

11. Explain why the male testes must be outside the human body.
12. Label this diagram of the female reproductive system:

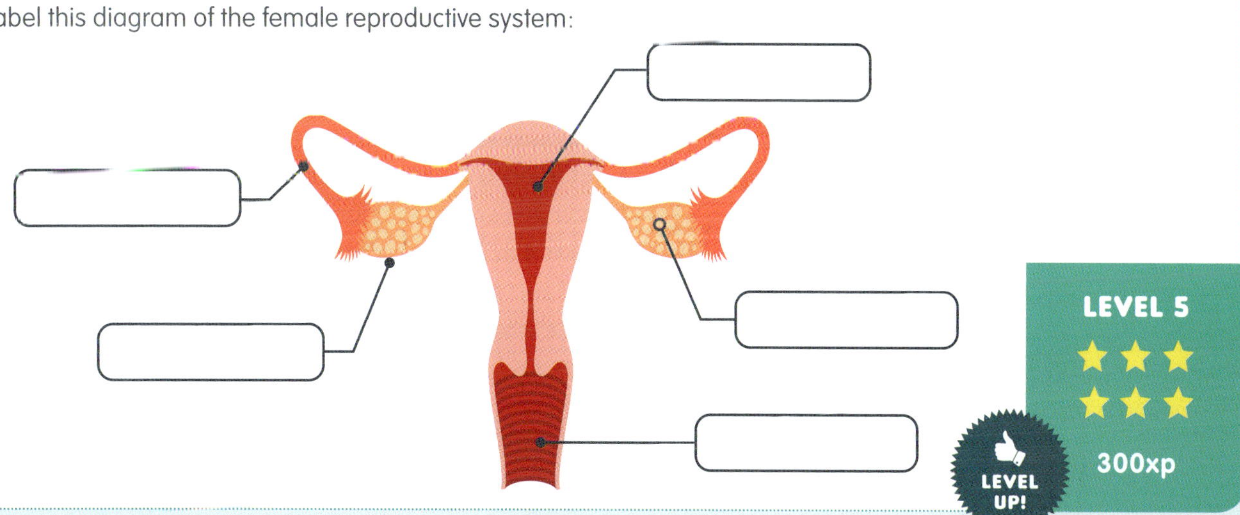

LEVEL UP!

LEVEL 5

★★★
★★★

300xp

BIOTECHNOLOGY

16

Developments in science and technology happen every day. Many new tools and devices improve our lives in countless ways. Discoveries shape our knowledge and opinions, paving the way for even more advances, especially in medicine and biotechnology.

What scientific and technological discoveries are you most excited about? The search for 'super-Earths' (rocky planets twice the size of ours)? The progression of artificial intelligence and robotics? Growing replacement organs in laboratories? Gene mapping and editing? The future of science and technology seems limited only by our imaginations.

1 LEARNING LINKS

What do you already know about biotechnology?

How do social and environmental influences affect product design?

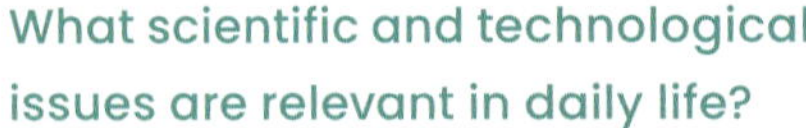

What scientific and technological issues are relevant in daily life?

What attitudes do people hold towards science and technology?

2 SEE-KNOW-WONDER

List three things you can **see**, three things you **know** and three things you **wonder** about this image.

3 CRITICAL + CREATIVE THINKING

B-A-R: What would you make *bigger*, *add* and *replace* for a bionic leg?

Commonality: Write as many points of commonality as you can between the spread of disease and an organ transplant.

The ridiculous! Try to substantiate (prove/support) the following statement: 'The world would be better without sanitation and waste treatment'.

4 THE LONGEST!

The longest organ transplant chain in history involves over 100 kidney donors and recipients. It started in 2013 in Alabama, in the USA, The average wait for a kidney is more than five years, so this long transplant chain helps many people.

In the USA, an organ transplant chain begins with one person who decides to donate an organ because they want to help others. The person who receives the organ may have a relative who wanted to help them but they weren't the right match. Instead, this relative then donates their organ to another patient and the chain goes on.

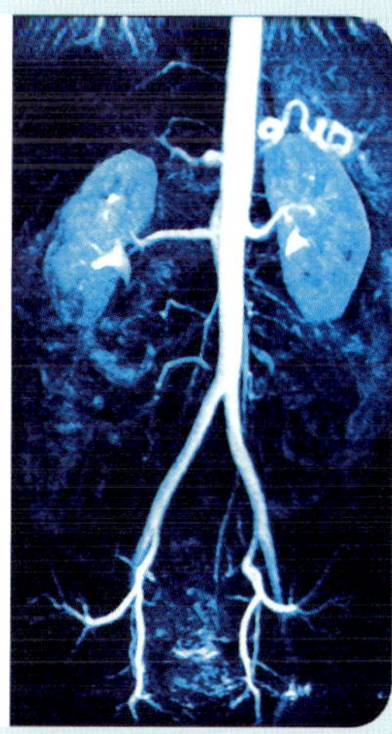

16.1 TREATING DISEASES

At the end of this lesson I will be able to:

- **describe** at least one example of how changes in scientific knowledge have contributed to finding a solution to a human health issue.

KEY TERMS 

bacteria
tiny, single-celled organisms that can live in a range of environments

immune
resistant to a particular illness or disease

transmissible
able to be passed from one person to another

virus
a tiny infectious agent that multiplies in its host

LITERACY LINK 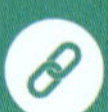

Write an opinion piece about the ethics of Edward Jenner testing his smallpox vaccine on an eight-year-old boy.

NUMERACY LINK 

10 people are infected with a transmissible disease, and 4 more are infected every day.

Write an equation to model the spread of the disease, and calculate how many people in total are infected after 17 days.

Nothing has caused more loss of human life than **transmissible** diseases. Historically, during some disease outbreaks, 10% of the world's population was wiped out.

Scientists have been vital in treating diseases, in some cases completely eliminating them. They have made vaccines and antibiotics, and educated people about how diseases spread, saving millions of people from suffering and death.

1 Transmissible diseases are passed from person to person

Have you ever been home sick with 'the flu'? Sometimes people say that when they just have a bad cold, but the real flu is a disease called influenza. This is a transmissible disease – one that passes from person to person. Diseases are transmitted in different ways – some transmit through skin contact, others through tiny droplets in the air, or several other ways.

There are different types or strains of influenza, and some are stronger and more dangerous than others. In 1918, a strain called the Spanish flu killed 3–5% of people on Earth. That was between 50 and 100 million people! One of the reasons so many people tragically died was that there was no vaccine available to make people **immune** to influenza.

What is a transmissible disease?

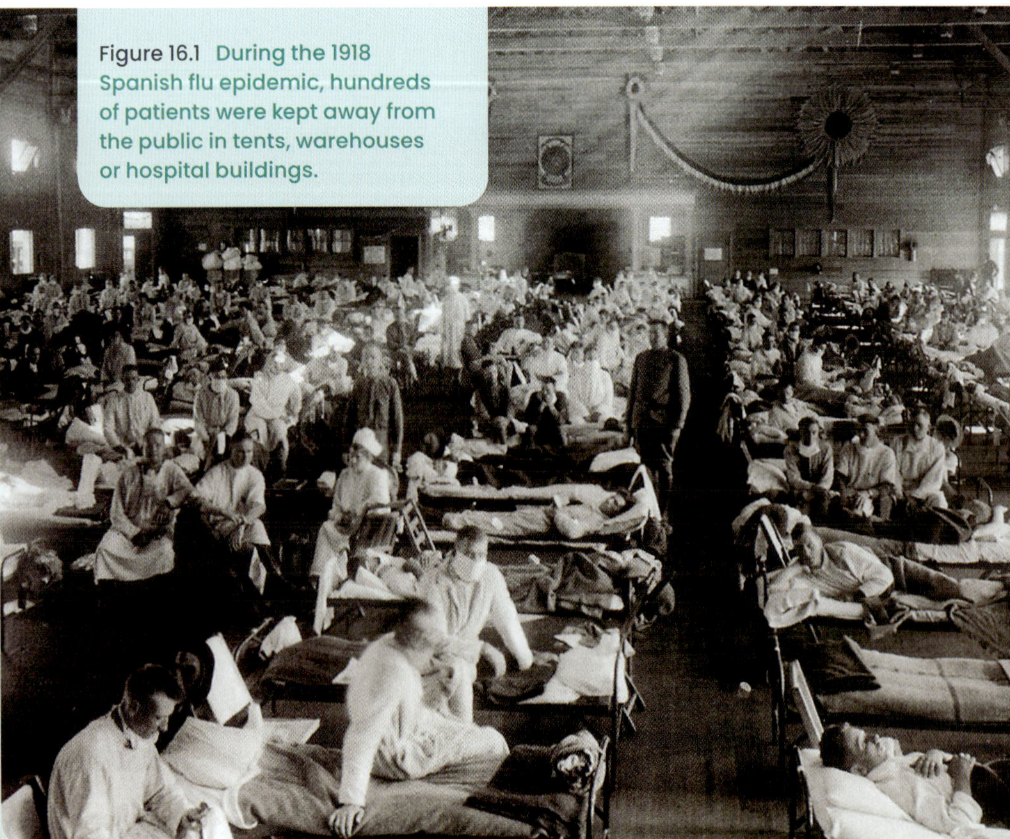

Figure 16.1 During the 1918 Spanish flu epidemic, hundreds of patients were kept away from the public in tents, warehouses or hospital buildings.

2 Vaccines can prevent people from catching diseases

Vaccines are medicines that can prevent people from catching diseases in the first place. The vaccine is similar to the disease itself, so when a person is injected with it or swallows it, their immune system acts to fight it off. The person's body then usually 'remembers' the disease and becomes immune to it. If the person comes in contact with the disease again, they can remain healthy.

The first vaccine was discovered and developed at the end of the 18th century by English doctor Edward Jenner. He lived at a time where a disease called smallpox was killing millions of people all around the world.

Jenner noticed that people who had been exposed to cowpox – a very similar but much less dangerous disease – were immune to smallpox. One day he borrowed the eight-year-old son of his gardener and experimented on him. He rubbed pus from cowpox blisters into an open wound on the boy's arm. Jenner's experiment worked, and the boy became immune to smallpox!

How was the vaccine for smallpox developed?

INVESTIGATION 16.1
The contagion game

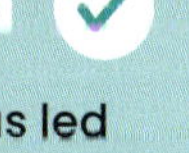

3 Antibiotics treat diseases caused by bacteria

Have you have ever cut yourself and then noticed the wound is red and hot? It may have been infected with **bacteria**. Medicines called antibiotics can treat many infections like this. Antibiotics are also able to treat illnesses such as tuberculosis and pneumonia, which in the past have killed millions of people.

The earliest antibiotics used types of fungus, such as mould, to fight bacteria. There are records showing the use of mould in treating infections in Ancient Greece and Egypt. People at that time may not have known the science behind it, but they knew that it worked. British scientist Alexander Fleming was the first person to work out why moulds could treat infections, and he invented the first antibiotic medicine, penicillin from a fungus, in 1928. Australian scientist Howard Florey and British scientist Ernst Chain did more research to make this medicine in large amounts, so that it could treat many people.

Antibiotics have limitations, because they only work on bacteria. Diseases such as colds and influenza are caused by **viruses**, tiny infectious agents even smaller than bacteria, so antibiotics have no effect on them. Sometimes antibiotics are used too much or when they shouldn't be, which can cause them to lose their effectiveness. Many doctors suggest avoiding the use of antibiotics unless they are absolutely necessary.

Figure 16.2 The pale green bacteria cannot grow near the small discs of test antibiotics.

What can be treated with antibiotics?

CHECKPOINT 16.1

1 Scientific knowledge has led to many improvements in human health. Identify at least two of the ways that now exist to fight disease.

2 Explain in your own words how vaccines work.

3 There are many types of diseases that humans can have.
  a With a partner, list as many different diseases as you can.
  b Separate the diseases from part **a** into two columns: 'transmissible' (those you can catch from someone or something) and 'non-transmissible'.

4 Explain why antibiotics are not useful to treat influenza.

CHALLENGE

5 Imagine you are the Prime Minister of Australia. Devise a plan to ensure the safety of all Australians in case of a disease outbreak. What are the three most important aspects of your plan?

SKILLS CHECK

- I can describe at least one way that scientific knowledge has contributed to solving a human health issue.

16.2 THE IMPORTANCE OF CLEAN WATER

At the end of this lesson I will be able to:

- **explain** how evidence from a scientific discovery has changed understanding and contributed to solving a real world problem, e.g. hygiene, sewage treatment or biotechnology.

KEY TERMS 

hygiene
ways of doing things that support health and prevent disease, often through cleanliness

sanitation
facilities and services to ensure safe treatment of human waste (urine and faeces)

sewage
wastewater and human waste (urine and faeces)

waterborne
carried or existing in water

LITERACY LINK 

Create a poster that gives tips on good hygiene for primary-aged students. Ensure your poster is easy to understand.

NUMERACY LINK 

A chemist tests 11 samples of bottled water for contaminants and records the following percentages:

0, 0, 0, 0, 1, 1, 2, 2, 3, 4, 4, 11

Remove the outlier from this data set, then calculate its mean, median and mode.

We use water for many things: to produce food, to make goods in factories, and for cleaning and drinking. Clean water is essential for health.

The biggest cities in the world began in places with easy access to water. People living downriver often had health problems, because people upstream were polluting the water with waste. **Waterborne** diseases such as cholera still cause the death of millions of people each year in places that do not have effective ways to get rid of wastes.

❶ Sanitation programs keep water safe and clean

Sanitation is about keeping the environment – especially water – clean and free of disease. As well as removing and treating wastes such as sewage, it includes making sure water is safe for people to drink. Without good sanitation, people can be exposed to waterborne diseases, viruses, bacteria and parasites such as intestinal worms.

Good sanitation programs educate people about the importance of **hygiene**. Do you wash your hands after using the bathroom? Do you have regular showers or baths? Do you help your family by doing the dishes and wiping surfaces? All of these questions promote good hygiene, keeping people healthy and preventing the spread of disease.

In the 19th century, English doctor John Snow identified that the disease cholera came from contaminated water. Previously, it was thought that the disease was due to 'bad air'. His research led to important findings in germ theory, disease transmission and sanitation, and improve health around the world.

What does sanitation involve?

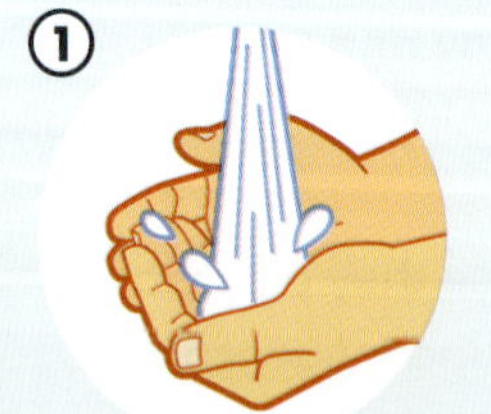

Figure 16.3 Washing your hands maintains health and hygiene, not just for yourself but for everyone and everything you contact.

❷ Sewage is treated before it is released to the environment

Can you imagine life without a toilet? In the 19th century, many people would throw their waste straight out of the window onto the street, or directly into waterways. Aside from smelling really bad, this is dangerous! Wastes such as urine and faeces can contain viruses, bacteria and parasites. These cause serious diarrhoea, particularly in children. This leads to millions of deaths each year in developing countries where sanitation is poor.

In the late 19th century, British chemist Edward Frankland did some experiments to treat **sewage** with chemicals. He wanted to make the sewage safe to release into the environment. He found that by treating the sewage and then letting it sit for a while, he was able to make it safe to release.

Frankland's discoveries led to the development of large sewage treatment plants, which are used throughout Australia. The sewage is treated in a number of steps, starting with removing objects such as nappies, plastics and cotton buds. Air is pumped in to encourage bacteria to start doing the hard work of breaking everything down.

How is sewage treated to make it safe?

Figure 16.4 Sewage treatment plants, like this one near Sydney, treat wastewater so that it can be used for non-drinking purposes such as flushing toilets and watering gardens.

INVESTIGATION 16.2
Treating muddy water

CHECKPOINT 16.2 ✓

1 What kinds of contaminant can be found in sewage and wastewater?

2 What is an example of a waterborne disease?

3 Explain how the work of John Snow and Edward Frankland led to important improvements in human health.

4 Explain why good hygiene is critical to ensuring good health.

5 Many people in developing countries still lack basic sanitation. Explain how you would go about assisting these communities to improve their sanitation practices.

6 Explain the difference between hygiene and sanitation.

7 Figure 16.4 shows a sewage treatment plant. Write three things you can see when you look at the image, three things you know and three things you wonder.

CHALLENGE

8 Create a six- or eight-panel cartoon strip that describes some of the main steps of sewage treatment.

SKILLS CHECK

- I can explain how the understanding of sanitation has improved ways of waste management such as sewage treatment.

16.3 ORGAN TRANSPLANT TECHNOLOGY

At the end of this lesson I will be able to:

- **describe**, using examples, how developments in technology have contributed to finding solutions to a contemporary issue, e.g. organ transplantation.

KEY TERMS 

embryo
an early stage of life where the parts of an unborn organism are still being developed

ethical
relating to principles about what people think is 'wrong' and 'right'

Nobel Prizes
world-famous awards given each year for academic, cultural and scientific advances

stem cell
a cell that can produce a different type of cell and divide to make more cells

LITERACY LINK 

Defend or criticise this statement: 'Embryonic stem cells should not be used to create organs for organ transplantation'.

NUMERACY LINK 

The chance of Ruben's body accepting his new kidney is 84%. What is the chance of his body rejecting the transplant?

Organ transplantation allows organs, such as the heart, kidneys and skin, to be given to someone who needs an organ because theirs is no longer working.

The biggest medical challenge in organ donation is an organ being rejected after someone receives it. Developments in technology mean that donations are now more successful.

1 Technology has improved organ transplant medicine

Transplanting organs from one person to another has been tried again and again over hundreds of years, but not very successfully. As doctors and scientists learnt more about the body, organ transplants became more successful.

One problem for organ transplants is keeping the organs fresh and healthy after being removed from the donor. Currently, organs are kept in cold storage during transfer, but this can damage them. In recent medical technology, scientists have made a system that copies the conditions inside the human body, keeping organs fresh and capable of 'surviving' as they're moved from the donor to the recipient.

How has technology led to more successful organ transplantation?

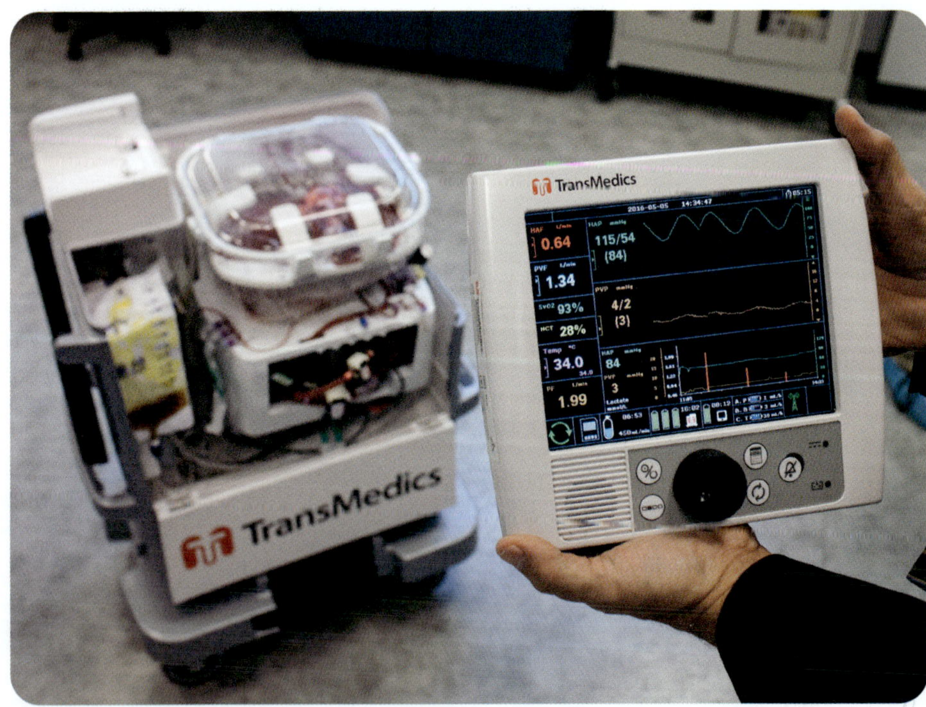

Figure 16.5 Instead of freezing organs, the TransMedics Organ Care System copies the conditions within the human body, keeping the organ alive.

❷ The human body tries to reject transplanted organs

As with any surgery, things can go wrong during organ transplants. The major cause of problems during transplants is actually the human immune system. The human body is always on the lookout for things that shouldn't be there – that's how our immune systems protect us. Unfortunately, these things include cells from something that would be helpful – in this case, a donor organ.

To avoid this, doctors try to find a very close match between donor and recipient, and anti-rejection medication can be used. The medication tries to stop the body from attacking the new donor organ.

Peter Medawar, a British scientist, worked out why a person's body rejects an organ. This led to the first anti-rejection medication, and he received a **Nobel Prize** for his work in 1960.

Why does the body try to reject donor organs?

❸ Stem cells may allow new organs to be grown

Current research into the use of some types of cell is exciting for the future of organ transplant. **Stem cells** are cells that can produce any other type of cell. They exist in some places in an adult human body, but some of the most powerful stem cells come from **embryos**.

The benefits are huge – imagine if you could grow a new beating heart out of your own stem cells! You wouldn't have to worry about organ rejection either. The new heart would have the same DNA as you, so your body would identify the heart as yours.

There are **ethical** concerns, though. If embryonic stem cells are used, the embryo is destroyed afterwards. Some people consider this to be loss of life, because the embryo could have developed into a baby.

How could stem cell science change organ transplantation?

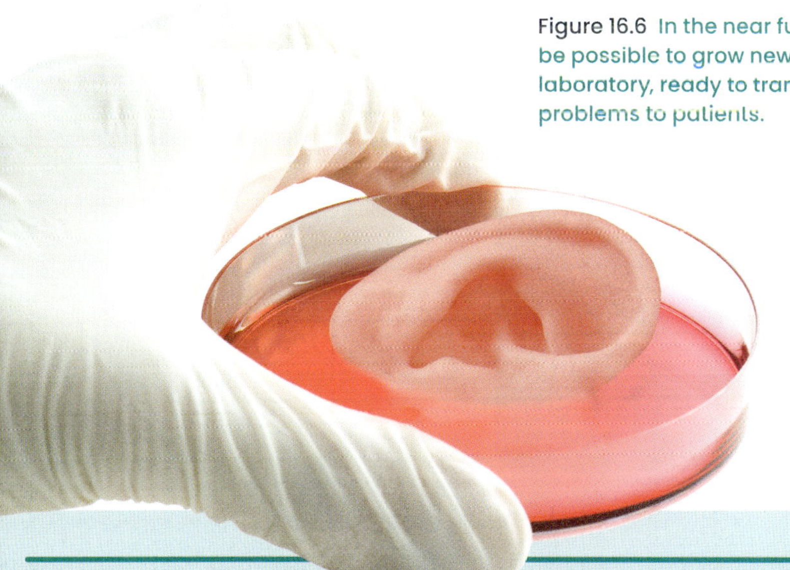

Figure 16.6 In the near future, it may be possible to grow new organs in a laboratory, ready to transplant without problems to patients.

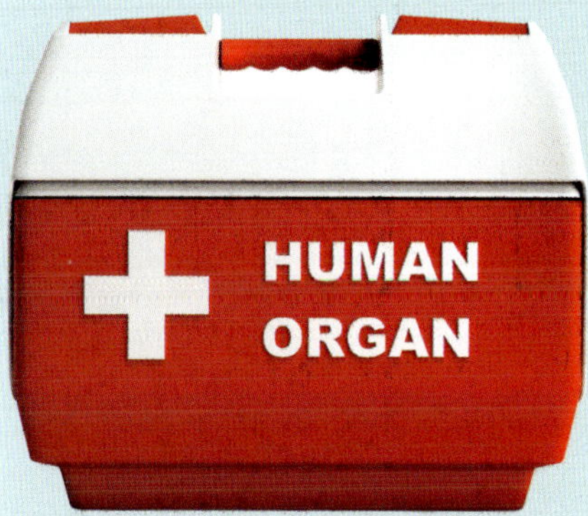

CHECKPOINT 16.3 

1 List some of the organs that can be transplanted.

2 Describe how organs are currently transported for organ transplantation.

3 Describe some of the risks of organ transplantation.

4 Describe at least two ways that developments in technology have led to improvements in organ transplantation.

5 Explain why the immune response of trying to kill foreign cells is usually an important and useful thing for the body to do.

6 What do you think the job of the immune system is in the human body? Give an example with your explanation.

7 Explain how stem cells could be used to make improvements in organ transplantation.

CHALLENGE

8 Would you donate your organs? Conduct research to create a list of pros and cons of organ donation.

SKILLS CHECK

- I can explain how developments in technology have led to improvements in organ transplantation.

16.4 OPINIONS ABOUT ORGAN TRANSPLANTS

At the end of this lesson I will be able to:

- **give examples** to show that groups of people in society may use or weight criteria differently in making decisions about the application of a solution to a contemporary issue, e.g. organ transplantation.

KEY TERMS 

controversial
creating public disagreement and debate

ethical
relating to principles about what people think is 'wrong' and 'right'

humane
compassionate and sympathetic

LITERACY LINK 

Create a pamphlet aimed at encouraging people to consider organ donation. Include information about the importance of organ donation and what is involved in becoming a donor.

NUMERACY LINK 

22 467 people responded to a recent survey on their opinions on organ transplants.

Round 22 467 to the nearest hundred.

Choosing to be an organ donor, or to receive a transplanted organ, isn't always a straightforward decision. Different groups in society have different beliefs and opinions about organ transplantation.

There are more people waiting for transplants than there are organs available, so animal organs are sometimes used. Also, some people sell organs illegally.

1 Organ transplantation raises concerns

There is a huge gap between how many organs are available and how many people need them. In NSW, only about 17 in every million people donate an organ. This raises several **ethical** questions. How do we best select who receives an organ? Should we use organs from children?

Some people – living donors – choose to donate an organ such as a kidney while they are still alive. Some decide that their organs may be donated to others after they die. For potential living donors, there can be pressure to donate an organ, and this can cause stress and relationship issues. The family of a person who has died may be too distressed to agree to the organ donation planned by that person.

What are some of the issues with organ transplantation?

2 Organ trafficking is a worldwide problem

When there is a need for organs, this is a chance for the illegal sale of human organs. This is also known as organ trafficking or trade.

Organ trafficking can affect disadvantaged people who may need money for themselves and their families – they may sell an organ such as a kidney and become sick or die. Cases of kidnapping and stealing organs have been reported all around the world. The price for some organs has been reported to be as high as $150 000, or even more.

Why does organ trafficking happen?

Figure 16.7 The illegal buying and selling of human organs is a major problem in some parts of the world.

3 The use of animal organs is controversial

There are many more people needing organs than there are donors. Some people support the use of animal organs, like those of pigs or baboons, for transplants. This is a field of medical research called *xenotransplantation*. The Greek prefix *xeno* means 'foreign' or 'different'.

Using animal organs in transplantation has never been entirely successful, because the human body rejects them. The use of animals in science and medicine has always been **controversial**. People concerned about the **humane** treatment of animals may not agree with animal organ transplants.

Why is the use of animal organs controversial?

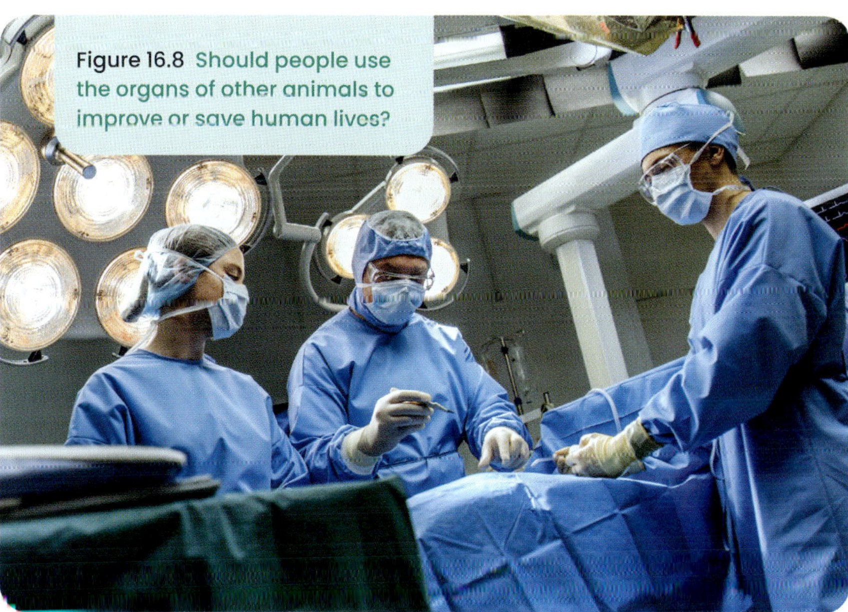

Figure 16.8 Should people use the organs of other animals to improve or save human lives?

4 Some cultures and religious groups have concerns about organ transplants

Different cultures and religious groups have different attitudes about many things, including medicine. They may have different ways of thinking about the value of organ transplantation.

In many religions there is no objection to organ donation and organ transplantation. Some groups support it, seeing organ donation as an act of generosity or compassion. For some there are moral or spiritual concerns. For example, some people believe that the body of a person who has died must remain complete, in readiness for an afterlife. Others may accept an organ donation if the organ no longer contains blood from the donor.

How can culture or religion affect decisions about organ transplantation?

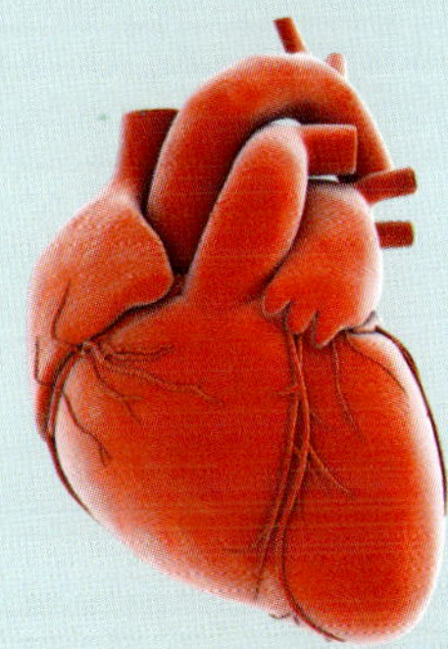

CHECKPOINT 16.4 

1 Describe at least three ethical considerations about organ transplantation.

2 Explain why the illegal trade of organs exists.

3 Explain why these groups may be hesitant to undergo organ transplantation.
   a Indigenous Australians
   b Romani people
   c those of the Jehovah's Witness faith

4 Explain why the use of animal organs can be controversial.

5 Create a list of positives and negatives in relation to organ transplantation.

6 Some people consider the use of pig organs for transplantation to be more acceptable than the use of organs from primates such as baboons, even though primate organs are a closer match to our own. What is your opinion? Defend or criticise the use of pig organs over primate organs.

CHALLENGE

7 Research organ trafficking and prepare a short report on the illegal trade of organs.

SKILLS CHECK

- I can describe how different groups in society make decisions about organ transplantation.
- I can give examples of the concerns of at least two groups of people in society.

CHAPTER SUMMARY

Biotechnology
Transmissible diseases are passed from person to person. They can be extremely dangerous, but modern medical science has developed ways to combat them.

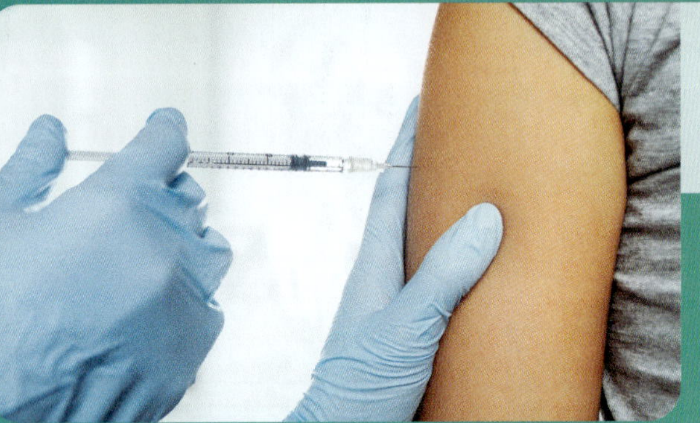

◀ **Vaccines** prevent people from catching diseases.

Antibiotics ▶ treat diseases caused by bacteria.

▼ Waterborne diseases like cholera still cause deaths in places that lack adequate sanitation.

▼ Sewage is treated to reduce its environmental impact. Bacteria break down the organic waste in sewage water so that the water can be recycled for non-drinking purposes.

▼ Many different organs and tissues can be transplanted, from hearts and eyes to bone marrow and skin. Modern organ transplant technology allows donor organs to be preserved and kept healthy.

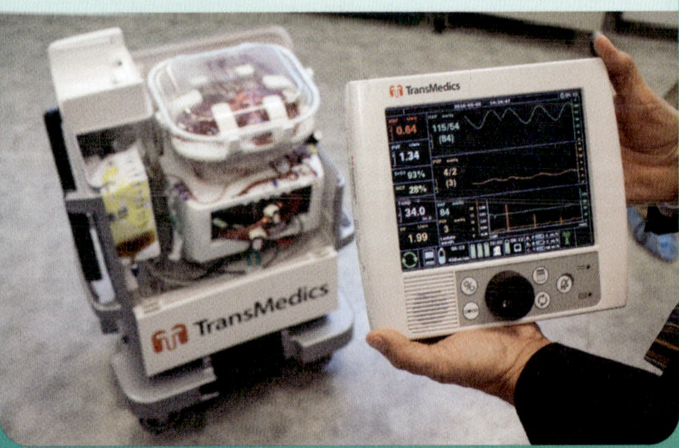

The illegal buying and selling of human organs is a major problem in some parts of the world. Some cultures and religions have concerns about organ transplants, while other people may have ethical concerns. ▶

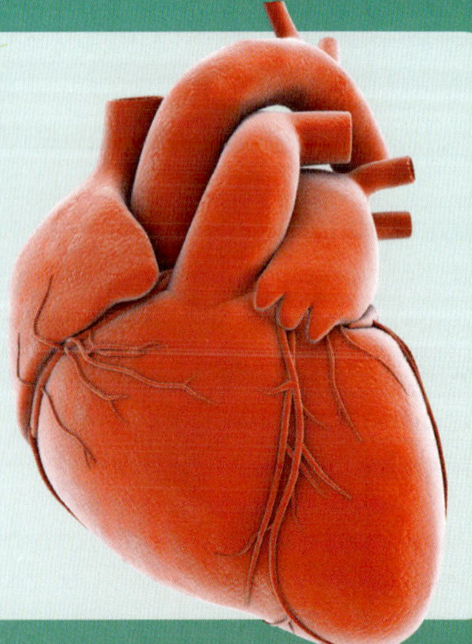

★ FINAL CHALLENGE ★

1. Give a definition of biotechnology in your own words.
2. Complete the following sentence: sanitation is not only ensuring that human waste like urine and faeces is safely disposed of, but also…
3. Give some basic advice for avoiding transmissible diseases.

4. Match the terms below to their definitions.

Transmissible	Cells that can produce other kinds of cells and keep dividing.
Immune	Practices that support health and stop the spread of disease.
Sewage	Capable of being passed from one person to another.
Stem cells	Resistant to a particular illness or disease.
Hygiene	Waste water and excrement.

5. Explain why antibiotics will not help someone who has a cold.
6. Explain how the discovery of the vaccine has improved human life.
7. Describe the difference between a vaccine and antibiotics.

8. Explain how waste is treated and made safe in Australia.
9. The Spanish flu wiped out around 3–5 % of the world's population; suggest how this was possible.
10. Describe some of the historic problems with organ transplants that are now being addressed or improved due to improvements in technology.
11. Explain stem cell science, including what stem cells are and how scientists hope to use them in the future.

12. Describe some implications or considerations of organ transfer under the following sub-headings:
    a ethical
    b cultural
    c legal

17 ECOSYSTEMS

An ecosystem is a biological community of living things, such as plants and animals, and non-living things, such as air and water. The way that living things interact can be shown using food chains and webs.

Many people struggle to share their home with spiders, but killing them can have a big impact on an ecosystem. Spiders eat insects and pests that might otherwise damage natural plant environments and ruin crops. They are a food source for animals such as birds, so removing them from an ecosystem can be devastating.

1 LEARNING LINKS

What do you already know about ecosystems?

How do animals and plants interact?

What are the roles that different organisms play in a habitat?

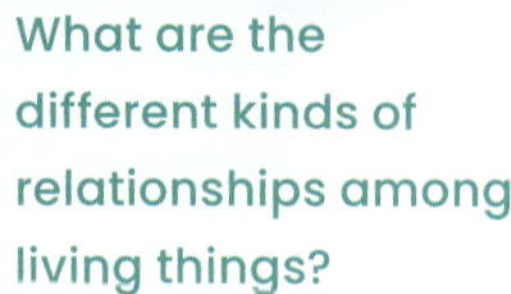

What are the different kinds of relationships among living things?

How do natural changes in the environment affect relationships between plants and animals?

2 SEE-KNOW-WONDER

List three things you can **see**, three things you **know** and three things you **wonder** about this image.

3 CRITICAL + CREATIVE THINKING

Variations: How many ways can you think of to control a fire?

What if ... there were no more decomposers on Earth?

Mismatches: Solve the problem of deforestation using a robot, 20 donkeys and a cabbage.

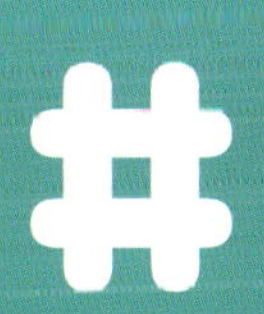

4 THE SCARIEST!

Not all relationships in an ecosystem are nice ones. Parasites are organisms that live in or on another living thing in order to survive, causing damage or even death. A parasitic, bloodsucking fish lives in the Amazon River – and it's not a piranha. It is called a candiru, also known as the vampire or toothpick fish.

The candiru finds its way behind the gills of other fish. Once there, it sticks out its spines so that the victim bleeds and provides it with a meal. Nasty!

17.1 FOOD CHAINS AND FOOD WEBS

At the end of this lesson I will be able to:

- **construct and interpret** food chains and food webs, including examples from Australian ecosystems.

KEY TERMS 

consumer
an organism that gains energy by consuming other organisms

decomposer
an organism that gains energy by breaking down dead material

ecosystem
a community of living and non-living things

food chain
a path of energy through an ecosystem

food web
a system of interlocking food chains

producer
an organism that produces energy at the start of a food chain

LITERACY LINK 

Copy and complete this sentence: 'The difference between food chains and food webs is ...'

NUMERACY LINK 

A local ecosystem has $\frac{42x}{7}$ food webs.

Simplify the expression $\frac{42x}{7}$: ___________

All living organisms need energy to grow, repair damage and reproduce. Without energy, an organism cannot survive. Plants make their own food and energy, but other organisms – including humans – must feed on other organisms to gain energy.

The feeding relationships between organisms, and the energy transfers in an **ecosystem**, can be shown in food chains and food webs.

1 Food chains show the path of energy in an ecosystem

Food chains are diagrams that show one possible path that energy can move through an ecosystem. Energy is transferred between organisms when they eat or are eaten.

Food chains always begin with a **producer** – a plant or microorganism that can produce energy. This is because they provide the glucose for all other organisms to feed on. Every ecosystem includes producers.

When you draw a food chain, you draw arrows between each organism to show the direction that energy moves. In a food chain for an Australian grassland ecosystem, a grasshopper gains energy from grass, so you would draw an arrow from the grass to the grasshopper.

Figure 17.1 Each organism in a food chain provides energy for the next organism.

The other organisms in the food chain – those that feed on organisms to gain energy – are **consumers**. They can be described in terms of the organisms they feed on.

Table 17.1 Types of consumer and how they obtain their energy

Consumer	Obtains energy by ...	Examples
Herbivore	... feeding on producers	Koala, grasshopper
Omnivore	... feeding on producers and other consumers	Sea star, most humans
Carnivore	... feeding on other consumers	Kookaburra, brown snake
Decomposer	... breaking down dead matter	Bacteria, fungi

Organisms can also be described by their position in a food chain.

Table 17.2 Organisms and their positions in a simple food chain

Organism	Position in food chain
Grass	Producer
Grasshopper	Primary consumer
Blue-tongue lizard	Secondary consumer
Eastern brown snake	Tertiary consumer
Wedge-tailed eagle	Quaternary consumer

INVESTIGATION 17.1
Modelling a pond ecosystem

What type of consumer is a herbivore?

2 Food chains interact to form food webs

Organisms are usually part of more than one food chain. This is because most organisms don't feed on a single food source, or get eaten by a single consumer. Having more than one food source is important so that a species can survive in a changing environment. If a food source runs out, a species is more likely to survive if it has another food supply.

Food webs show all the feeding relationships in an ecosystem. These types of diagram show a much bigger picture of how each organism feeds and interacts. An organism may be a tertiary consumer in one food chain, while being a secondary consumer in a different food chain within the ecosystem.

When organisms feed, they only gain a small amount of energy from their food source. They use a large amount of their energy to move, grow, repair cells and reproduce.

What is the difference between a food chain and a food web?

Figure 17.2 This New South Wales bushland food web shows how different food chains can connect.

CHECKPOINT 17.1

1 Define the term *producer*.
2 What do the arrows in a food chain show?
3 Use the food web in Figure 17.2 to:
  a draw one food chain
  b identify the secondary consumer in your food chain.
4 Explain why is it important for an organism to have a variety of food sources.
5 Explain how an organism can be both a secondary and tertiary consumer.

CHALLENGE

6 Imagine you're a tiny carbon atom stored in the leaf of a eucalypt tree. Describe a possible journey you could take as you cycle through an ecosystem.

SKILLS CHECK

- I can explain what a food chain is and construct a simple food chain independently.
- I can explain what a food web is and construct a simple food web independently.

17.2 HOW ORGANISMS INTERACT IN AN ECOSYSTEM

At the end of this lesson I will be able to:

- **describe** interactions between organisms in food chains and food webs, including producers, consumers and decomposers.

KEY TERMS 

carnivore
an organism that eats only meat

decomposition
the process of rotting and decay

herbivore
an organism that eats only plants

omnivore
an organism that eats both plants and meat

parasite
an organism that lives in or on another organism, causing it harm

LITERACY LINK 

Create a mind map for these terms: *producer, consumer, decomposer, predator, prey, competition.*

NUMERACY LINK 

Selina records the number of predators active in eight different local ecosystems: 4, 7, 9, 9, 14, 15, 18, 20.

Calculate the mean, median and mode of this data set.

All organisms within an ecosystem, including humans, depend on interactions with each other for energy, nutrients and survival. Food chains and webs show the feeding relationships between producers, consumers and decomposers.

There's more to an ecosystem than just who eats whom. Interactions between organisms can be helpful or harmful.

1 Organisms compete for resources

Competition often happens when organisms share the same limited resource, such as food, in the same ecosystem.

Consumers compete with each other for the same food. This can happen between different species or members of the same species. Depending on the amount of food available, organisms may be harmed, may starve or have to find a new food source.

Producers also compete for resources, but not food. Plants compete for space, light, water and nutrients – the things they need to produce energy. Some plants will grow and survive, while others will die.

Name a resource that plants compete for in an ecosystem.

2 Predators and parasites get nutrients from other organisms

Predators are consumers that kill and feed on another animal, their prey. Predators in a food web include **omnivores** and **carnivores**, both of which eat meat. Carnivores are always predators within a food web, but omnivores are predators in some food chains and prey in others. Predators benefit by gaining energy and nutrients from the prey, which is killed.

Parasites are organisms that live in or on another organism, the host. Parasites harm the host, but they don't kill it. Examples of parasites include hookworms, which live in the intestines of their human hosts, feeding on nutrients from digested food.

What is the main difference between predators and parasites?

Figure 17.3 Ticks are parasites that eat the blood of other animals.

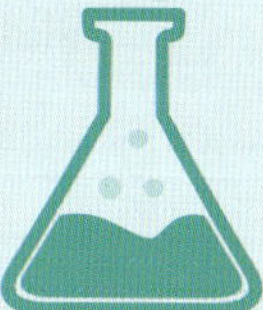

❸ Some plants benefit from being eaten

Herbivores are consumers that only eat plants, which are the producers in a food chain. Some of them only eat the reproductive parts of a plant – the fruits and flowers. This can benefit the plant, because the animal plays a part in the plant's reproductive process.

When a herbivore feeds on nectar, pollen is released from the flower. The pollen is transferred from one plant to another, possibly on the body of the animal. It then fertilises the next plant, resulting in a fruit being produced.

Other herbivores and omnivores feed on the fruit produced by plants. The seeds within the fruit pass through an animal's digestive system and are spread out away from the parent plant, to grow into a new plant. This seed spreading helps the plant because it reduces competition for space and light within the ecosystem.

Figure 17.4 When a rainbow lorikeet feeds on the nectar of red bottlebrush flowers, it transfers pollen from one plant to another.

How can plants benefit from animals eating their fruits and flowers?

❹ Bacteria and fungi break down dead matter

The process of breaking down dead matter is called **decomposition**. If decomposition didn't happen, the dead remains and waste materials of organisms would be everywhere, and the nutrients in them would stay trapped inside, never to be used again.

Most bacteria and fungi are neither producers nor consumers, but decomposers. They interact with every other organism in a food web by breaking down dead matter and waste in an ecosystem. This is how they gain their energy to grow, reproduce and survive.

The nutrients of dead organisms are released into the soil to be recycled, making them available to new food chains. Plants take up these nutrients, along with water, through their roots. This allows consumers to obtain these nutrients when they feed. All other organisms, including humans, rely on decomposers to survive.

Figure 17.5 The decomposers breaking down this dead tree gain its stored energy and recycle its nutrients.

What are two types of decomposer found in all ecosystems?

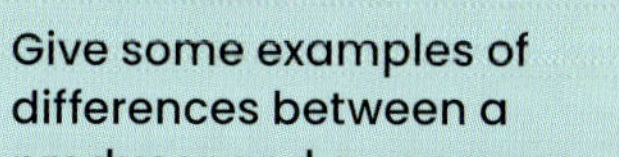

INVESTIGATION 17.2
Observing ecosystems

CHECKPOINT 17.2 ✓

1 Give some examples of differences between a producer and a consumer.

2 Give an example of an Australian predator.

3 Write definitions for the terms *decomposer* and *competition*.

4 Identify one way a plant can benefit from being eaten by a herbivore.

5 What resources do plants compete for in an ecosystem?

6 Venus fly traps are both producers and consumers. These plants trap and kill flies and other insects to gain energy and nutrients. Identify the interaction between Venus fly traps and flies.

CHALLENGE

7 Use the food web in Figure 17.2 to answer these questions.
   a Which organism is the producer? How do you know it is a producer?
   b Identify one example of each of:
      i competition
      ii predation.
   c Suggest how competition and predation would change if all the crickets were eaten so they were no longer a food source in this food web.

SKILLS CHECK

- I can explain what a producer, consumer and decomposer are and give an example of each.
- I can describe ways that producers, consumers and decomposers interact in an ecosystem.

17.3 HOW MICRO-ORGANISMS AFFECT ECOSYSTEMS

At the end of this lesson I will be able to:

- **describe** examples of beneficial and harmful effects that microorganisms can have on living things and the environment.

KEY TERMS 

bacteria
microscopic, unicellular organisms

microorganism
an organism only visible under a microscope

pathogen
a microorganism that can cause disease

protist
a microorganism that is not a fungus or bacteria

unicellular
made of one cell

LITERACY LINK 

Find a news article about harmful or beneficial microorganisms. Read the article and carefully summarise it into one paragraph.

NUMERACY LINK 

Roscoe tests for microorganisms in a pond that is x metres long and y metres wide. The area of the pond is 360 square metres.

Write an algebraic expression to represent the area of the pond, and determine x if $y = 40$ m.

Microorganisms are microscopic organisms that are too small to see with the naked eye. They're found at all feeding levels of food webs, in every ecosystem on Earth.

Microorganisms have different roles in ecosystems. Some of these roles benefit other organisms and the environment, while others are harmful and can lead to major changes.

1 Bacteria, fungi and protists are microorganisms

Microorganisms exist all around you. They live in your intestines and on your skin, in the soil in your backyard and in the water you drink. They also live at the bottom of the ocean, in volcanic vents and in the air.

There are three main types of microorganisms: bacteria, fungi and protists. Most are **unicellular** organisms.

Bacteria are extremely small unicellular organisms. They usually exist as large colonies of thousands of organisms. The different sizes and shapes of bacteria can be used to identify them. Bacteria that live in water environments often have hair and tail-like structures, called cilia and flagella, to help them move.

Protists are larger than bacteria and they exist alone, rather than in colonies. Every type of protist has a unique structure that helps it survive. Some protists carry out photosynthesis, and act as producers in ecosystems, while others are consumers and decomposers.

Fungi range in size from unicellular yeasts to multicellular mushrooms. Although they are often grouped with plants, fungi are not producers in an ecosystem because they cannot photosynthesise.

What are the three main types of microorganism in ecosystems?

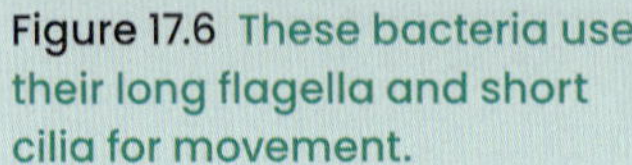

Figure 17.6 These bacteria use their long flagella and short cilia for movement.

Figure 17.7 Freshwater protists have eyespots that can detect light. They use their flagella to move towards the light to carry out photosynthesis.

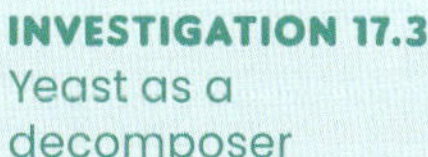
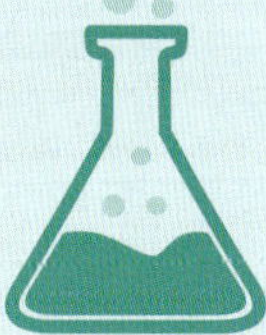

2 Some microorganisms are helpful

Many microorganisms have benefits to ecosystems. Producer microorganisms provide the energy source for food chains and webs. Consumer microorganisms break down food into products that other organisms can use. Decomposer microorganisms recycle nutrients and break down harmful waste products.

Here are some other ways that microorganisms benefit the natural world, or are used by humans:

- Penicillin is an antibiotic produced by the *Penicillium* fungus. It stops the growth of harmful bacteria, and so is useful as a medicine. Many other medicines come from fungi.
- Bacteria in the intestines of animals, including humans, help digestion. They break down food to release nutrients, make vitamins and remove wastes.
- Bacteria are used as decomposers in sewage treatment plants. Under controlled conditions, they break down sewage into non-toxic substances. The nutrients in the non-toxic substances are used as fertilisers, and clean water is released to waterways.

What is one beneficial microorganism used in medicine?

3 Some microorganisms are harmful

Some bacteria, fungi and protists are harmful to ecosystems. When these organisms are harmful, they are called **pathogens**.

Bacteria are responsible for many diseases. Food poisoning is a common illness caused by toxins made by *Salmonella* bacteria. This pathogen enters the stomach in contaminated food and can cause diarrhoea, vomiting and stomach cramps.

Protists can also cause illness. Malaria is caused by a protist carried by mosquitoes. When the mosquito feeds, the pathogen enters the animal's bloodstream and infects red blood cells. This causes fever, aches, vomiting and sometimes death.

In water environments, when nutrient levels are too high, some protists and bacteria can reproduce much faster than they are eaten. The water becomes covered in a 'bloom' of algae. This stops light and oxygen from entering the water, which can kill some organisms and allow harmful bacteria to increase.

What is a pathogen?

Figure 17.8 Phytoplankton are microorganisms that photosynthesise in oceans. They are producers, supplying marine food webs with energy.

INVESTIGATION 17.3
Yeast as a decomposer

CHECKPOINT 17.3

1 Write a definition for the term *microorganism*.

2 Are all microorganisms beneficial to an ecosystem? Why or why not?

3 Describe one difference between bacteria and protists.

4 Describe one way that decomposers are beneficial to all ecosystems.

5 A student wrote this incorrect sentence:

'Producers are always plants. They are the only organisms that are able to carry out photosynthesis in an ecosystem.'

Explain why the student's statement is incorrect.

6 Give one example of how microorganisms can be beneficial and one example of how they can be harmful.

CHALLENGE

7 Some people eat or drink probiotics, which are live bacteria yoghurts. These contain thousands of helpful bacteria. Explain how taking a probiotic can be beneficial to humans.

SKILLS CHECK

- I can explain what a microorganism is and list the three main types.
- I can describe ways in which microorganisms can be harmful and beneficial.

17.4 HUMAN IMPACTS ON FOOD WEBS

At the end of this lesson I will be able to:

- **predict** how human activities can affect interactions in food chains and food webs, including examples from Australian land or marine ecosystems.

KEY TERMS 

biodiversity
the variety of organisms in an ecosystem

deforestation
the removal of trees to make land suitable for other uses

non-biodegradable
does not break down in the environment

pesticide
chemicals used on farms to protect crops by killing pests

urbanisation
the creation of urban areas such as cities

LITERACY LINK 

Write a speech that argues for significantly decreasing levels of deforestation in Australia.

NUMERACY LINK 

After an ecosystem is cleared, a property developer builds 6 houses on it. They sell for: $400 000, $550 000, $620 000, $835 000, $910 000 and $1 080 000.

Calculate the mean and median of the property prices.

Human activities often use fossil fuels and **non-biodegradable** materials to meet demand for food, energy and technologies. This can damage food webs, such as by polluting air and removing forests.

Even a seemingly small change in the environment can have a dramatic impact on the feeding relationships in an ecosystem. The balance between producers, consumers and decomposers in a food web can be easily disturbed.

1 Urbanisation and deforestation damage food webs

Urbanisation is the replacement of natural ecosystems with urban areas such as cities and suburbs. This is due to the constant need for more space for increasing human populations. As cities grow, the surrounding land is cleared and many of the organisms in the ecosystem die.

Deforestation is the removal of large trees to make space for urban landscapes and farms. Big trees are the main producers in forest ecosystems, as well as homes and food sources to many herbivores and pollinators. All organisms within food webs are affected by deforestation, because they all depend on the original energy source from these producers.

Removing trees can increase soil erosion and water pollution. This is because the roots of the trees no longer hold the soil together or filter the water flowing through the ecosystem.

What is deforestation?

Figure 17.9 The land covered by this new housing development previously had large trees and an ecosystem for a variety of organisms.

2 Removing or introducing organisms affects biodiversity

Biodiversity is the variety of species within an ecosystem. The more species there are in a food web, the better the chances each type of organism has of survival if the environment changes. If there are fewer organisms, there are fewer food chains and less variety of food sources.

Loss of biodiversity causes major problems. These problems can happen in ecosystems on land or in water, for example during commercial fishing. If a consumer relies on a single food source, and this source is lost, the consumer will be affected. It will decline, either by individuals moving to a new area or dying. This has a flow-on effect to the consumers that feed on it, and causes greater competition for the reduced food sources available. Humans take organisms out of natural ecosystems for their own use, such as for food. Biodiversity can be affected if too many organisms are taken, because they don't get a chance to repopulate.

Another way humans damage biodiversity is by introducing new organisms into an environment. These species , such as rabbits and blackberries, often thrive and take over food sources and habitat of other organisms.

What is the name for the variety of different species in an ecosystem?

Figure 17.10 The practice of overfishing – taking too many fish from an area – can completely remove some species of fish from food webs.

3 Farming, fossil fuels and plastics can damage the environment

Pesticides are chemicals used on farms to kill pests, including unwanted plants and insects, to protect crops. These chemicals may be passed on to other animals in a food chain that feed on the pest. This can cause a build-up of these chemicals to toxic amounts in an ecosystem, and harm organisms (including humans) that have consumed the pesticide.

Algal blooms can happen when large amounts of nutrients from farms are released into streams. The algae in the streams feed off the nutrients and grow quickly, forming a bloom that covers the surface of the water. This prevents light reaching other producers, which die.

The burning of fossil fuels is a cause of global warming, which has a big impact on the habitat and food sources in food webs. These fossils fuels are also used to make plastics, which are non-biodegradable and enter food webs when animals accidentally consume tiny plastic particles.

How do farming practices and fossil fuels affect ecosystems?

CHECKPOINT 17.4 ✓

1 What is meant by the terms *urbanisation* and *deforestation*?

2 Explain how urbanisation can affect interactions in food chains and food webs.

3 How can an introduced species affect other organisms within an ecosystem?

4 Which has higher biodiversity – a tropical rainforest or a wheat farm? Give evidence to support your answer.

5 Suggest why organisms are more likely to survive in an ecosystem that has high biodiversity.

CHALLENGE

6 Research coral bleaching and outline how human activities have contributed to it.

SKILLS CHECK

- I can suggest how human activity can affect interactions within food webs and chains.
- I can give an example of the effect of human activities in Australia on natural environments (on land or in water).

17.5 MANAGING BUSHFIRES

At the end of this lesson I will be able to:

- **explain**, using examples, how scientific evidence and/or technological developments contribute to developing solutions to manage the impact of natural events on Australian ecosystems.

KEY TERMS 

biodiversity
the variety of organisms in an ecosystem

germinate
grow and put out shoots

intensity
how much heat is released from a fire front

LITERACY LINK 

Research and create a bushfire action plan for people living in areas with thick vegetation. Include advice on what to do in case of a nearby bushfire.

NUMERACY LINK 

In March 2018, a bushfire in the South Coast region of NSW burned through approximately 1250 hectares of land. Convert 1250 hectares into square metres.

1 hectare = 10 000 square metres

Australian ecosystems are often affected by unpredictable natural events that harm ecosystems and reduce **biodiversity**. These events include droughts, floods and bushfires.

Scientists can't always predict when and where these events will happen, but they can use research and technology to manage the impact on ecosystems and biodiversity.

1 Bushfires happen every year in Australia

Bushfires are common natural events in Australia due to the hot, dry summer climates of many regions. These conditions create the perfect environment for bushfires.

Lightning strikes are the most common natural cause of bushfires. When lightning hits a dry or dead tree, the heat of the strike makes the tree catch fire and break apart, scattering burning wood that then ignites other plants. Bushfires are also caused by human activity – often due to accidents, but sometimes deliberately.

The **intensity** and duration of a bushfire depend on the plants in the ecosystem, because they provide the fuel needed for the bushfire to burn.

What is the most common natural cause of bushfires?

2 Low-intensity bushfires are helpful in some ecosystems

Low-intensity bushfires usually don't last very long and don't destroy many plants. In fact, they can be helpful for some ecosystems as a whole. These fires remove older or dead plants, making more space for seeds to **germinate** and grow. Fire can even trigger germination in some plants. The ash from the fire returns some nutrients to the soil, and the ecosystem recovers relatively quickly.

Lighting low-intensity bushfires has been an Indigenous Australian practice for thousands of years.

What is one advantage of a low-intensity bushfire to an ecosystem?

Figure 17.11 The Warlpiri people in the Northern Territory burn spinifex during winter, so that the grasses won't catch fire during summer.

❸ High-intensity bushfires usually harm ecosystems

High-intensity bushfires cause a lot more damage to ecosystems. Tall, thick, compact grasses, and fallen leaves and branches, can provide a bushfire with a large fuel supply. They create high-intensity fires that take a long time to burn. The bushfire burns entire trees, right up to the top leaves.

High-intensity bushfires are more likely to happen in areas that haven't had a fire for many years. There is a large amount of dry plant material, such as fallen branches and dead trees at the ground layer. This provides the fire with fuel to burn for a long time.

What provides the fuel for a high-intensity fire?

❹ Bushfires can be managed and controlled

Some scientists develop ways to manage bushfires in Australia. They gather scientific evidence by monitoring fires and learning from previous fires. They can then develop techniques and technology to manage bushfires and better understand how they behave.

The Commonwealth Scientific and Industrial Research Organisation (CSIRO) has a national bushfire research facility in Canberra called the Pyrotron. It is used to safely carry out bushfire experiments. The Pyrotron is a 25-metre-long wind tunnel with different sections and compartments. Fuel sources, such as different types of leaves, are placed in the tunnel and set on fire while a fan blows wind through the tunnel. Sensors and cameras measure how the fire spreads, while observers can watch through fireproof windows.

The Pyrotron improves our understanding of how bushfires spread in different conditions. It also helps firefighters train, be better prepared and make better decisions during a fire.

A computer model can also be used to predict the movement, intensity and duration of a bushfire. The CSIRO's bushfire modelling system, Spark, simulates a bushfire. It uses the information it receives to predict the direction, speed and intensity of a bushfire. Spark's models are based on records of previous fires, vegetation and soil types. It also considers conditions such as wind and soil moisture.

Figure 17.12 Dry eucalypt leaves are being used as a fuel source in this experiment at the Pyrotron.

How can technology help firefighters when they are in action?

CHECKPOINT 17.5

1 Explain how computer modelling systems can assist with the management of bushfires.

2 What are the benefits of a controlled burn?

3 Why does a fire need plant material to burn?

4 Suggest what plant characteristics you would see in an ecosystem that is prone to low-intensity bushfires.

5 Give two reasons why Indigenous Australian land management practices include regular burning of areas of vegetation.

6 Give an example of how technology is helping manage the impact of fires on Australian ecosystems.

CHALLENGE

7 Research how a different natural event, such as a cyclone or drought, affects Australian ecosystems, and how science and technology are used to manage its impact.

SKILLS CHECK

- I can name at least three natural events that affect Australian ecosystems.
- I can suggest how improvements in scientific knowledge and technology have resulted in better management of bushfires.

17.6 IMPROVING CROP SUSTAINABILITY

At the end of this lesson I will be able to:

- **describe** how scientific knowledge has influenced the development of practices in agriculture, e.g. crop cultivation, to improve yields and sustainability.

KEY TERMS 

agriculture
the science or practice of farming

biodiversity
the variety of organisms in an ecosystem

cultivation
preparing and using land for crops or gardening

deforestation
the removal of trees to make land suitable for other uses

irrigation
supplying water to land or crops

monoculture
the practice of cultivating a single crop in a given area

selective breeding
breeding organisms with desirable traits

Farming crops for food production is called **cultivation**. This is part of the practice of **agriculture**, where farmers use land to grow crops and animals for food and other products. Agriculture often has negative impacts on ecosystems. Farming is one of the main causes of **deforestation**, and it often reduces **biodiversity**.

Scientists work with farmers and governments to come up with ways of cultivation that improve crop yields and cause less harm to ecosystems.

1 Monocultures lead to a loss of biodiversity

Some farms grow only a single crop, such as canola or wheat, that covers a very large area of land. These crops are called **monocultures** – other species are removed so that they don't compete for nutrients or feed on the crop. One problem with this loss of biodiversity is that if a disease or pest affects one plant in a monoculture, it can damage all of them because they are all the same.

After a crop is harvested, the soil is ploughed (turned over) to remove the remaining roots of the crop from the soil. This increases seed germination and removes weeds, but it also increases soil erosion and removes nutrients stored in decomposing plants.

Many farmers are starting to use agricultural methods that allow nutrients to remain in the soil. If the plant matter is left in the soil, without being ploughed, it decomposes, leaving carbon and other nutrients in the soil. This reduces the amount of fertiliser that needs to be added and it is a cheaper method.

What is a monoculture?

2 Chemicals can build up in ecosystems

To give better yields, crops needs nutrients. Traditionally, chemical fertilisers have been added to the soil – but it runs into waterways when there's heavy rain or too much is used. This affects other ecosystems; for example, it can cause toxic algal blooms.

Pesticides and herbicides are used in agriculture to remove pests and weeds from crops. These chemicals can build up in ecosystems, affect other organisms and enter waterways. They can be passed along food chains and cause harm to other organisms.

LITERACY LINK 

Write three questions about this section for a classmate. Swap questions and complete each other's answers.

Figure 17.13 Planting canola seeds into soil without ploughing after the harvest causes less erosion.

Reducing and improving chemical use is important in agriculture. Organic fertilisers such as compost and manure add nutrients without causing the same harm as chemical fertilisers. When chemicals are used, they must be registered and controlled. Farmers are required to record their use, including the date, type of chemical and the amount being used.

What is an example of an organic fertiliser?

Figure 17.14 Micro-irrigation systems are replacing traditional spraying methods for some crops.

3 Irrigating crops uses large amounts of water

Have you ever seen a jet of water shooting up high to spray downwards on a crop at a farm? These types of **irrigation** waste a lot of water – some of the water evaporates or doesn't reach the plant, so a large amount needs to be used to soak into the soil.

Micro-irrigation systems are replacing some traditional irrigation sprayers. They are water pipes sitting just above the crops, allowing water to drip slowly on the plants.

Scientists also use scheduling tools to work out the best time to irrigate a particular crop based on rain forecasts, plant stress and temperature. This gives better plant growth and prevents water from being wasted when irrigation isn't needed.

What is irrigation?

4 Selective breeding is used to develop new crops

Some scientists study different species of crop to increase yield and find varieties that can resist drought, pests and diseases. This can be done as part of **selective breeding** programs.

When a disease destroys a crop, a few plants usually survive because they are naturally resistant. Scientists can then breed more of those plants to create an entire crop that's resistant to the disease.

Blackleg is a fungus that damages canola crops in New South Wales and Victoria. Farmers usually use chemical sprays to kill and prevent the spread of blackleg, but the chemicals also affect the ecosystems. Scientists have studied varieties of canola that can resist blackleg. Growing these varieties of canola is the best way to control the effects of blackleg in Australia.

What is selective breeding?

CHECKPOINT 17.6 

1 Define agriculture in your own words.

2 Give two examples of crops grown in Australia.

3 Give one example of how scientific knowledge has influenced agriculture.

4 Explain why chemicals are often used in agriculture.

5 Suggest why some farmers are starting to use organic fertilisers.

6 Why do farmers usually grow monocultures rather than a variety of crops?

7 Describe why deforestation is necessary for agriculture.

CHALLENGE

8 Undertake some research to suggest one way farmers could improve the productivity of wheat in times of drought.

SKILLS CHECK

- I can explain what agriculture is with at least two examples.
- I can describe how improvements in scientific knowledge have led to better agricultural practices.

CHAPTER SUMMARY

An **ecosystem** is a biological community made up of living things like plants and animals, as well as non-living things like air and water.

Food chains show the path of energy in an ecosystem

A **food web** is a system of interlocking food chains.

Dingo
Red-bellied black snake
Kangaroo
Sheep
Possum
Lizard
Tree frog
Moth
Cricket
Grass
Nectar

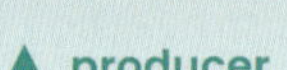

producer
an organism that produces energy at the start of a food chain.

decomposer
an organism that gains energy by breaking down dead material.

consumer
an organism that gains energy by consuming other organisms.

CONSUMER	OBTAINS ENERGY BY
Herbivore	feeding on producers
Omnivore	feeding on producers and other consumers
Carnivore	feeding on other consumers
Decomposer	breaking down dead matter

Some microorganisms are helpful in ecosystems, breaking down dead matter or acting as a food source.

Other microorganisms are harmful, spreading disease or consuming resources needed by consumers.

Human activities can damage the environment. Our actions can destroy ecosystems or reduce biodiversity.

Human activities can also help the environment, such as using science to fight bushfires or changing the way we grow crops.

★ FINAL CHALLENGE ★

1. Which type of organisms always occupies the start of a food chain?
2. Name the process that producers carry out to obtain energy.
3. Koalas feed on eucalyptus trees. In this feeding relationship, the koala is a ______________ while a eucalyptus tree is a ______________.

LEVEL 1
★☆☆
☆☆☆
50xp
LEVEL UP!

4. In your own words, write a definition for each of the following terms:
   a microorganism
   b deforestation
   c biodiversity
   d monoculture
5. Give an example of an unpredictable natural event that can cause harm to Australian ecosystems.

6. Explain the difference between a food chain and a food web.
7. Copy and complete the table by filling in the empty spaces to compare the four different types of consumers.

Type of consumer:	Herbivore	Omnivore		Decomposer
Energy is obtained by:		Feeding on both producers and other consumers		Breaking down decaying matter
An example is:	Grass		Kookaburra	

LEVEL 3
★★★
☆☆☆
150xp
LEVEL UP!

8. Suggest why is it important for organisms to have more than one food source?
9. Predict what you might see if you came upon an area that had been impacted by a low-intensity bushfire the previous year.

10. Draw and fully label a food chain for an Australian bushland ecosystem that consists of four trophic levels.
11. Outline the role of decomposers in the cycling of nutrients within an ecosystem.

SKILLS AND INVESTIGATIONS

SCIENCE SKILLS

1. Questioning, predicting and planning

2. Collecting and using data

3. Writing investigation reports

4. Safety in science

5. Laboratory equipment

INVESTIGATIONS

Chemical World

Earth and Space

Investigations labelled TD are Teacher Demonstration only.

SCIENCE SKILLS 1

QUESTIONING, PREDICTING AND PLANNING

KEY TERMS

controlled variables
all the things that need to stay the same during an investigation

dependent variable
the thing that will be measured and is altered by the independent variable

experiment
an investigation carried out under controlled conditions, to test a hypothesis

fair test
an investigation in which only one factor is changed and all other variables are kept the same

fieldwork
an investigation conducted in the natural environment, not a laboratory

hypothesis
a scientific statement that can be tested

independent variable
the thing that is purposely changed during an investigation

reliable
provides consistent results when repeated

research
gather data and information in an organised way to inform a hypothesis or investigation

valid
measures what is intended to be measured

Science is all about investigating – asking questions, looking at data and drawing conclusions about how things work. A scientist is like a detective, but instead of investigating a crime, they're investigating the world. To be useful, a good scientific test needs to follow certain principles.

1 Good science needs to be valid and reliable

When scientists design investigations, they ask themselves 'is this *good science*?'

To figure out if something is good science, you need to check that it's both **valid** and **reliable**. If a test is reliable, you can do the test over and over again and get very similar results. If a test is valid, it measures what it is supposed to measure.

Imagine you design a catapult that launches marshmallows and decide to test it against a friend's design. Just as your friend is firing the catapult, a massive gust of wind blows their marshmallow further than yours – that's not fair, right? It's not a valid outcome because the wind caused the increased distance, not the catapult. The test didn't measure what you wanted it to measure (the power of the catapult); it measured the power of the catapult *and* the power of the wind. It's not reliable because, if you did the test again, the wind might be weaker, stronger or not there at all.

Why does good science need to be valid and reliable?

2 A fair test needs to be controlled

Fair tests are essential for good science. A **fair test** is one in which only one variable is changed and all other variables are kept the same. Variables are the things that can be controlled, changed or measured during an investigation or experiment. There are three main types of variable: independent, dependent and controlled variables.

The **dependent variable** is what you are measuring in an investigation, and is what is altered by the independent variable. Examples include time in seconds or mass in grams. The **controlled variables** are all the things you will keep the same. Examples of controlled variables are temperature, mass, equipment, location and volume.

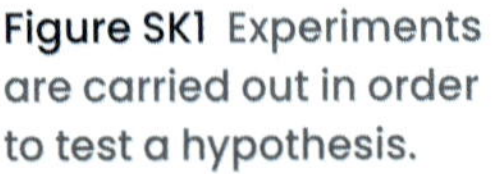

Figure SK1 Experiments are carried out in order to test a hypothesis.

The **independent variable** is the one thing you want to change in an investigation. If you change more than one thing, the investigation probably won't be a fair test anymore.

Let's say you decide to put three plants in three different amounts of sunlight to see which plant grows the most. You would make sure the plants were the same size, health and species, and only change the amount of sunlight the plant is getting – this is the independent variable. The dependent variable would be your measurement of the plants' growth (which could be their weight or their size) and the controlled variables are all the other factors.

What are the three types of scientific variable?

3 A hypothesis is a prediction of the outcome

A **hypothesis** is a prediction made to test something. A good hypothesis involves some reading and research so that scientists can make an informed decision about what they think will happen, before testing it in an investigation. A hypothesis can be supported (found to be correct) or rejected (found to be incorrect).

You use the independent and dependent variables when writing a hypothesis, so the first step is always to identify these. The general rule to use when writing a hypothesis is:

independent variable → If I [do this], then [this] will happen. ← dependent variable

Even though this rule has the word 'I' in it, that's not how you write the hypothesis! You should always write it formally and in the third person (don't use *I*, *we*, *you*, etc.).

What is a hypothesis?

Figure SK2 These scientists are doing fieldwork to test water samples.

ELEMENTS OF AN INVESTIGATION

Think again about the investigation that involves plants in different amounts of sunlight to see which plant grows the most. The elements of this investigation are:

- *hypothesis*: If a plant is placed in direct sunlight, then it will grow more than a plant in indirect or no light.
- *independent variable*: amount of direct sunlight (one plant is put in a dark cupboard, one is put outside in direct sunlight and one is put near a window)
- *dependent variable*: growth of the plant, in millimetres
- *controlled variables*: species of plant, starting size of plant, health of plant, amount of water given to plant.

TYPES OF INVESTIGATIONS

Scientists do many different types of investigation, depending on their area of science and the information they need to gather.

Fieldwork happens when information and data are collected outside of the laboratory or usual setting. Environmental scientists often do fieldwork, such as collecting water samples from streams to study the water quality or counting the number of species of plants and animals in an area.

Experiments are usually carried out to test a hypothesis. Experiments in science include those in chemistry, physics, earth science, and with living things in biology.

Research informs a hypothesis before it is created. Scientists often share their research so they can build scientific understanding and discoveries over time.

SCIENCE SKILLS 2

COLLECTING AND USING DATA

KEY TERMS 

inference
an educated guess or judgement based on observations

observation
something you see and know to be true

prediction
a statement about the future based on observation and evidence

primary data
first-hand data, from your own investigation

qualitative
written descriptions and observations

quantitative
numerical information and data

secondary data
second-hand data, from someone else

CALCULATING THE MEAN

To calculate the mean (also known as the average) of a group of numbers, add all the numbers together and then divide them by how many numbers you added together.

For example, to calculate the average growth of plant 3 in the investigation about plants and sunlight (Table 1), you would add 2, 3 and 5, then divide by 3.

The mean would be (2 + 3 + 5) ÷ 3 = 3.33mm.

Data is like evidence – you need it to draw your conclusion. Scientists collect and analyse data to test their hypotheses.

1 Scientists collect different types of data

One way to describe data is that it can be **qualitative** or **quantitative**. Quantitative data relates to quantities – that is, numbers. Quantitative data can include the number of something, the volume, the length, time ... anything that scientists can physically measure or count. Qualitative data relates to the qualities of something – that is, written descriptions or observations about data.

Another way to describe data is as **primary data** or **secondary data**. Primary data is first-hand data that you collect yourself through scientific investigation. Secondary data is second-hand data, gathered by someone else and given to you.

To make sure secondary data is valid and reliable, you need to check that it comes from a reliable source. It's also important to make sure the data is accurate. If it is a survey, was the sample size large enough or did it only involve a small number of people? Are the results from just one country or population group?

How are qualitative and quantitative data different?

2 Data needs to be carefully collected and recorded

To ensure that their data is valid, scientists record observations and measurements very carefully. They might do this in a logbook or table for quantitative data. Qualitative data might be recorded in a journal or workbook. Sometimes data is visual and can be recorded with a camera.

When taking measurements and recording quantitative data, use the appropriate units for physical quantities. This table shows some common metric units for physical quantities.

Physical quantity	Measurement and unit	Conversion
Length	Millimetre (mm)	10 mm = 1 cm
	Centimetre (cm)	100 cm = 1 m
	Metre (m)	1000 m = 1 km
	Kilometre (km)	
Mass	Milligram (mg)	1000 mg = 1 g
	Gram (g)	1000 g = 1 kg
	Kilogram (kg)	
Volume	Millilitre (mL)	1000 mL = 1 L
	Litre (L)	
Temperature	Celsius (°C)	

Why is it important to use the correct units when measuring?

3 Organising data makes it easier to understand

Collecting data isn't the end of the process – the data needs to be analysed and considered. That means it must be well organised and clearly presented, or else it will be difficult to understand.

One of the best ways to arrange and present scientific data is in a table. Design and rule out your table before you start your investigation. This ensures you are ready and organised to collect the correct data, and that you don't forget to collect important data.

MAKING A GOOD SCIENTIFIC DATA TABLE

1 Use a ruler so that your table is clear and easy to read.
2 Give your table a descriptive and useful title and include the table number in case you want to refer to it in your investigation report.
3 Include the units in the column headings where needed (e.g. mm).

Clear descriptions in titles.

The table has a number and title.

Units are given at top of column.

Lines are ruled and easy to follow.

TABLE 1 EFFECT OF SUNLIGHT ON PLANT GROWTH

Plant environment	Initial plant height (mm)	Growth after 1 week (mm)	Growth after 2 weeks (mm)	Growth after 3 weeks (mm)
Plant 1: no sunlight	181	0	−1	−3
Plant 2: indirect sunlight	175	1	2	4
Plant 3: direct sunlight	178	2	3	5

Another way to present data is to create a graph, using the data from your table. Graphs are an excellent visual way to illustrate data.

Why are tables used to organise data?

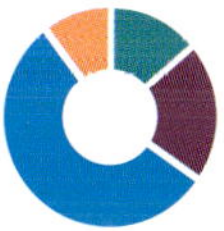
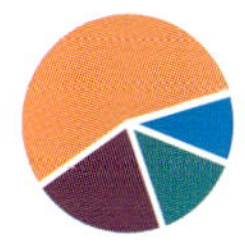
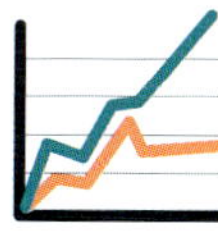

Figure SK3 Graphs can be used to present and explain data.

OBSERVATIONS, INFERENCES AND PREDICTIONS

An **inference** is something you think might be the case, but you don't know for sure. An **observation** is something you see and know is definitely true. A **prediction** is what you think will happen in the future.

In science, an observation can often lead to an inference. You could observe that your cactus is dying, and then infer that this was because it was overwatered. You could then stop watering it and observe it again – this could re-inform and change your inference.

Figure SK4 Recording your observations is a great way to gather first-hand data.

SCIENCE SKILLS 3

WRITING INVESTIGATION REPORTS

Writing an investigation report is a key skill in science, and one you will use many times during scientific study. By writing a clear, consistent report at the end of your investigation, you ensure that other people will understand your work.

1 An investigation report has a consistent structure

Your investigation reports should typically have a similar structure to the one shown here.

The title should be clear and in plain language. Many scientists write their title as a research question.

DOES THE AMOUNT OF DIRECT SUNLIGHT AFFECT PLANT GROWTH?

Use your research question/title to write your aim. Make sure it starts 'To investigate ...'

AIM

To investigate whether the amount of direct sunlight affects growth of a certain species of plant

Write the variables in an investigation report. They can help you to write your hypothesis.

- Independent variable: amount of direct sunlight
- Dependent variable: growth of the plant (in millimetres)
- Controlled variables: species of plant, health of plant, amount of water given to plant

If your investigation is an experiment, then you should include a hypothesis. It should should refer to your independent variable and your dependent variable. Remember to write in the third person.

HYPOTHESIS

A plant in direct sunlight will grow more than a plant of the same species in indirect light or no light.

List all materials and equipment with amounts and sizes as simple bullet points.

MATERIALS

- 3 plants of the same species and of similar size
- 250 mL beaker

The method provides clear, step-by-step instructions.

Remember to number the steps of your method, and to write in the past tense and third person. Methods should be written like a cooking recipe – imagine that someone from another school has to follow your method.

METHOD

1 Each plant was labelled 1, 2 or 3 and measured at the start of the investigation in order to get a starting height for each. The heights were recorded in a simple table.
2 Plant 1 was placed in a dark cupboard.
3 Plant 2 was placed near a window where it could receive indirect sunlight.
4 Plant 3 was placed outside in direct sunlight.
5 The height of the plants was measured every week for three weeks and the growth was recorded in the results table.
6 The plants were watered the same amount every three days.

RESULTS AND DISCUSSION

As the results in Table 1 show, the plant that had the most growth was the plant in direct sunlight (plant 3). The plant in direct sunlight had the highest growth, with 2 mm after the first week, 3 mm in the second week and 5 mm in the third week, compared to the plant in no sunlight (plant 1) which had no growth, then shrank and lost growth in week 2 and 3.

Use exact figures from your table and compare them to others, to show you have analysed the data.

This is because sunlight is used in photosynthesis – the process by which plants make energy for growth.

Describe your results (referring to the table or figures) in the discussion and link them to your understanding of science.

One error that could have occurred is in the amount of water given to each plant. The method could be improved by measuring the exact volume of water (in mL). There may have also been some errors to do with accurately measuring the plants. The method could be improved by including photos of the plants against the same ruler backdrop, to improve the accuracy of measurements.

Identify any potential errors here and suggest improvements to the method to try to control these. Discussion questions can be answered here too.

TABLE 1 EFFECT OF SUNLIGHT ON PLANT GROWTH

Plant environment	Initial plant height (mm)	Growth after 1 week (mm)	Growth after 2 weeks (mm)	Growth after 3 weeks (mm)
Plant 1: no sunlight	181	0	−1	−3
Plant 2: indirect sunlight	175	1	2	4
Plant 3: direct sunlight	178	2	3	5

CONCLUSION

The results of this investigation show that the amount of direct sunlight does impact on the growth of this species of plant. The investigation supported the hypothesis that if this species of plant is put in direct sunlight, it will grow more than a plant of that species in indirect or no light. This is due to more sunlight being available for photosynthesis, which is how plants grow.

Your conclusion summarises the investigation by responding to the aim.

Mention whether the results supported or rejected your hypothesis, and briefly summarise the investigation, but don't introduce any new information.

REFERENCES

BBC, 2019. What is photosynthesis? [online] Available at: https://www.bbc.com/bitesize/articles/zn4sv9q. [Accessed 5 January 2019]

References show the source of any information you used that was not your own.

This is particularly important when you use secondary data.

SCIENCE SKILLS 4

SAFETY IN SCIENCE

Safety is critical in science, because all investigations carry some degree of **risk**. From using a Bunsen burner to dissecting a sheep heart, things can go wrong if we don't plan and act in a safe manner.

1 Get to know good safety practices

Whenever you step into the lab for an investigation, it's important to identify **hazards** and use good safety practices to control the risk of them causing harm.

How many safe practices can you identify in the image below? Use each practice that you identify to make a safety rule for working in science.

It's important to know about these safety tools and materials in the science lab:

- *Safety glasses/shields* protect your eyes from fumes, particles and irritants.
- *Lab coats and aprons* protect your clothes and body from chemical spills, flames and other dangers.
- *Gloves and hand protectors* protect your hands when handling chemicals, biological materials and sharp objects.
- *Fire extinguishers* use dry chemicals (not water) to put out the flames if there is a fire.
- *Eye wash stations* are used if you get something in your eye. Never rub your eye – use the eye wash station to wash it instead.
- *Laboratory hoods* are like exhaust fans in your bathroom or kitchen, removing gases and fumes.

Name and describe three pieces of laboratory safety equipment.

Figure SK5 Safety glasses and rubber gloves reduce risk in the laboratory.

❷ Get to know Bunsen burner safety

A **Bunsen burner** is a common piece of scientific equipment and is most often used for heating things in the science laboratory. Learning how to set up and safely use a Bunsen burner is extremely important. Misuse can result in severe burns or gas leaks.

A Bunsen burner can operate two main flames: the **safety flame** and the **heating flame**. The safety flame is an orange flame and reaches temperatures of about 300°C. That sounds like a lot, but the heating (blue) flame reaches temperatures of about 1500°C!

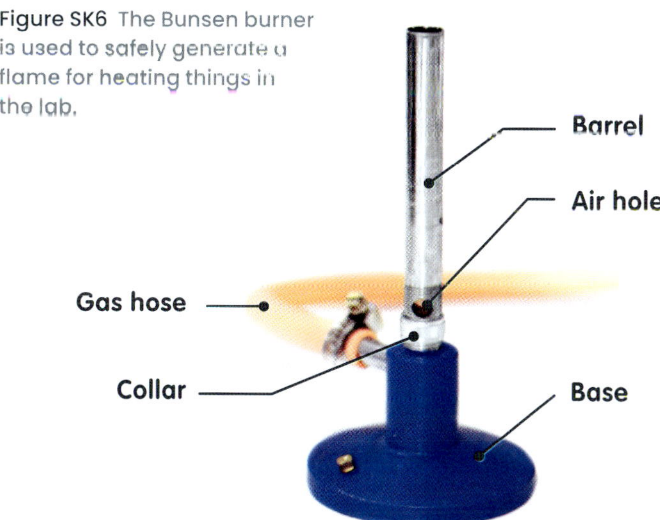

Figure SK6 The Bunsen burner is used to safely generate a flame for heating things in the lab.

SETTING UP AND LIGHTING A BUNSEN BURNER

1 Place a heatproof mat on the lab bench, place the Bunsen burner on top and connect the gas hose tightly to the gas tap.
2 Turn the collar of your Bunsen burner so that the air hole is closed.
3 Time for the safety check! Make sure long hair is tied back, safety glasses are on, you know the location of the fire extinguisher and blanket, and then check the gas hose for any cracks, holes or tears.
4 Ask your teacher to check the set-up of your Bunsen burner and provide any feedback.
5 The lighting of the Bunsen burner is best done by two people. Ask a lab partner to get ready to turn the gas tap on.
6 Light your match or taper before the gas tap is turned on and position it over the top of the barrel.
7 Ask your lab partner to turn the gas tap on.
8 Your Bunsen burner should now be lit. Move away from the burner and extinguish the match.

How hot is the orange flame of a Bunsen burner, and how hot is the blue flame?

KEY TERMS 

Bunsen burner
a piece of equipment used in science that produces a single open gas flame

hazard
something that can harm living things, objects or the environment

heating flame
the blue (very hot) flame of a Bunsen burner (approx. 1500°C), used for heating substances

risk
the chance that a hazard will cause harm

safety flame
the orange (cooler) flame of a Bunsen burner (approx. 300°C), used between heating substances

AFTER YOUR BUNSEN BURNER IS ALIGHT, CALMLY TAKE THE MATCH OR TAPER AWAY FROM THE BURNER.

SHAKE OR BLOW THE FLAME AWAY FROM THE BURNER. IF YOU BLOW THE FLAME OUT NEAR THE BUNSEN BURNER, YOU RISK BLOWING OUT THE BURNER FLAME TOO.

IF THIS HAPPENS, CALMLY TURN OFF THE GAS VALVE AND START AGAIN.

SCIENCE SKILLS 5

LABORATORY EQUIPMENT

1 Get to know common laboratory equipment

By learning the names and uses of laboratory equipment, you can select and use the correct equipment for any investigation. Some of the most common equipment is shown here.

Name and describe five pieces of common laboratory equipment.

❷ Get to know your light microscope

The most common type of microscope in the school science laboratory is a light microscope. It allows you to study samples by shining a bright light through an extremely thin slice of material. The image is magnified by the microscope's lenses, which you look through.

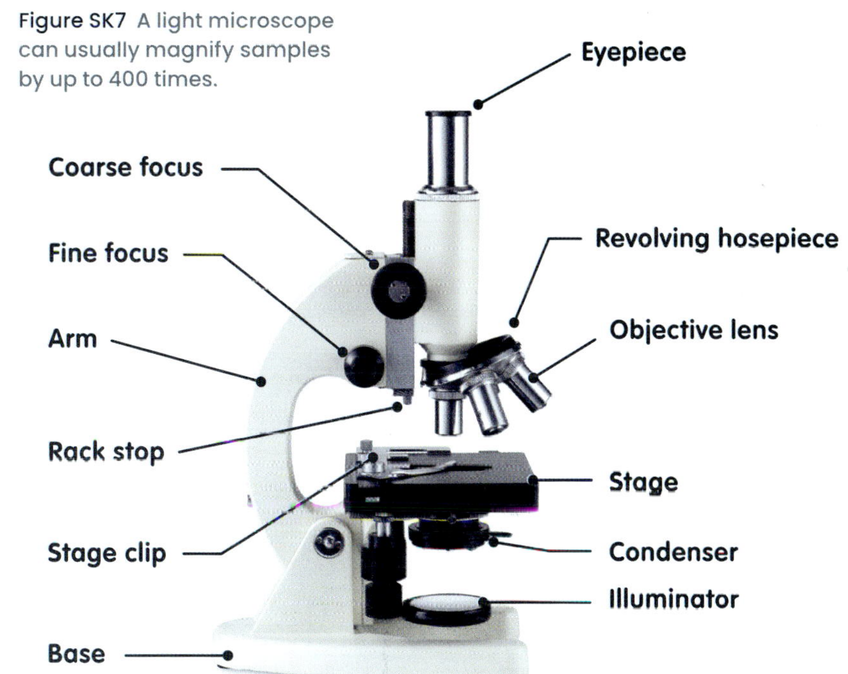

Figure SK7 A light microscope can usually magnify samples by up to 400 times.

SETTING UP AND USING A LIGHT MICROSCOPE

1 Place your microscope on the bench or table, making sure that it's not too close to the edge and that the arm is facing you.
2 Plug your microscope in and turn it on. The illuminator (light) will come on.
3 Lower the stage as far as it can go, using the coarse focus knob.
4 Consider each of the rotating objective lenses. Often they are different colours and have the magnification written on them.
5 Start with the lowest magnification – this is usually 4× (four times). Your microscope eyepiece already has a 10× magnification on its own, so when coupled with the 4× eye piece, what you are looking at will be 40 times the actual size.
6 Carefully insert the microscope slide onto the top of the stage and hold it firm under the stage clip. Position it so the object you need to see is in the middle of the stage.
7 Look through the eyepiece and slowly bring the stage upwards towards you, using the coarse focus knob. This can take some time and everything will look bright and fuzzy until you get a glimpse of the slide as it comes into focus.
8 When the slide is roughly in focus, use the fine focus to turn it into a clear image.
9 Increase the magnification by changing the objective lens to 10×, 20× or 40×.
10 40× is usually the highest available magnification and can be tricky to find and focus on. It can also bring the lens extremely close to the slide, enough to crack and break it – monitor this carefully.

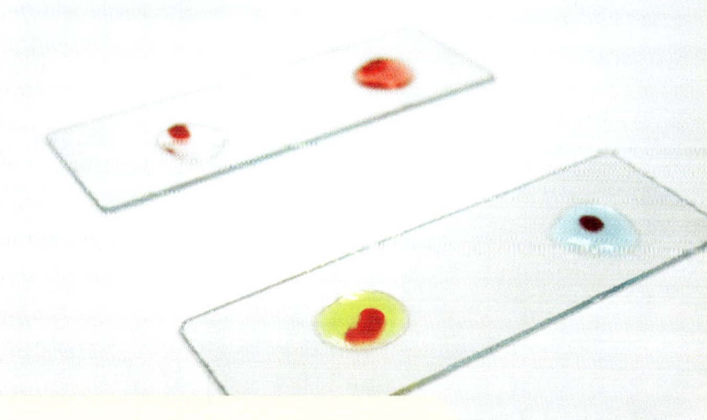

Figure SK8 Microscope slides are usually made of glass and are very fragile – treat them carefully.

INVESTIGATION 1.1

Compressing liquids and gases

AIM

To investigate the compressibility of liquids and gases

MATERIALS

- plastic syringe
- water
- small beaker

METHOD

1 Draw some water into the syringe so that it is about half full.
2 Hold your finger over the nozzle so that water cannot come out, then try to push in the plunger. Are you able to compress the water? Record your observations.
3 Empty the water from the syringe and pull the plunger back so that the syringe is half full of air.
4 Again, holding your finger over the nozzle, try to push in the plunger. Are you able to compress the air? Record your observations.

DISCUSSION

1 What do the results tell you about the compressibility of gases?
2 What do the results tell you about the compressibility of liquids?
3 Use the particle model to explain your results.
4 How could you investigate the compressibility of solids using an ice cube?

CONCLUSION

Copy and complete:
'The results show that: (*respond to the aim*)'.

INVESTIGATION 1.2

Heating materials

AIM

To investigate how quickly different materials can heat up

MATERIALS

- 3 large spoons (one metal, one wooden, one plastic)
- wax
- water
- 500 mL beaker
- Bunsen burner, heatproof mat, tripod, gauze mat and matches OR kettle

METHOD

1 Set up the materials as shown.

2 Scoop a small amount of wax on to each spoon.
3 If using a kettle, half fill the beaker with hot water from the kettle. Place all three spoons into the water, handle side down.
4 If using a Bunsen burner, set it up on the heatproof mat, with the tripod and gauze mat. Half fill the beaker with water, then place all three spoons into the water, handle side down. Ignite the burner and heat the water.
5 Observe how quickly the wax melts on each spoon. Record your observations.

DISCUSSION

1 Which material was able to conduct heat the best?
2 What evidence led you to this answer?
3 If you were to design a cooking utensil that ensured heat was not transferred to the person using it, what materials would you consider best to use and why?

CONCLUSION

Copy and complete:
'The results show that: (*respond to the aim*)'.

AN OPEN FLAME IS A HAZARD. TAKE CAUTION. IF YOU BURN YOURSELF, TELL YOUR TEACHER IMMEDIATELY AND PLACE THE BURNT AREA UNDER COLD RUNNING WATER FOR 20 MINUTES.

INVESTIGATION 1.3

Expanding gases

AIM

To investigate how the volume of a gas changes when heated

MATERIALS

- balloon
- conical flask
- 2 beakers (large enough for the conical flask to fit inside)
- ice water
- hot water from tap
- string
- ruler

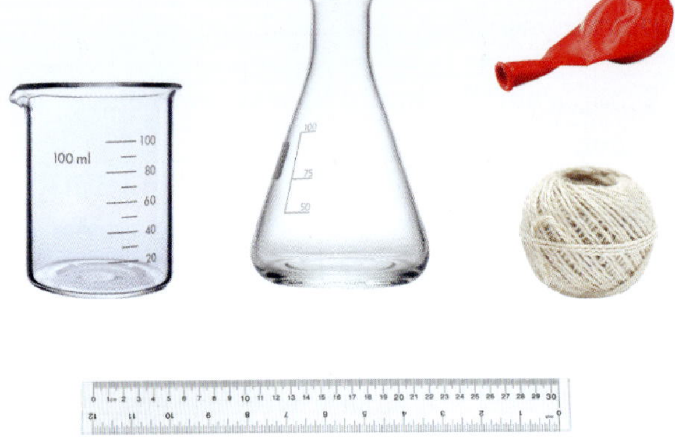

METHOD

1. Inflate and deflate the balloon a couple of times to stretch it out.
2. Blow up the balloon so that it has a diameter of about 10 cm. Place the balloon over the neck of the conical flask.
3. Wrap the string around the widest part of the balloon and mark the string where the two parts touch. Unwrap and measure the string with a ruler. Record the circumference of the balloon at room temperature.
4. Half fill one of the beakers with ice water and place the conical flask inside the beaker. Let it stand for 5 minutes.
5. Use the string to measure the circumference of the balloon after the air has been cooled by the water and record the measurement.
6. Transfer the conical flask to the beaker containing hot water. Let it stand for 5 minutes.
7. Use the string to measure the circumference of the balloon after the air has been warmed by the water, and record the measurement.

DISCUSSION

1. What effect did heating have on the circumference of the balloon?
2. Use the particle model to explain why this happened.
3. What would you expect to happen to the circumference of the balloon if the water was heated further?
4. How could you improve the reliability of your results?

CONCLUSION

Copy and complete:
'The results show that: (*respond to the aim*)'.

INVESTIGATION 1.4

Exploring melting points

AIM

To investigate the melting and boiling points of different household substances

MATERIALS

- household liquids (e.g. water, vinegar, dishwashing liquid, vegetable oil, apple juice, milk, candle wax, soft drink)
- ice-cube tray (to make ice cubes that fit inside the test tubes)
- test tubes (one for each substance, large enough to fit ice cubes)
- thermometer
- large beaker
- Bunsen burner
- heatproof mat
- tripod
- gauze mat
- matches

RESULTS

TABLE I1.4

Sample	Melting point	Boiling point

METHOD

1. The day before the investigation, pour each substance into a separate ice cube mould. Place the trays in a freezer overnight.
2. Collect your solid, frozen samples.
3. Copy the results table into your notebook, adding a title and rows as needed.
4. Set up the Bunsen burner, tripod and gauze mat.
5. Half fill a large beaker with water. This will act as your water bath for the test tube samples.
6. Remove one of the samples from the ice-cube tray and place it into a test tube. Place the test tube into a large beaker. Continue to place test tubes into the beaker if space permits.
7. Set up the apparatus.
8. Light the Bunsen burner beneath the beaker and heat the sample until it has completely melted. Use the thermometer to measure the melting point of the sample and record in your table.
9. If time permits, continue to heat the sample until it begins to boil. If you are boiling several samples together, make sure that there is some space between the test tubes.
10. Use the thermometer to measure the boiling point of the sample and record in your table.
11. Repeat steps 6–10 for the rest of your samples.

DISCUSSION

1. Which substances had low melting points?
2. Which substances had high melting points?
3. Which substances had low boiling points?
4. Which substances had high boiling points?
5. Consider the classroom you are currently in. Can you identify something that would have a very high melting point? Justify your answer.

CONCLUSION

Copy and complete:
'The results show that:
(*respond to the aim*)'.

OPEN FLAMES, HOT LIQUID, WAX AND STEAM ARE HAZARDS. TAKE CAUTION. IF YOU BURN YOURSELF, TELL YOUR TEACHER IMMEDIATELY AND PLACE THE BURNT AREA UNDER COLD RUNNING WATER FOR 20 MINUTES.

INVESTIGATION 1.5

Exploring density

AIM

To investigate the relative densities of some household liquids and objects

MATERIALS

- honey
- corn syrup
- maple syrup
- dishwashing liquid
- water with food dye (to make it easier to see)
- vegetable oil
- isopropyl alcohol
- plastic building brick (e.g. Lego)
- die
- coin
- ping-pong ball
- measuring jug
- 1 L measuring cylinder (or any tall glass container)

METHOD

1. Use the measuring jug to measure out roughly equal amounts of each liquid to be poured into the measuring cylinder. Depending on the size of your cylinder or container, you may choose to use 100 mL of each liquid.
2. Pour the honey into the cylinder, followed by the corn syrup, maple syrup and dishwashing liquid. Carefully pour the water down the inside of the cylinder to ensure it doesn't splash. Add the vegetable oil and isopropyl alcohol in the same way. You should have a multi-layered cylinder of liquids, with the densest at the bottom.
3. Gently drop the plastic building brick into the measuring cylinder. It will sink through the liquids that are less dense than it, and float on any that are denser.
4. Predict where you think the coin, die and ping-pong ball will stop sinking in the cylinder. Add them, one by one, to check if you are correct.

QUESTIONS

1. Which object was the densest? Suggest how you could tell.
2. What was the role of the liquids in this investigation?
3. Can you think of an object that could float on top of all the liquids?
4. Rank the liquids from most to least dense.

CONCLUSION

Copy and complete:
'The results show that: (*respond to the aim*)'.

ISOPROPYL ALCOHOL IS FLAMMABLE. KEEP IT AWAY FROM SOURCES OF SPARKS AND FLAMES.

INVESTIGATION 1.6

Limitations of the particle model

TEACHER DEMONSTRATION

AIM

To investigate a limitation of the particle model

MATERIALS

- 250 mL of methylated spirits
- 250 mL of water
- 1 L measuring cylinder

METHOD

1 Before your teacher begins, predict the total volume of the water and methylated spirits if mixed together.
2 Your teacher will pour the methylated spirits and the water into the 1 L measuring cylinder. Observe and record the volume of the mixture.

QUESTIONS

1 Compare your predicted combined volume to the actual combined volume.
2 Imagine pouring a jar of sand into a jar of marbles. Would you expect the final volume to be equal to the sum of the two original volumes?
3 How can you use your answer to question 2 to explain the results of this investigation?

CONCLUSION

Copy and complete:
'The results show that: (*respond to the aim*)'.

METHYLATED SPIRITS IS FLAMMABLE AND TOXIC. KEEP AWAY FROM SOURCES OF SPARKS AND FLAMES. DO NOT DRINK IT.

INVESTIGATION 2.1

Comparing metals and non-metals

AIM

To investigate the properties of metals and non-metals

MATERIALS

- selection of metals and non-metals (e.g. sulfur, aluminium, iron, magnesium, carbon, plastic, wood, polystyrene)
- power supply
- light bulb
- 3 connecting wires
- alligator clips
- steel sewing needle

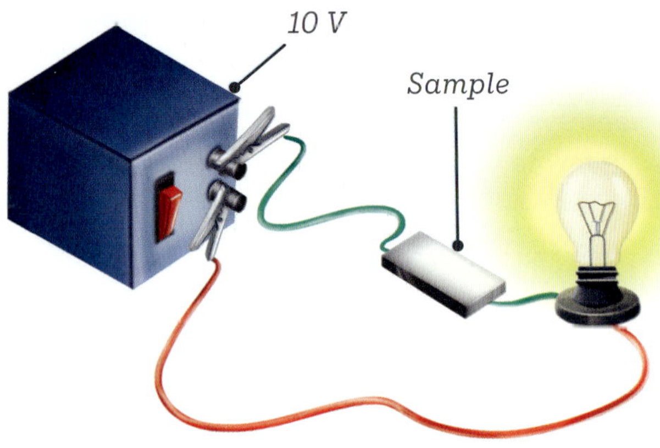

METHOD

1. Copy the results table into your notebook, adding a title and rows as needed.
2. Choose one of the samples and record your observations of its appearance (colour, shininess etc.)
3. Try to bend the sample to test for malleability.
4. Scratch the sample using the steel needle. Is the sample shiny underneath?
5. Connect the electric circuit as shown. Switch the power supply on to 2 V. Observe whether the light bulb turns on. If so, the sample conducts electricity.

DISCUSSION

1. What properties did all of the metals have in common?
2. Were there any properties common to both metals and non-metals?
3. Predict the properties of tungsten (metal) and iodine (non-metal).

CONCLUSION

Copy and complete:
'The results show that: (*respond to the aim*)'.

RESULTS TABLE I2.1

Sample	Metal or non-metal?	Appearance	Malleable?	Shiny?	Conducts electricity?

ELECTRICITY IS A HAZARD.
TAKE CAUTION.

INVESTIGATION 2.4

Separating a mixture

AIM

To investigate the process of separating a mixture of salt and water

MATERIALS

- 15 g of table salt
- 100 mL of water
- evaporating dish
- Bunsen burner
- heatproof mat
- tripod
- pipeclay triangle
- matches
- 150 mL beaker
- 100 mL measuring cylinder
- 10 mL measuring cylinder

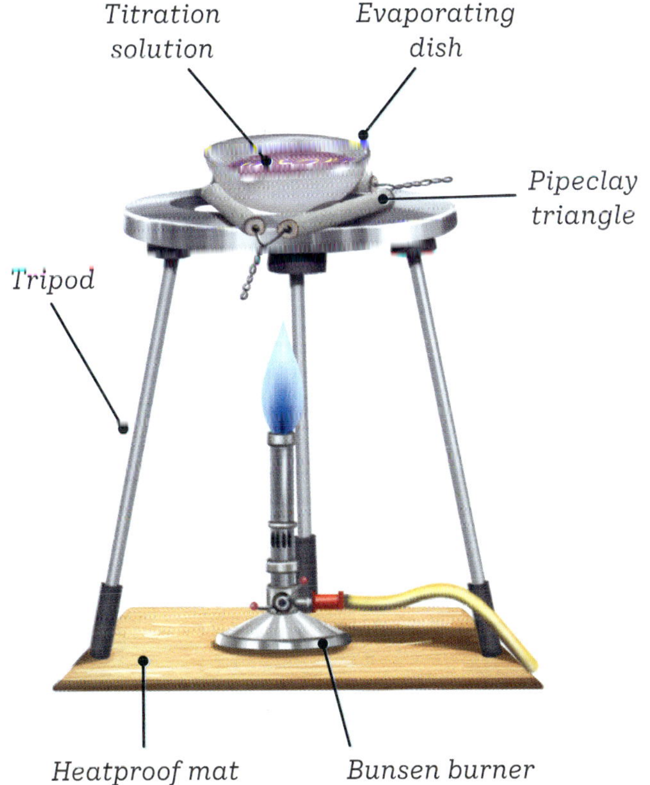

METHOD

1 Combine the water and salt in 150 mL beaker. Stir until all of the salt is dissolved.
2 Pour 10 mL of the salt solution into an evaporating basin.
3 Set up the apparatus as shown.
4 Heat the salt water solution for a few minutes until the water has all evaporated.
5 Observe the salt remaining in the dish. Record your observations.

DISCUSSION

1 Compare the appearance of the salt at the end of the investigation to the salt at the beginning. How is it different?
2 What happened to the water during the experiment?
3 How does this investigation show that salt water is a mixture?
4 Can you think of any other ways to separate a mixture of salt and water?

CONCLUSION

Copy and complete:
'The results show that: (*respond to the aim*)'.

AN OPEN FLAME IS A HAZARD. TAKE CAUTION. IF YOU BURN YOURSELF, TELL YOUR TEACHER IMMEDIATELY AND PLACE THE BURNT AREA UNDER COLD RUNNING WATER FOR 20 MINUTES.

INVESTIGATION 2.5

Properties of compounds

AIM

To investigate how the properties of a compound differ from those of the elements it is made of

MATERIALS

- 5 cm magnesium ribbon
- crucible and lid
- Bunsen burner
- heatproof mat
- tripod
- pipeclay triangle
- matches
- brass (crucible) tongs

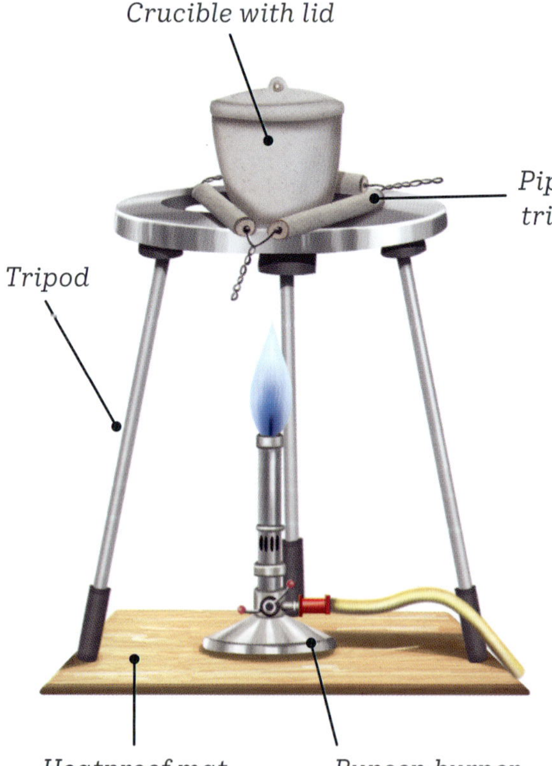

METHOD

1 Look closely at the magnesium ribbon. Record your observations (e.g. colour, shininess, feel).
2 Place the magnesium ribbon inside the crucible and put the lid on. Set up the apparatus as shown.
3 Using the Bunsen burner, heat the crucible containing the magnesium. Using tongs, lift the lid every 5 – 10 seconds to ensure enough air is in the crucible. *Do not look directly at the magnesium.*
4 The heated magnesium will combine with oxygen in the air to create magnesium oxide, a compound. After the magnesium has finished reacting, observe the magnesium oxide. Record your observations about its properties.

DISCUSSION

1 How did the appearance of the magnesium differ to that of the magnesium oxide?
2 What was the purpose of lifting the crucible lid occasionally?
3 Where did the oxygen in the magnesium oxide come from?

CONCLUSION

Copy and complete:
'The results show that: (*respond to the aim*)'.

AN OPEN FLAME IS A HAZARD. TAKE CAUTION. IF YOU BURN YOURSELF, TELL YOUR TEACHER IMMEDIATELY AND PLACE THE BURNT AREA UNDER COLD RUNNING WATER FOR 20 MINUTES.

MAGNESIUM RIBBON IS A HAZARD. IT IS FLAMMABLE AND REACTIVE. TAKE EXTREME CAUTION. NEVER LOOK DIRECTLY AT THE BURNING MAGNESIUM RIBBON.

INVESTIGATION 3.2

The Tyndall effect

AIM

To investigate three types of mixture: solutions, suspensions and colloids

MATERIALS

- 8 prepared mixtures of pure water and the following substances:
  1. milk; **2.** soil; **3.** soluble starch; **4.** food colouring; **5.** salt; **6.** sugar; **7.** flour; **8.** cooking oil
- 8 test tubes
- test-tube rack
- 10 mL measuring cylinder
- marker pen
- torch
- labelling tape
- stirring rod

METHOD

1 Copy the results table into your notebook, adding a title and rows as needed.

2 Using a marker, number the test tubes 1–8.

3 Measure 10 mL of each of the mixtures, using a measuring cylinder, and transfer into the appropriate test tubes.

4 Observe the mixtures. Record a description of each in your table.

5 After stirring, record which mixtures separate upon standing.

6 Make the room as dark as possible, or use an under-bench cupboard. Shine a torch on each mixture that does not separate upon standing. In your table, describe if the mixture exhibits the Tyndall effect – if light is scattered by the fine particles.

7 Classify each mixture as a solution, suspension or colloid.

DISCUSSION

1 The effect of scattering particles by light is known as the __________ ____________.

2 **a** If a mixture is left to stand and then separates out, it is a ________.

b If a mixture does not separate when left standing and the Tyndall effect is not seen, it is a ___________.

c If a mixture does not separate when left standing and exhibits the Tyndall effect, it is a _____________.

CONCLUSION

Copy and complete:

'The results show that (respond to the aim)'

RESULTS **TABLE I3.2**

Mixture	Brief description	Separates on standing?	Exhibits Tyndall effect?	Classification
Milk				
Soil				
Starch				
Food colouring				
Salt				
Sugar				
Flour				
Cooking oil				

INVESTIGATION 3.3

Purifying muddy water

AIM

To purify muddy water using decanting and filtration

MATERIALS

- soil and sand
- conical flask
- 250 mL beaker
- filter funnel
- filter paper
- glass stirring rod
- teaspoon
- wash bottle

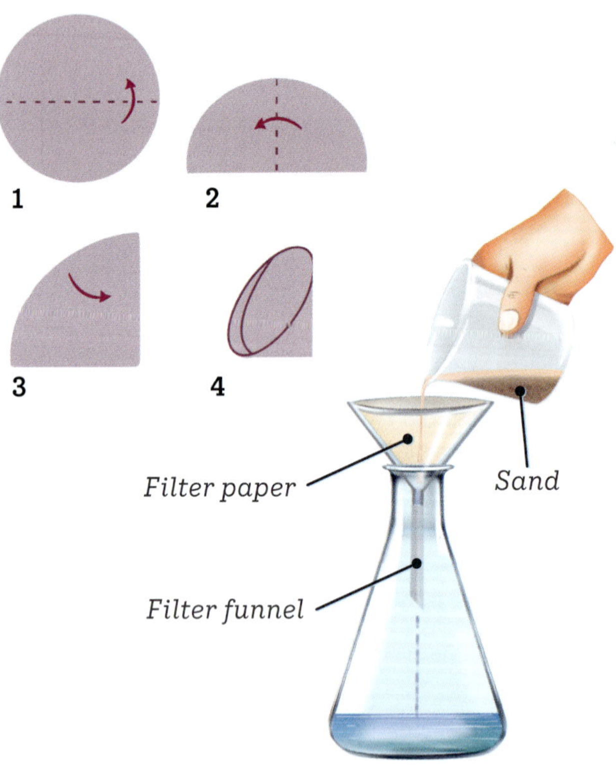

METHOD

1 Copy the results flowchart into your notebook.
2 Half fill the beaker with water.
3 Add 1 teaspoon of soil into the beaker and stir until a suspension forms.
4 Add 1 teaspoon of sand into the beaker and allow the sand to settle to the bottom.
5 Fold the filter paper as shown and place it in the funnel.
6 Set up the filtration apparatus as shown.
7 Carefully pour the mixture from the beaker into the conical flask by decanting, without disturbing the sand on the bottom of the beaker.
8 Record your observations in the flowchart.

RESULTS

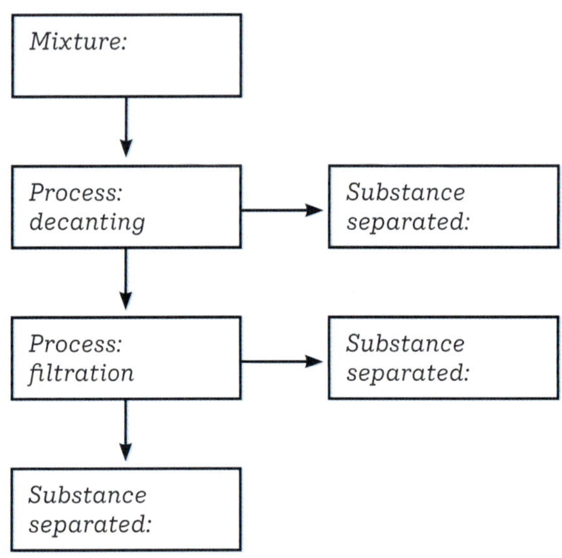

DISCUSSION

1 Explain the terms *suspension* and *sedimentation* with reference to the mud and sand used in the investigation.
2 Explain how decanting was used to separate sand.
3 Describe the physical properties that allowed you to separate the mud and the sand.

CONCLUSION

Copy and complete:
'The results show that: (*respond to the aim*)'.

INVESTIGATION 3.5A

Evaporating a solution

AIM

To investigate what happens when a solution evaporates

MATERIALS

- copper sulfate solution
- evaporating basin
- Bunsen burner
- heatproof mat
- tripod
- gauze mat
- matches

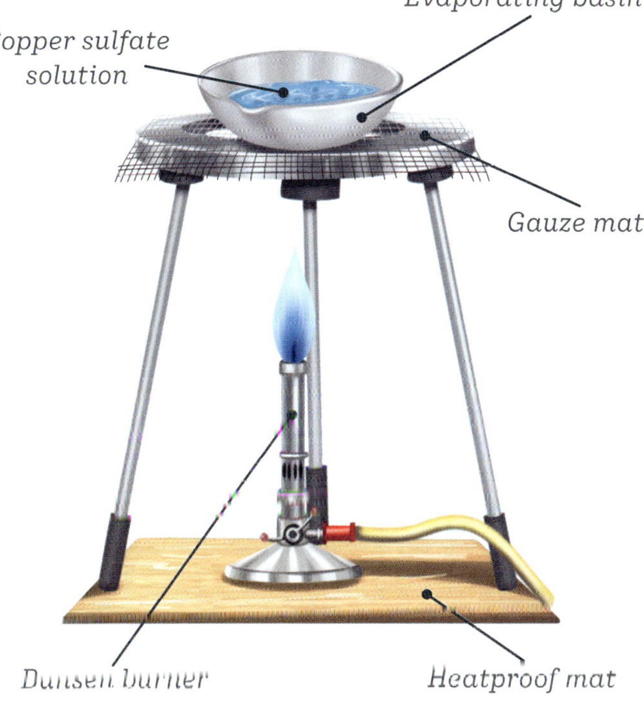

METHOD

1 Place a small amount (5 mL) of copper sulfate solution in an evaporating basin.
2 Set up the apparatus as shown and put on your safety glasses.
3 Place the evaporating basin on the gauze mat and heat the solution.
4 Record your observations as the water evaporates.
5 Turn off the Bunsen burner before the last few drops evaporate.

DISCUSSION

1 Describe the copper sulfate solution prior to the evaporation.
2 Describe the substance after evaporation had occurred.
3 Describe a safety precaution required for this investigation.
4 What happened to the water from the copper solution in this experiment?

CONCLUSION

Copy and complete:
'The results show that: (*respond to the aim*)'.

AN OPEN FLAME IS A HAZARD. TAKE CAUTION. IF YOU BURN YOURSELF, TELL YOUR TEACHER IMMEDIATELY AND PLACE THE BURNT AREA UNDER COLD RUNNING WATER FOR 20 MINUTES. COPPER SULFATE IS TOXIC. IF YOU COME INTO CONTACT WITH THE SOLUTION, TELL YOUR TEACHER IMMEDIATELY AND WASH IT OFF YOUR SKIN. ALWAYS WEAR SAFETY GLASSES.

INVESTIGATION 3.5B

Growing crystals

AIM

To investigate growing crystals at different temperatures

MATERIALS

- electronic balance
- 30 g of copper sulfate (solid)
- 2 × 100 mL beakers
- hot water from tap
- 2 stirring rods
- cotton
- 2 paperclips

METHOD

1 Add 30 g of copper sulfate to a beaker containing 40 mL of hot water. Stir with a stirring rod until the solid dissolves.

2 Pour half of the volume into another beaker. Cool this beaker by running cold water on the outside.

3 Tie a piece of cotton to each stirring rod and attach a paperclip at the other end. Hang the strings with paperclips inside each beaker.

4 Leave the beakers for several days to allow crystals to grow.

5 Describe what you can see inside each beaker.

DISCUSSION

1 Compare the sizes of the crystals in the two beakers. How could you make the crystals bigger?

CONCLUSION

Copy and complete:

'The results show that: (*respond to the aim*)'.

INVESTIGATION 3.5C

Distillation

TEACHER DEMONSTRATION

AIM

To investigate separating a salt solution using distillation

MATERIALS

- 150 mL beaker containing salt solution
- Bunsen burner
- heatproof mat
- tripod
- gauze mat
- matches
- Liebig condenser
- thermometer
- distillation flask
- marble chips
- retort stand with bosshead and clamp
- receiving flask

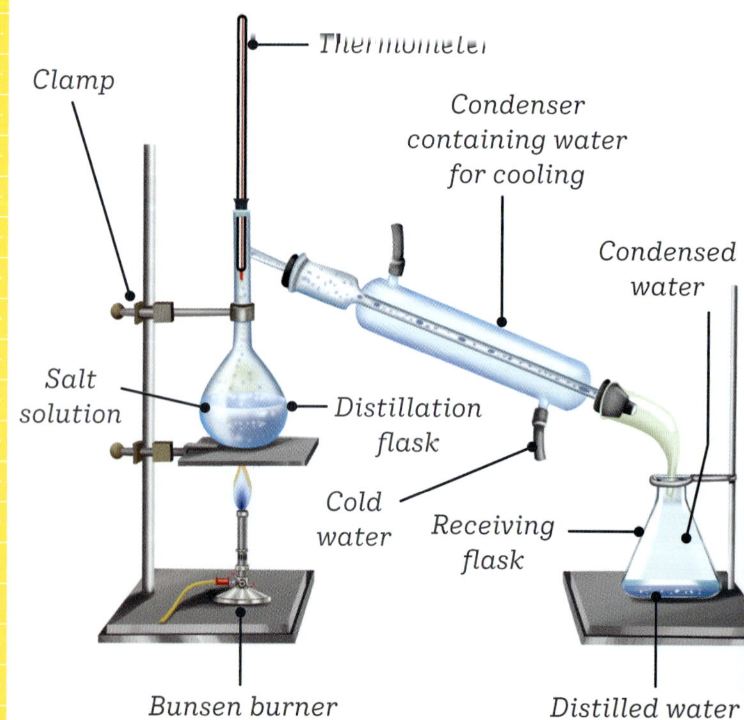

METHOD

1 Your teacher will set up the apparatus as shown.
2 They will pour the salt solution into the distillation flask, and then add a few marble chips to spread the heat evenly.
3 Your teacher will turn on the water through the condenser, boil the mixture in the flask and collect the water that comes out of the condenser in a small flask.
4 Identify the components separated as residue and distillate.

DISCUSSION

1 Explain why the two liquids could be separated.
2 Describe the changes of state that happened during the distillation.

CONCLUSION

Copy and complete:
'The results show that: (*respond to the aim*)'.

AN OPEN FLAME IS A HAZARD. TAKE CAUTION. IF YOU BURN YOURSELF, TELL YOUR TEACHER IMMEDIATELY AND PLACE THE BURNT AREA UNDER COLD RUNNING WATER FOR 20 MINUTES.

INVESTIGATION 3.7A

Separating colours using paper chromatography

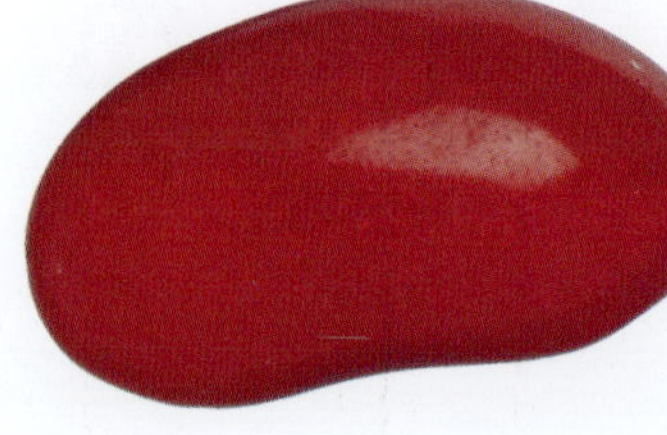

AIM

To investigate how chromatography can be used to separate colours

MATERIALS

- Coloured confectionery, e.g. jelly beans, Skittles, Smarties
- small beaker of water
- filter paper
- plastic pipette
- watchglass

METHOD

1 Place a sample of your chosen confectionery on a watchglass, then put 2 drops of water on the sample.
2 When the colour starts to dissolve, use an eyedropper to collect some coloured solution.
3 Add 1 drop of the coloured solution in the middle of the filter paper. The filter paper can be placed on an empty beaker.
4 Add 1 drop of water and observe what happens. Record your observations.
5 Repeat steps 1–4 using a different coloured lolly.
6 Describe what happens to the sample as the solvent (water) is added.

DISCUSSION

1 Which confectionery colours are single substances and which ones are mixtures?
2 How many different dyes are needed to make all the confectionery colours?

CONCLUSION

Copy and complete:
'The results show that: (*respond to the aim*)'.

NEVER EAT ANYTHING IN THE SCIENCE LABORATORY.

INVESTIGATION 3.7B

Separating two immiscible liquids

TEACHER DEMONSTRATION

AIM

To investigate using a separating funnel to separate a common mixture: oil and water

MATERIALS

- 2 × 25 mL measuring cylinders
- oil
- water
- separating funnel
- retort stand with bosshead and clamp
- retort ring
- measuring cylinder
- 2 × 100 mL conical flasks

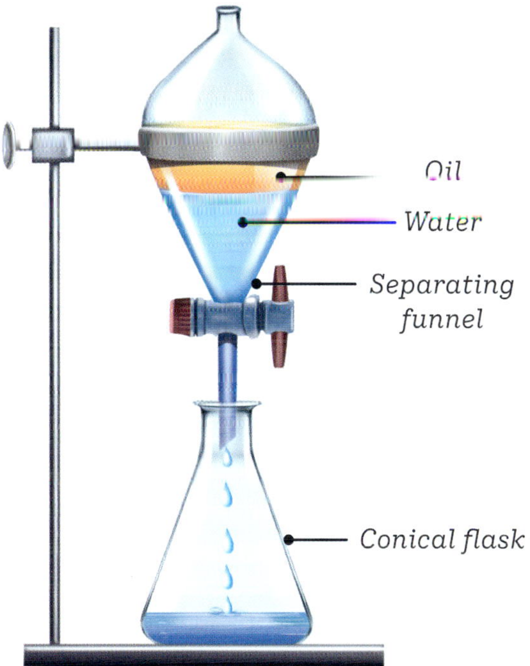

METHOD

1 Your teacher will set up the separating equipment as shown.
2 They will measure 20 mL of water and transfer it into the separating funnel.
3 Your teacher will measure 20 mL of oil and transfer it into the separating funnel.
4 They will stopper the funnel and shake it. Next, they will place the funnel back into the retort ring and observe the mixture separating. Record your observations.
5 After two obvious layers have formed, your teacher will slowly open the tap and collect the bottom liquid into a flask. Observe the transfer of the contents into a measuring cylinder and measure the volume of the liquid collected.
6 Once your teacher pours out the remaining liquid, measure the volume collected. Record your observations.

DISCUSSION

1 Describe what happened to the mixture when the separating funnel was shaken.
2 Explain why the mixture settled out after some time.
3 Consider the volumes you started with for each liquid and the volumes you recovered at the end. How would you describe the efficiency of separating a mixture using a separating funnel?

CONCLUSION

Copy and complete:
'The results show that: (*respond to the aim*)'.

INVESTIGATION 3.9

Froth flotation

AIM

To investigate the froth flotation process

MATERIALS

- kerosene
- liquid detergent
- mixture of 3 parts sand to 1 part iron filings
- water
- teaspoon
- large test tube
- rubber stopper

METHOD

1 Pour one teaspoon of the sand and iron mixture into the test tube.
2 Add water until the test tube is about half full, then insert the stopper and shake the test tube for 10 seconds. Record your observations.
3 Remove the stopper and add 3 drops of kerosene and 7–8 drops of detergent. Insert the stopper and shake the test tube for another 10 seconds. Record your observations.

DISCUSSION

1 What did you observe in the test tube after you added the kerosene and detergent?
2 Explain why the iron filings acted as they did.
3 Suggest how froth flotation could be used in industrial separation.

CONCLUSION

Copy and complete:
'The results show that: (*respond to the aim*)'.

KEROSENE IS A HAZARD. IT IS HIGHLY FLAMMABLE. NEVER USE IT NEAR MATCHES OR OPEN FLAME. NEVER INHALE OR INGEST IT. WASH YOUR HANDS IMMEDIATELY IF IT COMES INTO CONTACT WITH YOUR SKIN.

INVESTIGATION 4.1

Physical and chemical changes

TEACHER DEMONSTRATION

AIM

To investigate physical and chemical changes

MATERIALS

- Alka-Seltzer tablet
- vinegar
- bicarbonate of soda
- magnet
- mixture of sand and iron filings
- Petri dish
- 3 popsticks
- 5 cm magnesium ribbon
- 1 mol/L hydrochloric acid
- Bunsen burner
- heatproof mat
- tripod
- gauze mat
- matches
- evaporating basin
- 3 test tubes
- 3 × 10 mL measuring cylinder
- brass tongs
- water
- teaspoon

METHOD

1 Copy the results table into your notebook, adding a title.
2 Your teacher will set up the Bunsen burner on the heatproof mat, with the tripod and gauze mat.
3 They will use the Bunsen burner to burn the popsticks above the evaporating basin.
4 Observe as they add 1 teaspoon of bicarbonate of soda to a test tube and pour 5 mL of vinegar into it.
5 Your teacher will pass the magnet back and forth over the mixture of sand and iron filings (in the sealed Petri dish).
6 Your teacher will add one quarter of an Alka-Seltzer tablet to 5 mL of water in another test tube.
7 Your teacher will add the piece of magnesium ribbon to 5 mL of acid in a test-tube, using brass tongs. Observe closely. They will touch the bottom of the outside of the test tube.
8 Record all of your observations in your table. Explain why you categorised changes as physical or chemical.

DISCUSSION

1 What is the difference between a physical change and a chemical change?
2 From your results, list four properties you observed that indicated that the change was a chemical change.
3 Which change can be easily reversed, physical or chemical? Why?
4 What happens in all chemical changes?

CONCLUSION

Copy and complete.
'The results show that: (*respond to the aim*)'.

RESULTS **TABLE I4.1**

Substances	Observations	Type of change	Explanation
Popsticks and fire			
Baking soda and vinegar			
Magnet, sand and iron			
Alka-Seltzer and water			
Magnesium and acid			

INVESTIGATION 4.2A

Corrosion of iron

AIM

To identify the conditions that cause iron to rust (corrode)

MATERIALS

- 5 iron nails
- vegetable oil
- 10 mL of salt water solution
- distilled water
- tap water
- 5 test tubes
- 5 test-tube stoppers
- test-tube rack

METHOD

1. Set the test tubes in the test-tube rack.
2. Place nails and liquids in the test tubes as shown. The fifth test tube will not contain any liquid, only the nail and air.
3. Stopper all the test tubes.
4. Observe the test tubes for signs of rusting over the next week. Record your observations.

DISCUSSION

1. Rank the test tubes from containing the least rust to the most rust.
2. Suggest why some of the nails showed more signs of corrosion than others.
3. How could you make this investigation more reliable?

CONCLUSION

Copy and complete:
'The results show that: (*respond to the aim*)'.

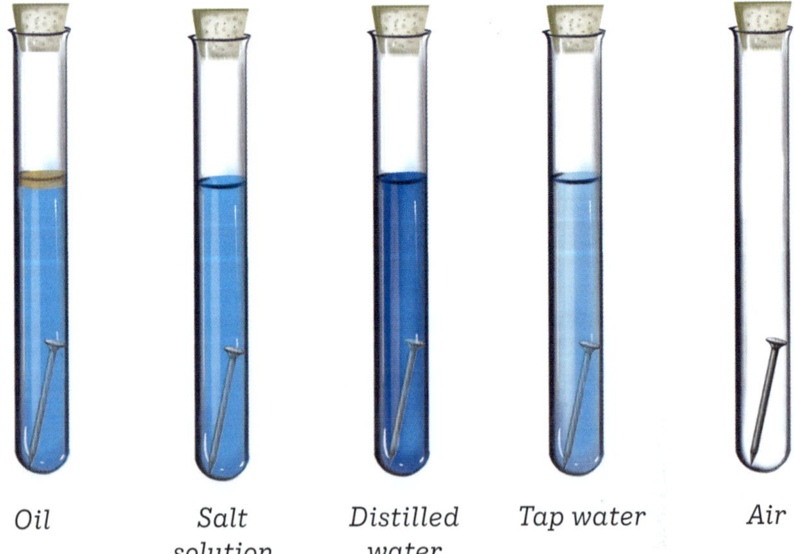

CHEMICALS OR SALT SOLUTION MAY GET INTO YOUR EYES. WEAR SAFETY GLASSES THROUGHOUT THE LESSON TO PREVENT THIS.

INVESTIGATION 4.2B

Preventing corrosion

AIM

To investigate which materials can prevent or slow corrosion

MATERIALS

- untreated nail
- galvanised nails
- painted nails
- nails sprayed with anti-rust spray
- 6 test tubes
- oil
- petroleum jelly

METHOD

1 Line your bench with some paper to protect the surface.
2 Prepare the nails so that you have a nail covered with oil, one with grease, a painted nail, a nail sprayed with anti-rust spray and leave one nail untreated.
3 Add each treated nail to a test tube. Fill each test tube with water so that the nails are covered.
4 Leave the test tubes aside for 2 days. Record your observations and determine which method best protects the nails from corrosion.

DISCUSSION

1 What was the purpose of putting the untreated nail in a test tube of water?
2 Which method prevented corrosion the best?
3 What other corrosion prevention substances could you potentially test?
4 List some ways of preventing rust.

CONCLUSION

Copy and complete:
'The results show that: (*respond to the aim*)'.

INVESTIGATION 4.3

Burning steel wool

TEACHER DEMONSTRATION

AIM

To investigate what occurs when steel wool is burnt

MATERIALS

- steel wool pad, cleaned with acetone
- watchglass
- brass tongs
- evaporating basin
- electronic balance
- 9 V battery

METHOD

1 Your teacher will place some of the steel wool in the watchglass on the electronic balance. Record the mass in your notebook.
2 Your teacher will place the steel wool in the evaporating basin. Record the total mass in your notebook.
3 Your teacher will touch the battery terminals to the steel wool, causing it to burn. This will continue until all of the wool is completely burned into ash or powder.
4 Record the current mass of the evaporating dish and steel wool. Calculate the current mass of the steel wool alone.

DISCUSSION

1 Does burning steel wool cause a physical or chemical change? Justify your answer.
2 What happened to the mass of the steel wool?
3 Use the particle theory to explain what happened to the steel wool.

CONCLUSION

Copy and complete:
'The results show that: (*respond to the aim*)'.

INVESTIGATION 4.6

Suitability of sports fabrics

AIM

To investigate which fabric is most suitable for sports clothing

MATERIALS

- 4 different types of fabric (cotton, polyester, nylon, spandex)
- 2 retort stands with bossheads and clamps
- 500 mL beaker
- long piece of string
- 4 stopwatches
- pegs
- warm water
- teaspoon
- salt
- fan

METHOD

1 Copy the results table into your notebook, adding a title.
2 Mix 500 mL of warm water and 1 teaspoon of salt together in a 500 mL beaker and leave aside.
3 Cut the four types of fabric into equal sized pieces.
4 Place the retort stands over a sink and tie a string from one retort stand to another.
5 Put the fan on, facing the retort stands.
6 Cover the prepared fabrics with salt water, take them out one at a time and peg them to the string, which serves as a clothesline.
7 Have four students time how long each fabric takes to dry. Record the times in your table and calculate an average time for each fabric.

DISCUSSION

1 Why do you think salt water instead of tap water was used in this investigation?
2 Why is it important for a sportsperson to wear clothing made from the right type of fabric?
3 Which fabric dried the fastest? Why do you think this was so?
4 Research the fast-drying fabric and write the properties that make it suitable for making clothes for sportspeople.

CONCLUSION

Copy and complete:
'The results show that: (*respond to the aim*)'.

RESULTS TABLE 14.6

Fabric	Time (min) (student 1)	Time (min) (student 2)	Time (min) (student 3)	Time (min) (student 4)	Average time (min)
Cotton					
Polyester					
Nylon					
Spandex					

INVESTIGATION 5.1

Modelling Earth's structure

AIM

To demonstrate how liquids of different densities interact and use their properties to model the different layers of the Earth.

MATERIALS

- 5 solutions of different densities (e.g. water, dishwashing detergent, vegetable oil, honey, light corn syrup) (100 mL of each)
- measuring cylinders (one each of 5, 10, 25 and 100 mL)
- 5 × 100 mL beakers
- electronic balance

METHOD

1 Copy the results table into your notebook, adding a title and rows as needed.
2 Use the electronic balance to find the mass of 10 mL of each liquid.
3 Use your knowledge of density to work out which liquid will represent each of Earth's layers. Check your order with your teacher. Add this to your table. (Hint: The densest solution will represent the inner core, while the least dense will represent the crust.)
4 Use the 25 mL measuring cylinder to measure 19 mL of liquid to represent the inner core. Pour this into the 100 mL measuring cylinder. Clean the 25 mL measuring cylinder.
5 Measure 35 mL of the liquid for the outer core. Carefully pour this into the 100 mL measuring cylinder so that it forms a layer on top of the first liquid. (Hint: Hold the measuring cylinder at an angle.) Clean the 25 mL measuring cylinder.
6 Measure 42 mL of the liquid for the lower mantle. Carefully layer this into the 100 mL measuring cylinder.
7 Measure 3 mL of the liquid for the upper mantle. Carefully layer this into the 100 mL measuring cylinder.
8 Measure 1 mL of the liquid for the crust. Carefully layer this into the 100 mL measuring cylinder.

DISCUSSION

1 Draw and label a diagram of your model.
2 How much thicker is the mantle than the inner core?
3 How much thicker is the lithosphere than the asthenosphere?
4 Explain how the model you have created represents Earth's layers and the density of the substances in each layer.
5 Models do not always accurately reflect what they are trying to show. Identify and explain one way that your model does not accurately reflect Earth's layering.
6 How could errors in measuring the density of the liquids have impacted this investigation?
7 What could you do to improve your method so that these errors are less likely to happen?

CONCLUSION

Copy and complete:
'The results show that: (*respond to the aim*)'.

RESULTS **TABLE I5.1**

Name of solution	Mass of 10 mL (g)	Density (g/mL)	Earth layer represented	Volume required for scale model (mL)

INVESTIGATION 5.2A

Observing minerals

AIM

To investigate the properties of some common minerals

MATERIALS

- selection of mineral samples (mineral kit)
- hand lens
- streak plate/unglazed white ceramic tile
- copper coin
- nail
- glass file/knife blade

METHOD

1 Copy the results table into your notebook, adding a title and rows as needed.
2 Complete steps 3–7 for each mineral sample.
3 Lustre: Determine if the mineral has a metallic or non-metallic lustre.
4 Colour: Describe the colour of the mineral.
5 Streak: Scratch the mineral along the streak plate. Describe the colour of the powder that is left on the plate. (Note: If your mineral has a measurement of more than 6.5 it will not make a streak. In this case write 'N/A' ('not applicable') in your table.
6 Hardness: Use your fingernail, copper coin, nail, glass file/knife blade and the streak plate along with Table 5.1 on page 77 to determine the approximate hardness of the mineral.

DISCUSSION

1 Which mineral was the softest and which was the hardest?
2 Which property do you think was the easiest to test? Why?
3 Is there one property that you could use on its own to tell the difference between each of the mineral samples you tested? Justify your response.
4 Explain why we observe both the colour and the streak of minerals.

CONCLUSION

Copy and complete:
'The results show that: (*respond to the aim*)'.

RESULTS TABLE I5.2A

Mineral name	Lustre	Colour	Streak	Hardness

INVESTIGATION 5.2B

Extracting copper

AIM

To investigate how chemical reactions are used to extract copper from a copper ore

MATERIALS

- 5 g of malachite or copper sulfate crystals
- zinc metal strip
- distilled water
- 100 mL beaker
- stirring rod

METHOD

1. Carefully dissolve 5 g of malachite or copper sulfate crystals in 50 mL of distilled water.
2. Add the zinc metal strip to the beaker so that it is covered by the solution.
3. Write the heading 'Start of the investigation' in your notebook and record your observations about the colour of the solution, colour of the zinc strip and any other observations.
4. Leave the beaker and solution overnight.
5. Write the heading 'End of the investigation' in your notebook and and record your observations about the colour of the solution, colour of the zinc strip and any other observations.
6. Include drawings or photos of the initial beaker and the beaker after 24 hours.
7. Describe the colour change to the solution in the beaker.
8. Describe the colour change to the zinc strip.

DISCUSSION

1. Describe the changes to both the zinc and the solution.
2. Research to find out the chemical reaction that occurred in this investigation. Write it as a word equation.
3. Where were the copper atoms located at the start of the investigation?
4. Where were the copper atoms located at the end of the investigation.
5. What does the investigation demonstrate about how chemical reactions can be used to extract minerals from ores?
6. What observations might you make to tell you that not all of the copper had been extracted from the solution?
7. How could you improve the method to extract all of the copper from the solution?

CONCLUSION

Copy and complete:
'The results show that: (*respond to the aim*)'.

INVESTIGATION 5.3

Modelling the rock cycle

AIM

To model the processes of the rock cycle using wax crayons

MATERIALS

- 2–4 wax crayons of different colours, cut into halves
- water
- butter or plastic knife
- cutting mat
- aluminium foil
- 2 large beakers
- brass tongs
- kettle
- camera (optional)

METHOD

1 Use the camera to take photographs of each step of this process.
2 Use the knife to cut up the crayons into very small pieces. Keep the colours separate for now.
3 Layer the different colours of crayon in the mortar.
4 Use the pestle to gently push down on your crayon layers so that they stick together.
5 Carefully take the crayon layers out of the mortar and place them on the cutting mat.
6 Use the rolling pin to further press down and combine the crayon layers.
7 Use the knife to carefully cut a section of the crayon layers that will fit into the crucible.
8 Set up the Bunsen burner, heat proof mat, tripod and pipeclay triangle.
9 Carefully place the crucible without its lid on the pipeclay triangle.
10 Light the Bunsen burner and turn to a blue flame.
11 Observe how the wax changes. Once all of the wax has melted turn off the Bunsen burner.
12 Remove the crucible with the tongs and allow the melted wax to cool.

QUESTIONS

1 What do the original crayons represent in this model?
2 Identify the rock cycle processes that are modelled by steps 2, 3, 4, 6, 11 and 12.
3 Identify the steps in the method that pose a safety risk. What did you do to reduce the risks?
4 Evaluate the method for its effectiveness in modelling the rock cycle. Could you make improvements? How could you improve the method?
5 Using the processes outlined in Figure 5.6, along with the photos you took, create your own rock cycle diagram.

HOT LIQUID, WAX AND STEAM ARE HAZARDS. TAKE EXTREME CAUTION. IF YOU SCALD (BURN) YOURSELF, TELL YOUR TEACHER IMMEDIATELY AND PLACE THE BURNT AREA UNDER COLD, RUNNING WATER FOR 20 MINUTES.

INVESTIGATION 5.4A

Cooling rate and crystal size

AIM

To investigate how cooling rate affects crystal size

MATERIALS

- saturated copper sulfate or alum solution
- solid copper sulfate or alum
- 3 test tubes
- 3 × 250 mL beakers
- 100 mL beaker
- ice
- cotton wool
- Bunsen burner
- heatproof mat
- tripod
- gauze mat
- matches
- spatula
- stirring rod
- magnifying glass

METHOD

1 Copy the results table into your notebook.
2 Place a test tube in each of the three 250 mL beakers. Surround one with ice up to the top of the beaker, one with cotton wool up to the top of the beaker and leave the other surrounded by air.
3 Set up the Bunsen burner, heat proof mat, tripod and gauze mat.
4 Add about 40 mL of saturated copper sulfate (or alum) solution to the 100 mL beaker.
5 Heat this solution using the Bunsen burner.
6 Add one spatula full of solid copper sulfate (or alum) to the solution at a time, stirring with the stirring rod until it dissolves. Keep adding solid until it no longer dissolves.
7 Carefully remove the solution from the Bunsen burner, pouring one third into each test tube.
8 Plug the top of each test tube with some cotton wool. Leave the test tubes to cool overnight.
9 Use the magnifying glass to help you observe the size of the crystals in each test tube and complete the results table.

DISCUSSION

1 Which tube cooled faster?
2 Describe the relationship that you observed between the rate of cooling and the size of the crystals formed. Was this expected? Why/Why not?
3 Which test tube represented the formation of an intrusive igneous rock? Justify your choice.
4 Which test tube represented the formation of an extrusive igneous rock? Justify your choice.
5 Molten rock can cool slower in some places than others. Suggest why.
6 Draw a labelled scientific diagram, with a scale showing the size and shape of the crystals that formed in each test tube.
7 With reference to the scientific method, explain why it was important to plug all three test tubes with cotton wool.

CONCLUSION

Copy and complete:

'The results show that: (*respond to the aim*)'.

AN OPEN FLAME IS A HAZARD. TAKE CAUTION. IF YOU BURN YOURSELF, TELL YOUR TEACHER IMMEDIATELY AND PLACE THE BURNT AREA UNDER COLD RUNNING WATER FOR 20 MINUTES.

RESULTS TABLE 15.4A

Beaker	Observations	Relative size of crystals (e.g. small, large)	Relative speed of cooling (e.g. slow, fast)
control			
ice			
cotton wool			

INVESTIGATION 5.4B

Observing igneous rocks

AIM

To observe key characteristics of some igneous rocks

MATERIALS

- selection of igneous rocks (e.g. basalt, granite, pumice, rhyolite, gabbro, obsidian)
- magnifying glass

METHOD

1 Copy the results table into your notebook, adding a title and rows as needed.
2 Carefully observe each sample, recording your observations in your results table.

QUESTIONS

1 Identify the extrusive igneous rocks. What feature enabled you to identify them?
2 Identify the intrusive igneous rocks. What feature enabled you to identify them?
3 You are walking through the bush and find a rock outcrop. What observations do you need to make to confirm that it is an igneous rock?

RESULTS TABLE I5.4B

Rock name	Approximate % of light and % dark minerals	Crystal size	Other observations	Rate of cooling

INVESTIGATION 5.5A

Modelling contact metamorphism

AIM

To investigate, using a model, the process of contact metamorphism using hot water and egg white

MATERIALS

- raw egg white
- salt
- water
- 100 mL beaker
- spatula
- Petri dish
- Bunsen burner
- heatproof mat
- tripod
- gauze mat
- matches
- beaker tongs
- camera (optional)

METHOD

1 Place the egg white in the Petri dish, making sure there is enough to cover the base of the dish.
2 Half fill the beaker with water and add a spatula full of salt.
3 Set up the Bunsen burner on the heatproof mat, with the tripod and gauze mat. Bring the water to the boil.
4 When the water is boiling, use beaker tongs to carefully remove the beaker from the tripod and place it in the centre of the Petri dish.
5 Observe what happens over 10 minutes. Record your observations. You may like to use photos to support your observations.

QUESTIONS

1 Describe the change that happened to the egg white.
2 What caused the change to the egg white?
3 This investigation modelled contact metamorphism.
   a Explain what the egg white represented in this process.
   b Explain what the beaker of hot water represented in this process.
4 Explain how the change to the egg white is similar to the changes to rocks that have undergone contact metamorphism.
5 Do you think that metamorphic rocks could be easily changed back into their original form? Justify your response.
6 Identify the steps in the method that pose a safety risk. What did you do to reduce the risks?

BOILING WATER, STEAM AND OPEN FLAME ARE HAZARDS. TAKE CAUTION. IF YOU BURN YOURSELF, TELL YOUR TEACHER IMMEDIATELY AND PLACE THE BURNT AREA UNDER COLD RUNNING WATER FOR 20 MINUTES.

IF YOU ARE ALLERGIC TO EGG WHITE, TELL YOUR TEACHER BEFORE PROCEEDING WITH THIS INVESTIGATION.

NEVER EAT ANYTHING IN THE SCIENCE LABORATORY.

INVESTIGATION 5.5B

Observing metamorphic rocks

AIM

To observe some samples of metamorphic rocks

MATERIALS

- selection of metamorphic rocks (e.g. quartzite, marble, slate, gneiss)
- magnifying glass

METHOD

1 Copy the results table into your notebook, adding a title and rows as needed.
2 Carefully observe each rock sample, recording your observations in your table.

QUESTIONS

1 Use your knowledge of metamorphic rocks to identify each sample as being formed by either contact or regional metamorphism.
2 What feature did you use to be able to identify the formation type of each sample?
3 You are walking through the bush and find a rock outcrop. What observations do you need to make to confirm that it is a metamorphic rock?

RESULTS TABLE I5.5B

Rock name	Size of crystals (mm)	Crystals in layers? (yes/no)	Other observations (e.g. colour)

INVESTIGATION 5.6

Modelling weathering of rocks

PART 1: MODELLING WEATHERING DUE TO TEMPERATURE CHANGE

AIM

To investigate how extreme temperature can weather rocks

MATERIALS

- piece of granite (approx. 20 × 20 × 20 mm)
- piece of quartzite (approx. 20 × 20 × 20 mm)
- 250 mL beaker
- Bunsen burner
- heatproof mat
- matches
- brass tongs
- paper towel

METHOD

1. Half fill the beaker with tap water.
2. Set up the Bunsen burner on the heatproof mat.
3. Using the tongs, heat the granite sample in the blue flame for 2 minutes.
4. Quickly put the sample in the beaker of water so it is covered. (This is called quenching.)
5. Record your observations.
6. Retrieve the sample from the water and pat dry with the paper towel.
7. Repeat steps 3–6 another four times.
8. Repeat the method for the quartzite sample.

QUESTIONS

1. Describe what happened to the granite sample after heating and quenching once. How did this change during the repetitions?
2. Describe what happened to the quartzite sample after heating and quenching once. How did this change during the repetitions?
3. What natural process is this investigation representing? Describe an environment where it might happen.
4. Granite is an igneous rock made of at least three different minerals, and quartzite is a metamorphic rock made of one mineral. Consider how this may affect how quickly they weather.
5. Is this investigation representing chemical or physical weathering?
6. Identify the steps in the method that pose a safety risk. What did you do to reduce the risk?

OPEN FLAME IS A HAZARD. TAKE CAUTION. DO NOT TOUCH THE HOT GRANITE - IT WILL BE EXTREMELY HOT.

FIRST AID: TELL YOUR TEACHER IMMEDIATELY ABOUT ANY INJURIES. TREAT BURNS OR SCALDS BY PLACING THE SKIN UNDER COLD, RUNNING WATER FOR 20 MINUTES.

TEACHER DEMONSTRATION

PART 2: MODELLING WEATHERING DUE TO ACID RAIN

AIM

To investigate how acid can react with rocks and chemically weather them.

MATERIALS

- small (less than 1 × 1 × 1 cm) pieces of each of limestone, sandstone and granite
- 1 mol/L hydrochloric acid
- 3 test tubes
- test-tube rack
- camera (optional)

METHOD

1. Your teacher will place one piece of each type of rock in three separate test tubes in the test tube rack.
2. They will then add hydrochloric acid to the test tubes to cover the rocks.
3. Record any initial observations.
4. The samples will be left until the end of the lesson or overnight before you consider them again to make final observations. You may like to include photos and drawings in your results and observations.

QUESTIONS

1. What evidence was there that chemical weathering happened?
2. Were your observations the same for all rock types?
3. Would you expect to observe the same reactions if you added water instead of acid? Justify your response.
4. Acid rain happens when pollution causes rain to be very acidic. What might you expect to happen to monuments and buildings made of limestone that are exposed to acid rain?

INVESTIGATION 5.7A

Modelling the formation of sandstone

AIM

To investigate, using a model, the formation of sandstone

MATERIALS

- separate samples of sands of different colours, with dry plaster of Paris mixed in
- water
- 2 clear disposable plastic cups
- spatula or spoon

METHOD

1 Place a layer of sand of one colour into one of the plastic cups.
2 Use the spatula or spoon to moisten (but not saturate) the sand with water.
3 Press the base of the second cup in to compress the layer.
4 Repeat steps 1–3 for sands of different colours.
5 Allow your 'sandstone' to dry before removing it from the plastic cup.

QUESTIONS

1 Identify the steps in the procedure that represent deposition, compaction and cementation.
2 Compare your 'sandstone' to a real sample of sandstone. What are the similarities? What are the differences?
3 What did the investigation demonstrate about the formation of sandstone?
4 What are the safety risks in this investigation?
5 What safety advice could you offer to address the safety risks?

INVESTIGATION 5.7B

Observing sedimentary rocks

AIM

To observe samples of sedimentary rocks

MATERIALS

- selection of sedimentary rocks (e.g. sandstone, conglomerate, mudstone, shale, limestone)
- magnifying glass

METHOD

1. Copy the results table into your notebook, adding a title and rows as needed.
2. Carefully observe each rock sample, recording your observations in your table.

QUESTIONS

1. Which rock contained the largest sediments?
2. Which rock contained the smallest sediments?
3. Large sediments are deposited in high-energy environments and small sediments are deposited in low-energy environments. Identify the rock that was deposited in the:
   a. highest energy environment
   b. lowest energy environment
4. Explain why some of your samples did not have obvious layers, even though they are still sedimentary rocks.
5. You are walking through the bush and find a rock outcrop. What observations do you need to make to confirm that it is a sedimentary rock?

RESULTS TABLE I5.7B

Rock name	Size of sediments (mm)	Obvious layers present? (yes/no)	Other observations (e.g. colour, presence of fossils)

INVESTIGATION 5.8

Making 'fossils'

AIM

To model the formation of a fossil

MATERIALS

- shell or other item
- plasticine
- plaster of Paris
- popstick
- plastic cup
- water
- plastic container (e.g. take-away food container)
- camera (optional)

METHOD

1 Carefully add water to the plaster of Paris and mix with popstick until it is the same consistency as toothpaste.
2 Press the plasticine into the base of the plastic container and smooth the top.
3 Press the shell or item into the plasticine to make an imprint, then remove it.
4 Carefully add a 1–2 cm layer of plaster of Paris on top of the plasticine. Tap the container down to remove any air bubbles.
5 Leave the plaster to dry.
6 Carefully separate the plaster from the plasticine.
7 Record your observations. You may like to photograph the plaster and the plasticine to support your observations.

QUESTIONS

1 Identify which of the plasticine and plaster is the mould and which is the cast. Propose a procedure you could carry out that would swap the two.
2 Use the terms *deposition*, *compaction* and *cementation* to compare the method you followed to the process of fossilisation.

INVESTIGATION 5.10

Mining for chocolate chips

AIM

To investigate which brand of chocolate chip biscuit contains the highest percentage of 'metal' in the 'ore'

MATERIALS

- chocolate chip biscuit (one of several different brands)
- toothpicks
- cupcake liner
- graph paper
- electronic balance

METHOD

MINING FOR METAL IN ORE

1 Copy the results table into your notebook, adding a title and rows as needed.
2 Place your biscuit in a cupcake liner. Set to zero mass on the electronic balance. Weigh and record the mass of the biscuit, then remove it and any crumbs from the liner.
3 Using the toothpick, carefully remove the chocolate chips from the biscuit, keeping them separate from those of other biscuits.
4 Place the chocolate chips in the cupcake liner. Set to zero mass on the balance. Weigh and record the mass.
5 Calculate the percentage of chocolate chips in the biscuit compared to the 'waste'.
6 Share your results with the rest of the class and calculate an average for all brands of biscuit.

DISCUSSION

1 Which brand contained the highest percentage of chocolate chips?
2 Which brand contained the lowest percentage of chocolate chips?
3 Compare your results with other groups – how did they differ?
4 Compare the class average results to the advertised percentage of chocolate chips – how did they differ?
5 What 'mining' strategies did you use? Could you have easily rehabilitated the biscuits after the chocolate chips were removed? How could you change them so they were more 'environmentally friendly'? How do you think this could impact any profits?

CONCLUSION

Copy and complete:
'The results show that: (*respond to the aim*)'.

RESULTS **TABLE I5.9**

Biscuit brand	Total biscuit mass (g)	Mass of chocolate chips (g)	% by mass of chocolate chips	Class average % by mass of chocolate chips	Labelled % of chocolate chips

INVESTIGATION 6.1

Modelling day and night

AIM

To investigate, using a model, how day and night are explained by the relative positions of the Sun and Earth

MATERIALS

- globe of Earth
- torch

METHOD

1 Have one person hold the torch. This is the Sun.
2 Have another person in charge of the globe. This is Earth. (The globe will have the correct tilt.)
3 Dim the light and aim the torch at the globe. Note that half of the globe is lit and half is in darkness.
4 Rotate the globe anti-clockwise and observe how the light travels over the continents.

QUESTIONS

1 How much of Earth is in daylight at any time?
2 How much of Earth is in darkness at any time?
3 When it is day in Australia, name a continent that has night.
4 Is it possible for New York and New Zealand to both have daylight at the same time?

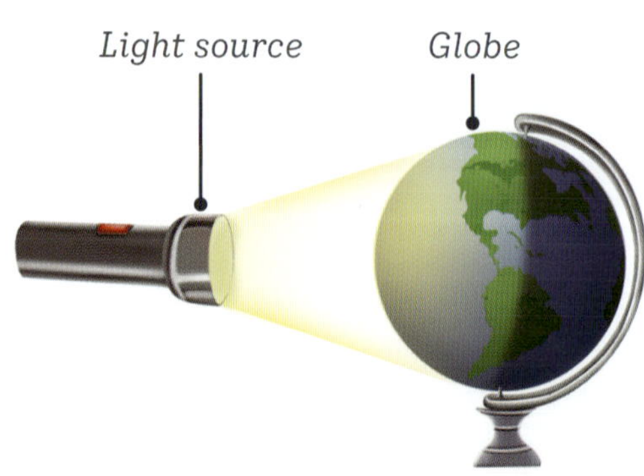

INVESTIGATION 6.2

Modelling the seasons

AIM

To investigate, using a model, how Earth's tilt affects how the Sun's rays strike Earth and cause seasons

MATERIALS

- backing board
- clips
- 7 sheets of graph paper
- 2 retort stands with bossheads and clamps
- torch
- pencil
- ruler
- protractor

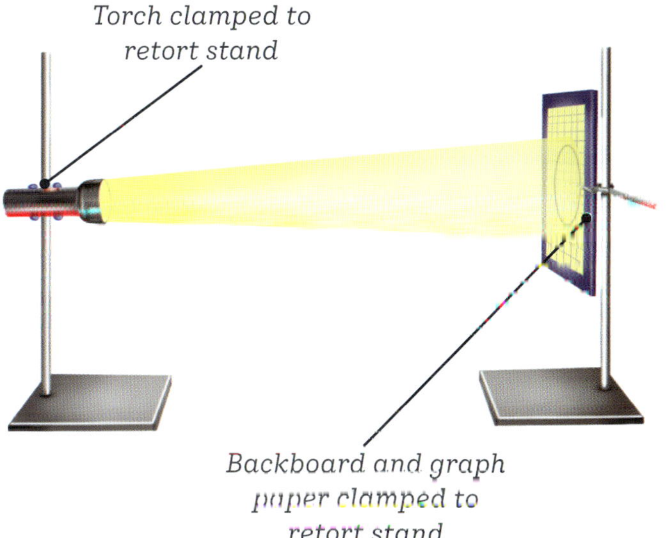

RESULTS

TABLE I6.2

Angle (°)	Area (number of lit squares)

METHOD

1. Copy the results table into your notebook, adding a title and rows as needed.
2. Set up the apparatus as shown.
3. Dim the lights so that the torch makes a clear circle on the graph paper.
4. Check that the board is parallel to the retort stand.
5. Measure and record the distance between the torch and the graph paper.
6. Use a pencil to trace the lit area. Be careful not to move the graph paper.
7. Remove the graph paper and label it 0°.
8. Place another sheet of graph paper on the board and tilt the board by 10° so that the top of the board is further away from the torch. Trace the lit area. Label this sheet with the angle of the board.
9. Repeat step 8 for each angle up to 60°.
10. Count the number of squares inside the traced areas and record them in the results table.
11. Graph your results, with the angle in degrees on the horizontal axis and the area in squares on the vertical axis. Remember to choose an appropriate scale.

QUESTIONS

1. Which angle do you think represents places near the equator? Why?
2. Which angle do you think represents Antarctica? Why?

INVESTIGATION 6.3A

Modelling a solar eclipse

AIM

To investigate, using a model, how solar eclipses are explained by the relative positions of the Sun, Moon and Earth

MATERIALS

- tape
- 2 cardboard tubes or roll of paper
- scissors
- aluminium foil
- coathanger wire (20 cm length)
- polystyrene ball (10–12 cm diameter)
- ping-pong ball or small polystyrene ball
- square of heavy card
- retort stand with bosshead and clamp
- torch

METHOD

1 Make a series of small (2 cm), even, vertical cuts around each end of one of the cardboard tubes.
2 Bend the cut pieces of both ends of the cardboard tube outwards and stand the tube upright. Tape one end of the tube to one of the pieces of heavy card to create a base for the model.
3 Tape the base to the polystyrene ball. This is the model of Earth.
4 Insert one end of the wire vertically into the Earth model.
5 Make two small holes in the ping-pong ball for the wire to pass through.
6 Cover the ping-pong ball with aluminium foil. This is the model of the Moon.
7 Bend the wire to secure the Moon as shown. The Moon's equator should be level with Earth's equator.
8 Adjust the torch using the retort stand so that it is in line with Earth's equator. The other cardboard tube can be used to direct the light.
9 Dim the lights.

To model the solar eclipse:
1 Stand facing the torch and swing the wire around until the Moon casts a shadow on Earth. The Moon is now positioned between Earth and the Sun, blocking Earth's sunlight.
2 Show that the shadow moves by slowly rotating the wire.
3 Bend the wire up so that the Moon only partially blocks the Sun.

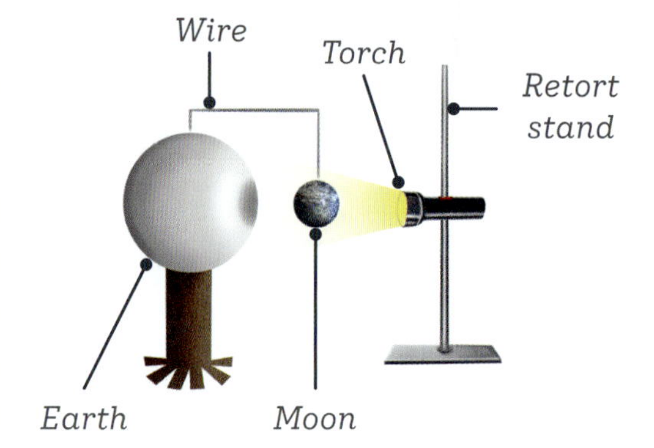

DISCUSSION

1 Which object is in shadow?
2 Which object is casting the shadow?
3 Does everyone in the world see a solar eclipse? Who can't see it?
4 Do you see solar eclipses at night or in the day?
5 During a solar eclipse, what would you see if you stood on the Moon and looked at Earth?
6 What is the phase of the Moon during a solar eclipse?
7 Do other planets have solar eclipses?

CONCLUSION

Copy and complete:
'The results show that: (*respond to the aim*)'.

INVESTIGATION 6.3B

Modelling a lunar eclipse

AIM

To investigate, using a model, how lunar eclipses are explained by the relative positions of the Sun, Moon and Earth

MATERIALS

- tape
- 2 cardboard tubes or roll of paper
- scissors
- aluminium foil
- coathanger wire (20 cm length)
- polystyrene ball (10–12 cm diameter)
- ping-pong ball or small polystyrene ball
- square of heavy card
- retort stand with bosshead and clamp
- torch

METHOD

1 Make a series of small (2 cm), even, vertical cuts around each end of one of the cardboard tubes.
2 Bend the cut pieces of both ends of the cardboard tube outwards and stand the tube upright. Tape one end of the tube to one of the pieces of heavy card to create a base for the model.
3 Tape the base to the polystyrene ball. This is the model of Earth.
4 Insert one end of the wire vertically into the Earth model.
5 Make two small holes in the ping-pong ball for the wire to pass through.
6 Cover the ping-pong ball with aluminium foil. This is the model of the Moon.
7 Bend the wire to secure the Moon as shown. The Moon's equator should be level with Earth's equator.
8 Adjust the torch using the retort stand so that it is in line with Earth's equator. The other cardboard tube can be used to direct the light.
9 Dim the lights.

To model the lunar eclipse:
1 Stand facing the torch and swing the wire around until the Moon is completely behind Earth. Earth is now positioned between the Moon and the Sun, blocking the Moon's sunlight .
2 Bend the wire so that only part of Earth's shadow falls on the Moon, to model a partial lunar eclipse.

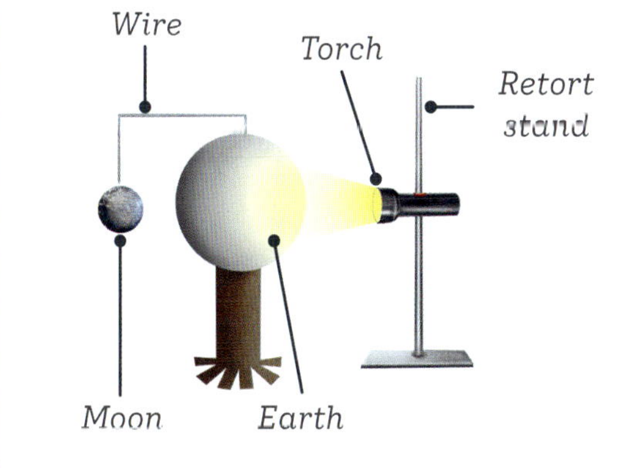

DISCUSSION

1 Which object is in shadow?
2 Which object is casting the shadow?
3 Does everyone in the world see a lunar eclipse? Who can't see it?
4 What is the phase of the Moon during a lunar eclipse?
5 Why don't we see a lunar eclipse during every full Moon?
6 Do other planets have lunar eclipses?

CONCLUSION

Copy and complete:
'The results show that: (*respond to the aim*)'.

INVESTIGATION 6.5

Making a simple telescope

AIM

To investigate how telescopes use lenses to magnify distant objects

MATERIALS

- metre rule
- Blu-Tack or plasticine
- retort stand
- strong convex lens (+14D) or magnifying glass
- weak convex lens (+2.5D) or lens from reading glasses
- greaseproof paper
- lamp

METHOD

1 Attach the weak lens to one end of the metre rule with plasticine or Blu-Tack.
2 Set the metre rule into the retort to hold it steady, with the weak lens pointing away from you.
3 Put the lamp on a table at the other end of the room and turn it on.
4 Turn the retort stand until the metre rule is aimed at the lamp.
5 Hold a piece of greaseproof paper just above the metre rule. Move it along the rule until you see a small image of the lamp on the paper. Have a partner hold it at that position.
6 Hold the strong lens at the other end of the metre rule. Using one eye, look through the lens at the image on the paper. Move the lens until the lamp image is magnified.
7 Have your partner take the paper away, but continue looking through the strong lens. You should still see the image of the lamp, although brighter than before.

QUESTIONS

1 Why does this telescope require two lenses, rather than just one?
2 Why is the image brighter once the paper is taken away?
3 How is this simple telescope different to a proper telescope?

CONCLUSION

Copy and complete: 'The results show that: (*respond to the aim*)'.

INVESTIGATION 7.1

Classifying resources used in the classroom

AIM

To investigate how the resources that are used in common classroom materials are made

MATERIALS

- selection of classroom items (e.g. paper, ruler, eraser, pencil case, calculator)

METHOD

1. Copy the results table into your notebook, adding a title and rows as needed.
2. Select 10 items in the classroom and include them in your table.
3. Identify the materials each item is made from. You may need to do some research.
4. Identify how the major material from each item is obtained.
5. Determine if the item is made from renewable or non-renewable resources.
6. Compare your list with those of others in your class.

QUESTIONS

1. Compare the number of renewable resources in your table to the number of non-renewable resources.
2. Describe the similarities between the items that are made from renewable sources.
3. Describe the similarities between the items that are made from non-renewable resources.
4. Where are most of the non-renewable resources you listed found in nature?

RESULTS TABLE I7.1

Item	Major resource in item	How is it obtained?	Renewable or non-renewable?
Paper	Paper	Cutting down trees	Renewable
Metal ruler	Aluminium	Mining from deposits in rocks	Non-renewable

INVESTIGATION 7.2

Investigating soil erosion

AIM

To investigate factors that influence soil erosion

MATERIALS

- several small seedlings
- soil
- small amount of mulch
- 6 empty soft-drink bottles (equal shape and size)
- piece of board or bench space (30 cm × 30 cm)
- glue or plasticine
- string
- 250 mL beaker
- water

METHOD

1 Cut a rectangle (approximately 10 cm × 20 cm) out of one side of three of the bottles.

2 Using glue or plasticine, position the bottles on their sides (holes facing up) on a table or board so that they will not move. The necks of the bottles should stick out over the edge.

3 Add an equal amount of soil to each of the three bottles.

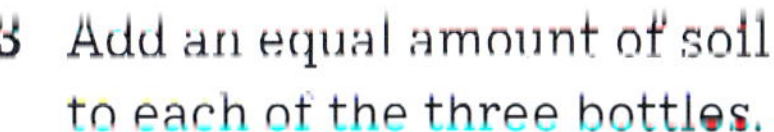

4 Plant your seedlings in the first bottle.

5 Cover the soil in the second bottle with a layer of mulch.

6 Leave the soil uncovered in the third bottle.

7 Cut the other three bottles in half and carefully pierce a hole on either side to thread the string through so that it can be suspended underneath the necks of the three bottles.

8 Use the beaker to measure 250 mL of water and carefully pour the water into each bottle containing soil, pouring into the bottom of the bottle, away from the neck.

9 When the water has drained through, record your observations of the differences in the water that collects in the buckets underneath the neck.

10 Repeat the watering process over several days, continuing to record your observations.

DISCUSSION

1 Compare the colour of the water in each of the collecting cups.

2 Compare the level of soil remaining in each of the three bottles.

3 Propose why the colours and levels of soil are different.

4 Farmers will often leave the roots of their old crop in the ground until they are ready to plant their new one. Justify why this is a useful strategy for preserving soil resources.

5 A farmer observed that soil will erode more from paddocks that are on a slope than in paddocks on flat ground. Design an experiment to test how the angle of a slope affects the amount of soil eroded.

CONCLUSION

Copy and complete:
'The results show that: (*respond to the aim*)'.

INVESTIGATION 7.3A

Designing a windmill to lift a weight

AIM

To investigate which design for a windmill can lift the heaviest weight

MATERIALS

- cotton reel
- string
- piece of dowel to fit through cotton reel
- block of polystyrene
- light cardboard, heavy cardboard, plywood and other materials suitable to make turbine blades
- skewers
- selection of fishing line sinkers or other small weights
- retort stand with bosshead and clamp
- fan

METHOD

1. Insert the dowel through the cotton reel and secure it so that the reel will not spin. This will be how you attach the string that will lift the weight.
2. Insert the other end of the dowel into the polystyrene block. This will be where you insert your turbine blades.
3. Carefully clamp the dowel into the retort stand and clamp. The dowel must still be able to spin freely.
4. Tape one end of the string onto the cotton reel. Thread the other ends through the cup that will be used to lift the weight.
5. Determine the size, shape and number of blades that your windmill will have. Construct these out of your chosen material.
6. Consider the angle (pitch) that your blades will have compared to the front of the windmill.
7. Put together your windmill from all of your components.
8. Attach different weights to the string. Use the fan to try to turn the windmill blades and lift each weight.
9. Record your observations and the heaviest weight your design could lift. Compare your findings with those of your classmates.

QUESTIONS

1. Which design was able to lift the heaviest weight?
2. Why is the angle of the windmill blades so important?
3. What materials (not provided) do you think would make an even stronger windmill?

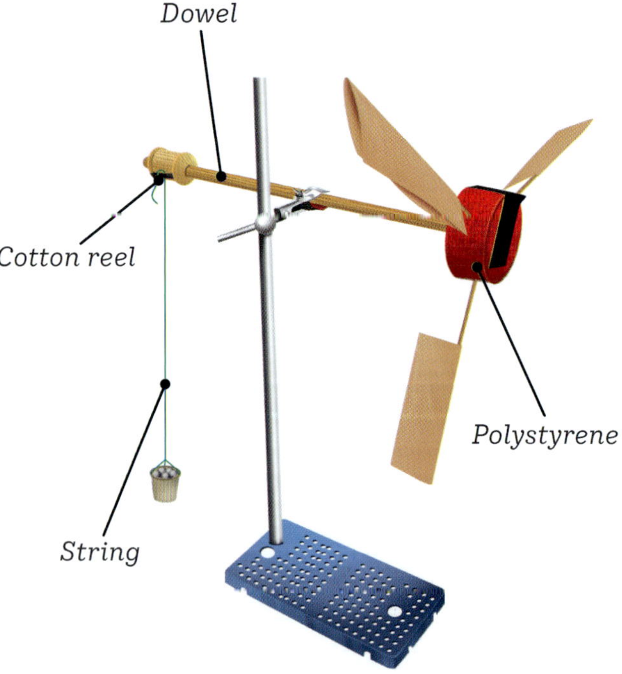

INVESTIGATION 7.3B

Making bioplastic

AIM

To investigate how plastic is made using cornflour

MATERIALS

- cornflour
- vinegar
- vegetable glycerine
- water
- aluminium foil
- 250 mL beaker
- hotplate
- stirring rod
- tablespoon
- teaspoon
- spatula
- selection of commercially produced plastics

METHOD

1. Copy the results table into your notebook, adding a title and rows as needed.
2. Set up the hotplate.
3. Add 1 tablespoon of cornflour, 1 teaspoon each of vinegar and glycerine and 4 tablespoons of water to the beaker. Stir with the stirring rod until combined.
4. Place the beaker onto the hotplate and switch on to low heat.
5. Continue stirring the solution until it thickens and starts to become translucent. This will not take long!
6. Carefully pour the plastic onto a sheet of aluminium foil and spread using the spatula.
7. Allow the plastic to completely cool. This will take several hours. If you wish to form the plastic into another shape, leave it spread out on the foil to cool for approximately 1 hour. It can then be carefully moulded into shape before it sets completely.
8. Compare the flexibility of your plastic with the samples provided. Record your observations.
9. Take a piece of your plastic and add it to water. Record your observations.
10. Take a piece of your plastic and heat it in a beaker over the hotplate. Record your observations.

DISCUSSION

1. Describe the properties of your bioplastic.
2. Compare these properties to the properties of some conventional plastics.
3. What happened when you added your bioplastic to water?
4. What happened when you reheated your bioplastic?
5. Explain the benefit of using a bioplastic, such as the one you have made, compared to a conventional plastic.

CONCLUSION

Copy and complete:
'The results show that: (*respond to the aim*)'

RESULTS **TABLE I7.3B**

Material	Flexibility	Behaviour in water	Behaviour when heated

TAKE CAUTION USING THE HOTPLATE. IF YOU BURN YOURSELF, TELL YOUR TEACHER IMMEDIATELY AND PLACE THE BURNT AREA UNDER COLD RUNNING WATER FOR 20 MINUTES.

INVESTIGATION 7.4

The sustainability game

AIM

To demonstrate how to use a resource so that it can continue to be used in the future

MATERIALS

- 60 marbles
- container

METHOD

The aim of the game is to have collected the most number of marbles at the end of eight rounds.

1 Count out 20 marbles into the container (your source), leaving the others as reserves.

2 Copy the results table into your notebook, adding a title and rows as needed.

3 To begin, decide how many marbles you will remove from the 'source'. The number of marbles you remove at each round can only vary by one. For example, if you remove three marbles in round 1, you can take two or four out in round 2.

4 At the end of each round, add one marble from the reserves for every four marbles left in the 'source'. For example, if there are 16–19 marbles left, four should be added. If there are 1–3 marbles left, none should be added.

5 Continue removing and adding marbles until you run out of marbles in the 'source' or make it through eight rounds. Record your results in your table.

6 Construct a graph of your results. Plot the number of marbles taken on the *y*-axis with the number of rounds on the *x*-axis.

QUESTIONS

1 Compare your results to those of other groups. Who collected the most marbles at the end of eight rounds? Who collected the least? Did anyone run out of marbles from the source before the end of eight rounds?

2 Compare the different strategies that were used. What strategies allowed for a sustainable use of the resource? What strategies were not sustainable? Why?

3 Did the marbles in this game represent a renewable or non-renewable resource? Justify your response.

4 How would the rules and outcome of the game have been different for the other type of resource?

RESULTS TABLE 17.4

Round	Number in source at start of round	Number removed from source	Number left in source	Number added to source at end of round

INVESTIGATION 8.1A

Investigating cohesion of water

AIM

To investigate cohesion in water using a 5-cent coin

MATERIALS

- 5-cent coin
- small amount of water in a 50 mL beaker
- pipette or eyedropper

METHOD

1 Copy the results table into your notebook, adding a title.
2 Use the pipette or eyedropper to practise making water drops of the same size.
3 Predict how many drops of water you will be able to fit onto the coin before it overflows. Record this in your notebook.
4 Carefully place one water drop on to the coin.
5 Keep adding water drops, counting how many you can fit on the coin before it overflows.
6 Dry the coin and repeat this process another three times.
7 Add up all four trial results and divide by four to calculate the average result.

RESULTS TABLE I8.1A

Trial	Number of drops before overflow	Average
1		
2		
3		
4		

DISCUSSION

1 Compare your prediction to your average result. Are they different? If so, why?
2 Compare your average result to others in your class. What might be the cause of differences, if any?
3 Would you expect to be able to fit the same number of drops of soapy water onto the coin? Design a procedure to test your hypothesis.
4 Research to find out more about the surface tension of water. How is this related to the cohesion between the water molecules?

CONCLUSION

Copy and complete:
'The results show that: (*respond to the aim*)'.

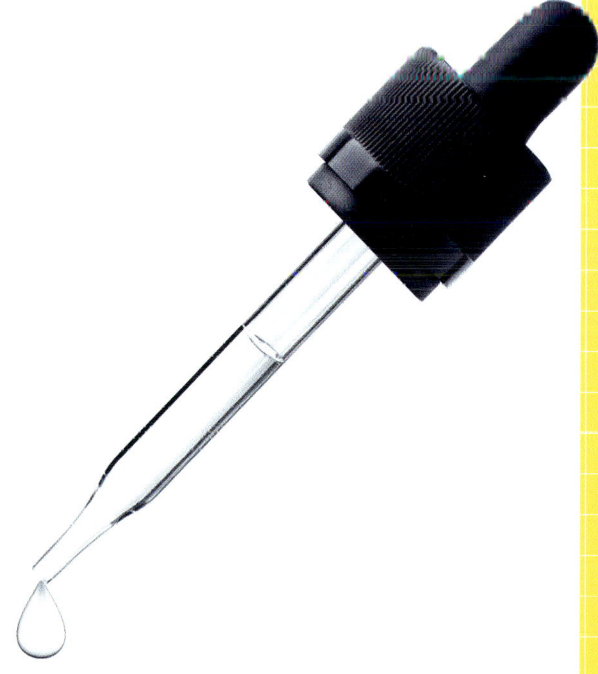

INVESTIGATION 8.1B

Observing capillary action

AIM

To investigate whether the thickness of a straw affects the movement of water by capillary action

MATERIALS

- 250 mL beaker
- food colouring
- water
- 3 clear straws or tubes of different widths, including one capillary tube (very thin tubing)
- marker pen
- ruler
- stopwatch

METHOD

1 Copy the results table into your notebook, adding a title.
2 Half fill the beaker with water and add some food dye so the water is a dark colour.
3 Arrange the straws so that their ends line up. Use the marker pen to draw a line across all of them about one-third of the way along.
4 Carefully hold the straws in the beaker of water, lining up the marker line with the top of the water. The straws should not touch the bottom of the beaker.
5 Start timing with the stopwatch. Hold the straws in place for 60 seconds.
6 Quickly and carefully mark where the water level in the straws rises to.
7 Remove the straws from the beaker and measure the difference between the original water level and the final water level. Record this in your table.
8 Repeat steps 4–7 another two times.

DISCUSSION

1 Describe the trend shown in your results.
2 Explain why you observed this trend. (You may need to do some research.)
3 Explain how both the cohesion and adhesion of water molecules allow water to travel up the straw. You may like to use a diagram to support your explanation.
4 Capillary action allows plants to transport water from their roots to their leaves through special transport vessels called xylem. Based on your results, do you think large trees would have narrow or wide xylem? Justify your response.

CONCLUSION

Copy and complete:
'The results show that: (*respond to the aim*)'.

RESULTS **TABLE I8.1B**

Trial	Straw/tube width (mm)	First distance (mm)	Second distance (mm)	Third distance (mm)
1				
2				
3				

INVESTIGATION 8.2

Modelling the water cycle

AIM

To investigate, using a model, the processes in the water cycle

MATERIALS

- large clear plastic bowl or container
- modelling clay
- water
- cling film
- small pebble
- elastic band
- food dye
- heat lamp
- camera (optional)

METHOD

1 Use the modelling clay to form an island that will sit in the middle of the container. Make a small depression in the top of the island that will represent a freshwater lake – do not make it too deep.
2 Add water to the container so that it comes about a quarter of the way up your 'island'. Add some food dye to colour the water.
3 Carefully cover the top of the container with cling film, securing it around the edges with an elastic band so that no water vapour can escape.
4 Place the pebble on the cling film.
5 Place the heat lamp over the container and switch it on. Leave it for a period of time and record your observations in your notebook. You may like to photograph your model to support what you have written.

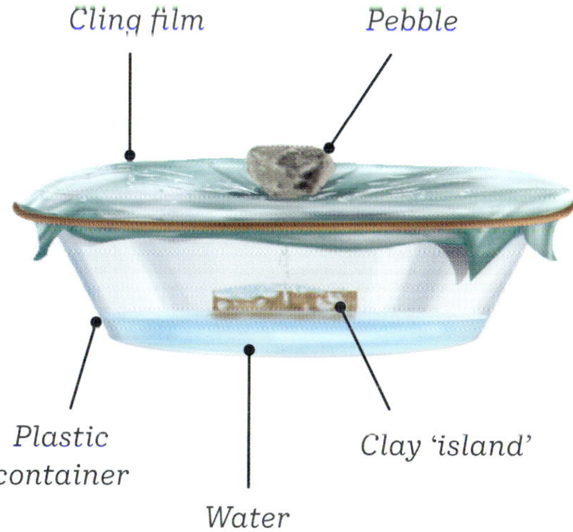

DISCUSSION

1 Identify the part of the model that simulated evaporation.
2 Identify the part of the model that simulated condensation.
3 Identify the part of the model that simulated precipitation.
4 Compare the colour of the water in the 'lake' with that in the ocean. Explain any differences.
5 Explain why water is considered to be a renewable resource.
6 This model does not include all parts of the water cycle. Which parts were not included? Suggest ways that the model could be altered to include them.

CONCLUSION

Copy and complete:
'The results show that: (*respond to the aim*)'.

TAKE CAUTION USING THE HEAT LAMP. IF YOU BURN YOURSELF, TELL YOUR TEACHER IMMEDIATELY AND PLACE THE BURNT AREA UNDER COLD RUNNING WATER FOR 20 MINUTES.

INVESTIGATION 8.3

Testing water quality

AIM

To investigate the quality of different water sources

MATERIALS

- water samples from different areas (e.g. dams, different points along a river)
- distilled water
- test tubes
- test-tube rack
- pipettes
- universal indicator and chart
- 0.1 mol/L silver nitrate in a dropper bottle
- 5% ammonia solution in a dropper bottle
- nitrate test strips

METHOD

Copy the results table into your notebook, adding a title and rows as needed.

PART 1: TESTING FOR PH

1 Add 2 cm of each water sample, including distilled water, into separate test tubes.
2 Add 1 drop of universal indicator to each sample.
3 Use the universal indicator chart to identify the pH of each sample. Record this in the results table.
4 Dispose of the water in the test tubes as per your teacher's directions.

PART 2: TESTING FOR SALINITY

1 Add 2 cm of each water sample, including distilled water, into separate test tubes.
2 Carefully add 1 drop of silver nitrate to each test tube and swirl to mix.
3 If salt is present in the water it will turn white and cloudy; the more salt, the cloudier it will be. Record your observations in your table.
4 Dispose of the water in the test tubes as per your teacher's directions.

PART 3: TESTING FOR NITRATES (NUTRIENTS IN FERTILISERS)

1 Add 2 cm of each water sample, including distilled water, into separate test tubes.
2 Carefully dip the reactive end of a nitrate strip in each water sample for the time shown on the directions.
3 Use the colour chart to determine if nitrates are present in your water sample. Record your observations in your table.
4 Dispose of the water in the test tubes as per your teacher's directions.

PART 4: TESTING FOR COPPER

1 Add 2 cm of each water sample, including distilled water, into separate test tubes.
2 Carefully add 5 drops of 5% ammonia to each test tube.
3 If copper is present in the water, a light blue colour will appear. Record your observations in your table.
4 Dispose of the water in the test tubes as per your teacher's directions.

RESULTS

TABLE I8.3

Sample	pH	Salinity	Nitrates	Copper

DISCUSSION

1 What was the purpose of testing distilled water?
2 Identify which tests were qualitative and which were quantitative.
3 Compare the results for each location.
4 Find out what different pH values can indicate about a water supply. Can you propose any reasons for the results you may have obtained?
5 Find out what the presence of salt, nitrates and copper might indicate about the water supply. Can you propose any reasons for the results you may have obtained?

CONCLUSION

Copy and complete:
'The results show that: (*respond to the aim*)'.

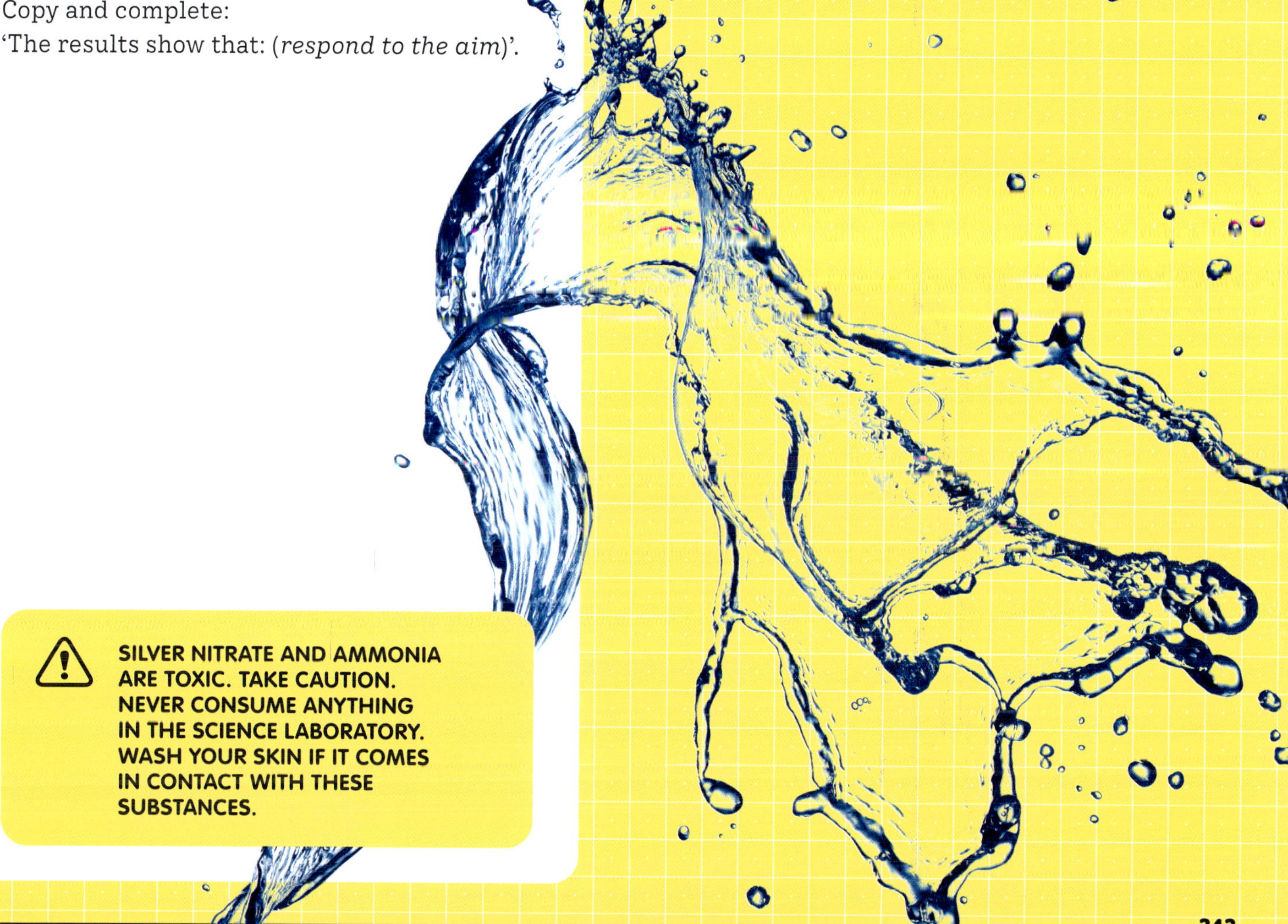

SILVER NITRATE AND AMMONIA ARE TOXIC. TAKE CAUTION. NEVER CONSUME ANYTHING IN THE SCIENCE LABORATORY. WASH YOUR SKIN IF IT COMES IN CONTACT WITH THESE SUBSTANCES.

INVESTIGATION 9.1

Push, pull or twist

AIM

To investigate the effects of forces on different objects

MATERIALS

- elastic band
- tennis ball
- plasticine or Blu Tack
- plastic ruler
- metal ruler
- A4 piece of paper

METHOD

1 Copy the results table into your notebook, adding a title.
2 Stretch the elastic band as far as it will go, then slowly let it go back to its normal shape.
3 Drop the tennis ball so that it bounces off the floor, then catch it before it falls again.
4 Squash the plasticine or Blu Tack into at least three different shapes.
5 Hold one end of the plastic ruler in each hand. Bend the ruler so that it flexes, then let it go back to its normal shape.
6 Repeat step 5 for the metal ruler, if possible.
7 Hold one corner of the paper in each hand. Bring your hands together so that the paper scrunches, then pull them apart until it tears.
8 For each object, record whether the forces you applied acted as a push, pull or twist. Consider all of the different forces that you applied to each object. Record your observations of what else happened when you applied forces.

RESULTS

TABLE I9.1

Object	Push, pull or twist	Effect of applying force
Elastic band		
Tennis ball		
Plasticine or Blu Tack		
Plastic ruler		
Metal ruler		
Piece of paper		

DISCUSSION

1 Were there any situations where you applied more than one force to an object? What was the effect of applying multiple forces?
2 Which objects changed their shape when you applied a force? Was this change temporary or permanent?
3 Did applying force affect the movement of any objects? If so, why?

CONCLUSION

Copy and complete:
'The results show that: (*respond to the aim*)'.

INVESTIGATION 9.2

Blowball

AIM

To investigate the effects of unbalanced forces on a ping-pong ball

MATERIALS

- table
- ping-pong ball
- 4 drinking straws

METHOD

1. Gather into groups of four. Each member of the group takes one straw and kneels at a different side of the table.
2. Place the ping-pong ball in the centre of the table.
3. Take turns to blow through your straws to move the ball around the table.
4. Have one group member blow the ball across the table while the others blow on it in the other direction.
5. Have one group member blow the ball across the table while the others blow on it from the sides.
6. Divide into two pairs, with each pair on a different side of the table. As a pair, try to blow the ball across the table and off the side while the other pair does the same.
7. Record observations in your book. In particular, detail how different sized forces were able to affect the motion of the ball.

QUESTIONS

1. When were the forces on the ball balanced?
2. What happened when you blew on the ball as it moved towards you?
3. What happened when you blew on the ball as it moved at a right angle to you?

INVESTIGATION 9.4

Crash cushions

AIM

To investigate how different materials absorb the impact of forces

MATERIALS

- dynamics cart or large toy car
- motion sensor that can measure force on impact (If this is not available, students can film the impact in slow motion and make observations based on the video footage.)
- inclined plane
- scissors
- tape
- construction paper, newspaper, cardboard boxes, straws, balloons, sponges, polystyrene and other materials

METHOD

1 Copy the results table into your notebook, adding a title and rows as needed.

2 Attach the sensor to the cart so that it measures the force of impact when the cart comes to a stop.

3 Set up the inclined plane as shown.

4 Allow the cart to roll down the plane, collide with the wall and come to a stop. Read the impact measurement from the sensor and record it in your table as 'trial 1'. Repeat the trial and record the second measurement.

5 Conduct at least three more trials. For each trial, create a 10 cm cushion out of different materials, place it next to the wall and move the inclined plane back so there is a 50 cm space between the plane and the cushion. Allow the cart to roll down the plane, hit the cushion and come to a stop. Record the impact measurements in your table.

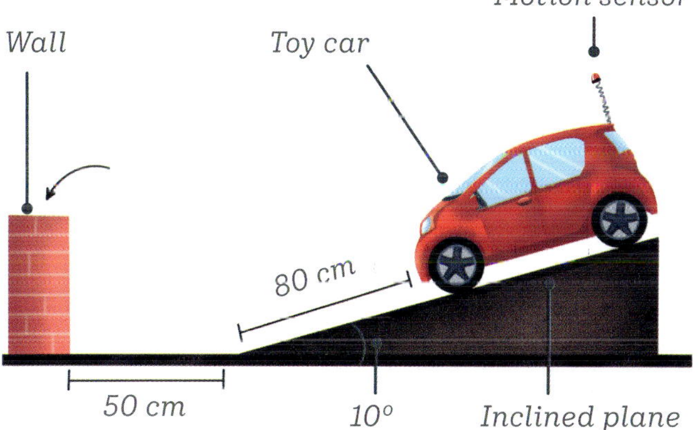

DISCUSSION

1 Which cushion was most effective at reducing the force of impact?

2 Suggest which materials were most useful for reducing impact, and explain why.

3 Suggest ways in which cushioning materials could be used in cars to reduce the impact of a crash.

CONCLUSION

Copy and complete:

'The results show that (*respond to the aim*)'

RESULTS **TABLE I9.4**

Trial	Cushion material	Force of impact (first run)	Force of impact (second run)
1	None		
2			
3			
4			

INVESTIGATION 9.5A

Investigating levers

AIM

To investigate the use of a lever to apply force

MATERIALS

- 30 cm ruler
- 2 paper cups
- tape
- approx. 30 marbles
- small cylinder (e.g. whiteboard marker)

METHOD

1. Copy the results table into your notebook, adding a title.
2. Tape the paper cups to the ends of the ruler so that they sit on top and can hold marbles. Label them A and B.
3. Set the ruler on the fulcrum (the cylinder) at the 15 cm mark.
4. Place 6–10 marbles in cup A.
5. Predict how many marbles you would need to put in cup B in order to lift cup A, and record your prediction in your table.
6. Place marbles one by one in cup B until cup A lifts. Record the number in your table.
7. Set the ruler on the fulcrum (the cylinder) at the 10 cm mark, and repeat steps 4–6.
8. Set the ruler on the fulcrum (the cylinder) at the 20 cm mark, and repeat steps 4–6.

RESULTS **TABLE I9.5A**

Fulcrum position (cm)	Marbles in cup A	Prediction for cup B	Actual marbles in cup B
15			
10			
20			

QUESTIONS

1. At which position did you need to place more marbles in cup B than in cup A?
2. At which position did you need to place fewer marbles in cup B than in cup A?
3. How accurate were your predictions?
4. What did this investigation demonstrate about the use of levers?

INVESTIGATION 9.5B

Investigating pulleys

AIM

To investigate the use of pulleys to apply force

MATERIALS

- spring balance
- suspension beam
- 2 single pulleys
- 2 double pulleys
- 4 m cord
- 1 kg weight
- 1 m ruler

METHOD

1 Copy the results table into your notebook, adding a title.
2 Set up the suspension beam and tie a single pulley to it. Run the 4 m cord through the pulley, then attach the 1 kg weight to one end and the spring balance to the other.
3 Pull the spring balance to lift the weight. Record how much effort is required.
4 Measure how far the cord had to be pulled in order to lift the weight by 1 m. Record the distance.
5 Calculate the mechanical advantage of the pulley by dividing the effort by the weight. Record it in your table.
6 Set up two single pulleys on the beam with the cord running through them, then repeat steps 3–5.
7 Set up two double pulleys on the beam with the cord running through them, then repeat steps 3–5.

QUESTIONS

1 Which number of pulleys gave the greatest mechanical advantage?
2 Which number of pulleys required you to pull the cord the longest distance?
3 Predict the mechanical advantage of an eight-pulley system, as well as the distance the cord would have to be pulled.

RESULTS TABLE I9.5B

Number of pulleys	Effort needed to lift weight	Distance load moves (m)	Distance effort moves (m)	Mechanical advantage
1		1		
2		1		
4		1		

INVESTIGATION 9.6

Heat from friction

AIM

To investigate the connection between friction and heat

MATERIALS

- microfibre cloth
- wire coathanger
- wooden skewer
- plastic rod
- cardboard roll

METHOD

1 Pick up each item to see how warm or cool it is.
2 Firmly rub the plastic rod 20–30 times with the cloth. Carefully feel it and record how its temperature has changed.
3 Repeat step 2 with each of the other items.
4 Record any other observations you make.

QUESTIONS

1 Which item had the greatest change in temperature?
2 Which item had the least change in temperature?
3 Explain how friction caused the change in temperature.

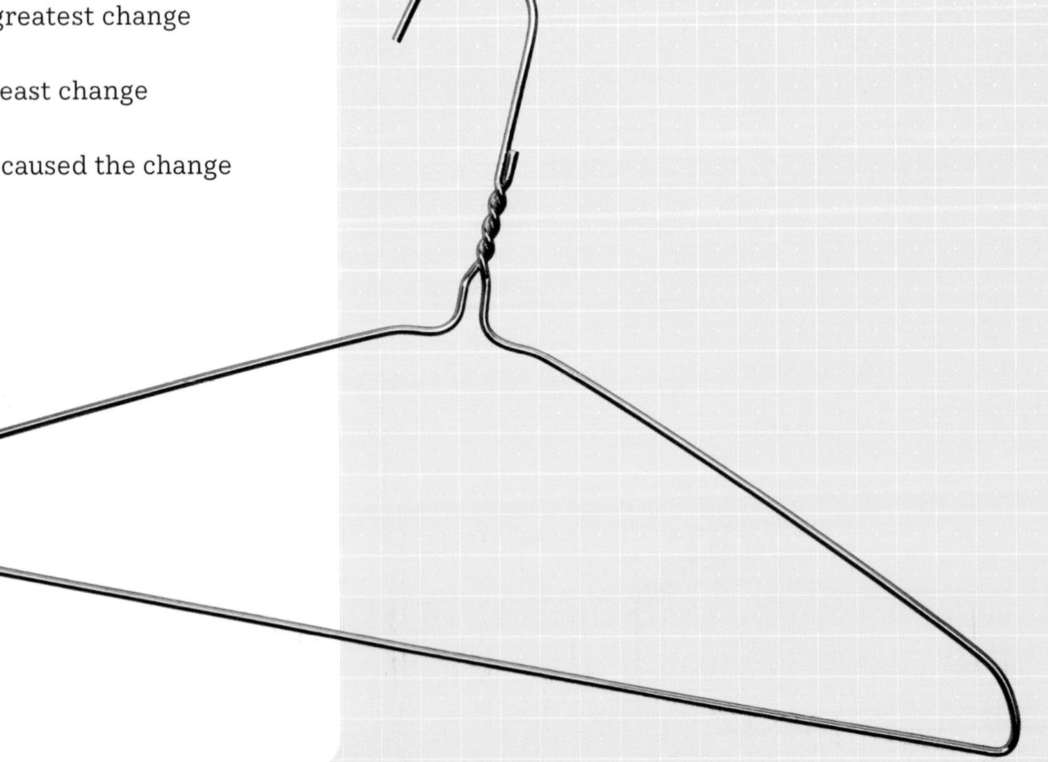

INVESTIGATION 9.7

Friction of materials

AIM

To investigate how different materials affect the application of friction

MATERIALS

- 5 × 10 × 10 cm block of wood
- 1 m wooden board with a smooth finish
- ruler
- stopwatch
- foil
- cotton cloth
- woollen cloth
- various other materials (if desired)
- cooking oil

METHOD

1 Copy the results table into your notebook, adding a title and rows as needed.
2 Prop the wooden board up on some books to create a ramp.
3 Set the wooden block at the top of the ramp. Release the block and time how long it takes to slide down the ramp and hit the floor. Repeat two more times, then calculate the average time.
4 Cover the surface of the board with foil, then repeat step 3.
5 Cover the surface of the board with the cotton cloth, then repeat step 3.
6 Cover the surface of the board with the woollen cloth, then repeat step 3.
7 Cover the surface of the board with any other materials you want to check, then repeat step 3.
8 Cover the surface of the board with cooking oil, then repeat step 3.

DISCUSSION

1 Which material gave the fastest time for the block?
2 Which material gave the slowest time for the block?
3 Explain why the different materials affected the block's slide, referring to the effects of friction.
4 Suggest a material that would have much more friction and much less friction than those you used in this investigation.

CONCLUSION

Copy and complete:
'The results show that: (*respond to the aim*)'.

RESULTS TABLE I9.7

Material	Time 1 (s)	Time 2 (s)	Time 3 (s)	Average time (s)
Wood		1		
Foil		1		

INVESTIGATION 10.2

Magnets and Matchbox cars

AIM

To investigate the behaviour of magnets

MATERIALS

- 2 bar magnets
- 2 Matchbox cars
- tape

METHOD

1 Tape a bar magnet to the roof of each Matchbox car, with the north pole pointing to the front.
2 Position the cars so that they are close together and facing each other, then release them. Record your observations.
3 Remove the bar magnet from one of the cars and replace it so that the south pole is pointing to the front.
4 Repeat step 2.

QUESTIONS

1 What happened when both cars had north poles at the front?
2 What happened when the cars had different poles at the front?
3 Explain the behaviour of the cars in terms of fields.

INVESTIGATION 10.3

Building an electric motor

AIM

To build and investigate a simple electric motor

MATERIALS

- insulated copper wire
- cylinder
- utility knife
- D-cell battery
- 2 metal paperclips
- electrical tape
- bar or round magnet

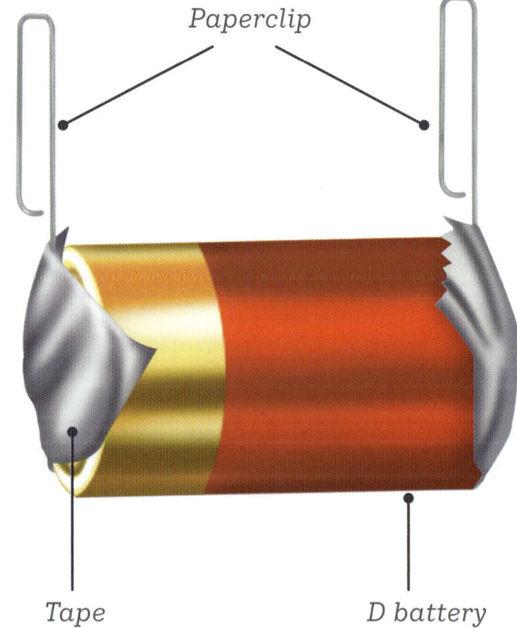

METHOD

1 Wrap the insulated copper wire around the cylinder several times to make a coil, leaving the ends sticking out.
2 Place the wire on a wooden board, and with extreme care scrape the plastic coating away from the wire.
3 Bend one end of each paperclip so that it sticks out. Tape the paperclips to the battery as shown.
4 Slide the coil of wire through the paperclips so that it is supported within the loops. Make sure the coil can spin freely.
5 Hold the magnet over the top of the coil and spin the coil.

QUESTIONS

1 What happened when you spin the coil?
2 Explain what happened to the coil in terms of magnetic fields.
3 Explain how a simple electric motor works, in your own words.

TAKE CAUTION USING THE UTILITY KNIFE. IF YOU CUT YOURSELF, TELL YOUR TEACHER IMMEDIATELY AND SEEK FIRST AID.

INVESTIGATION 10.5

Charging balloons

AIM

To investigate the behaviour of charged objects

MATERIALS

- 2 balloons
- plastic rod
- tissues
- piece of furry fabric
- string

METHOD

1 Tear the tissues into small squares of 1 × 1 cm.
2 Rub the plastic rod with the furry fabric vigorously for 1 minute.
3 Place the tip of the plastic rod close to, but not touching, the tissue squares. Record your observations.
4 Inflate the balloons and tie a 30 cm length of string to each one.
5 Rub the balloon against your hair for 1 minute. Record your observations.
6 Repeat step 5 with a partner, so that each of you now holds a charged balloon. Hold each balloon by the string, and move them close together (but not touching). Record your observations.

QUESTIONS

1 What did you observe happening to the tissue squares?
2 What did you observe happening to the balloons?
3 Explain the behaviour of the tissues and balloons in terms of electrostatic charge.

CONCLUSION

Copy and complete:
'The results show that: (*respond to the aim*)'.

INVESTIGATION 10.7

Measuring gravity

AIM

To investigate the strength of gravitational force

MATERIALS

- 5 × 50 g masses
- spring balance
- mass carrier

METHOD

1. Copy the results table into your notebook, adding a title and rows as needed.
2. Attach a 50 g mass to the bottom of the spring balance. The reading on the spring balance shows the weight of the mass. Record the reading on the spring balance in your table.
3. Repeat step 2, adding another 50 g mass each time, until there is a total mass of 250 g on the spring balance.
4. For each reading, divide the weight by the mass.

DISCUSSION

1. Is there a pattern in your answers?
2. What can you deduce about the force of gravity on Earth?
3. How would this investigation be different if you conducted it on the Moon?

CONCLUSION

Copy and complete:
'The results show that: (*respond to the aim*)'.

RESULTS TABLE I10.7

Mass (g)	Weight (N)	Weight ÷ mass

INVESTIGATION 11.1

Rolling balls

AIM

To investigate the relationship between gravitational potential energy and kinetic energy

MATERIALS

- balls or marbles with different masses
- electronic balance
- inclined plane
- ruler or tape measure
- milk carton or small, empty box

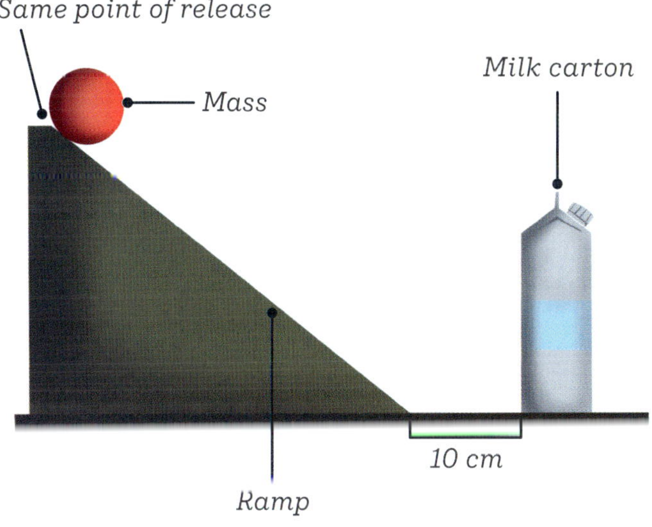

METHOD

1 Copy the results table into your notebook, adding a title and rows as needed.
2 Set up the ramp and the light box or milk carton as shown.
3 Record the mass of each ball.
4 Roll each ball down the inclined plane. Measure the distance that it moves the carton when the ball hits it.
5 Repeat the test five times with each ball.
6 Calculate the average of the five trials.

DISCUSSION

1 Which ball had the most gravitational potential energy?
2 What variable contributed to this?
3 What is the other variable that contributes to gravitational potential energy?
4 Design an experiment that would investigate the variable you identified in question 3 and its relationship to gravitational potential energy.

CONCLUSION

Copy and complete:
'The results show that: (*respond to the aim*)'.

RESULTS TABLE I11.1A

Ball mass (g)	Distance (cm) (trial 1)	Distance (cm) (trial 2)	Distance (cm) (trial 3)	Distance (cm) (trial 4)	Distance (cm) (trial 5)	Average distance (cm)

INVESTIGATION 11.2A

Conduction – heat energy transfer in a solid

AIM

To investigate how heat energy is conducted in a solid

MATERIALS

- metal bar
- wax beads
- retort stand with bosshead and clamp
- Bunsen burner
- matches
- tray

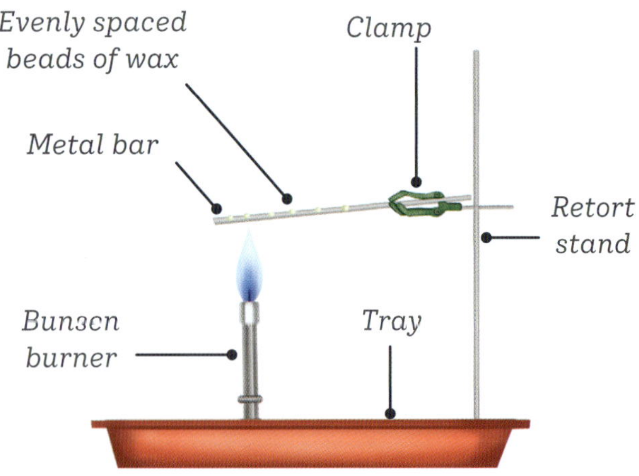

METHOD

1 Set up the apparatus as shown.
2 Light the Bunsen burner.
3 Observe the metal bar and wax beads, and record your observations.

DISCUSSION

1 Explain how heat is conducted in a solid, using your observations from this investigation.
2 Design an experiment to demonstrate that heat energy is transferred through different metals at different rates.
3 Design an experiment to determine which materials are good at conducting heat energy.

CONCLUSION

Copy and complete:
'The results show that: (*respond to the aim*)'.

AN OPEN FLAME IS A HAZARD. TAKE CAUTION. IF YOU BURN YOURSELF, TELL YOUR TEACHER IMMEDIATELY AND PLACE THE BURNT AREA UNDER COLD RUNNING WATER FOR 20 MINUTES.

INVESTIGATION 11.2B

Convection – heat energy transfer in a liquid

AIM

To investigate how heat energy is transferred in a liquid

MATERIALS

- 2 L beaker
- small conical flask
- cold water
- hot water
- red dye
- string or long-handled tongs

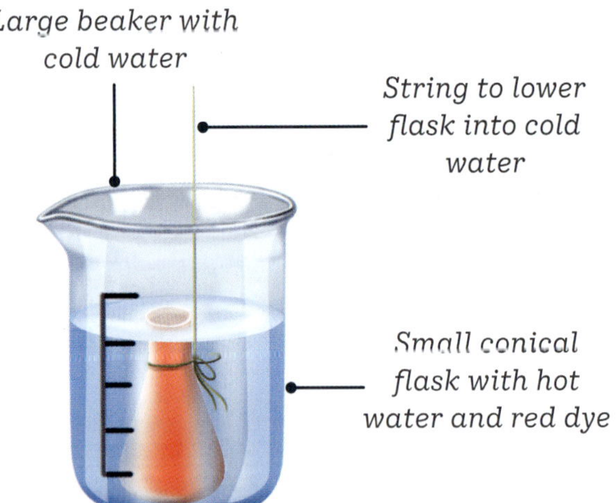

METHOD

1. Add cold water to the beaker until it is 3 cm from the top.
2. Tie the string around the neck of the conical flask.
3. Place a few drops of red dye into the conical flask, then fill it almost to the top with hot water.
4. Using the string and the long-handled tongs, carefully lower the small conical flask into the beaker. Take care not to spill the hot water or disturb the flask too much.
5. Record your observations.

DISCUSSION

1. Explain how convection works in a liquid, using your observations from this investigation.

CONCLUSION

Copy and complete:
'The results show that: (*respond to the aim*)'.

INVESTIGATION 11.2C

Radiation and colour

AIM

To investigate the relationship between colour and transfer of radiant heat energy

MATERIALS

- 6 test tubes
- test-tube rack
- coloured paper (black, white, red, blue and green)
- 6 thermometers
- scissors
- tape
- water
- heat source (sunlight or a heat lamp)

METHOD

1 Copy the results table into your notebook, adding a title and rows as needed.

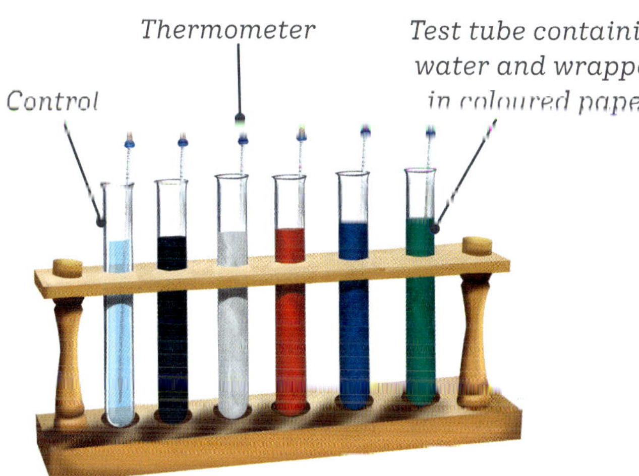

2 Tightly wrap the coloured paper around each test tube and secure with tape. Leave one test tube uncovered as a control.

3 Fill each test tube with the same volume of water and place in the test-tube rack.

4 Place a thermometer in each test tube and allow enough time for the temperature to stabilise.

5 Record this initial temperature in your table at time 0 minutes.

6 Leave the test tubes in sunlight, or near the heat lamp, and record the temperature for each colour every 5 minutes for half an hour.

DISCUSSION

1 Which colours transmitted the most radiant heat energy?

2 Which colours transmitted the least amount of radiant heat energy?

3 Explain what radiant heat energy is, in your own words.

4 Think of one improvement or addition you could make to the method of this investigation.

CONCLUSION

Copy and complete:

'The results show that: (*respond to the aim*)'.

TAKE CAUTION USING THE HEAT LAMP. IF YOU BURN YOURSELF, TELL YOUR TEACHER IMMEDIATELY AND PLACE THE BURNT AREA UNDER COLD RUNNING WATER FOR 20 MINUTES.

RESULTS **TABLE I11.2C**

Time (min)	Temperature (°C)					
	Control	Black paper	White paper	Red paper	Blue paper	Green paper

INVESTIGATION 11.4

Investigating conductors and insulators

AIM

To investigate which materials are electrical conductors and which are insulators in an electrical circuit

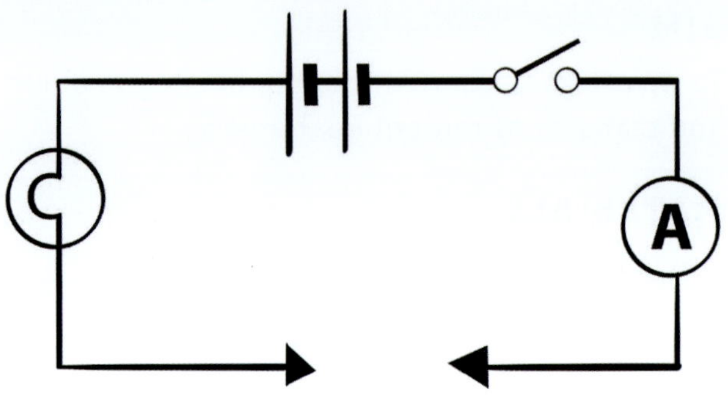

MATERIALS

- selection of materials (such as fabric, iron nail, strip of tin, icy pole stick, strip of copper, chalk)
- variable power supply
- ammeter
- switch
- 6 V, 10 W light bulb
- conducting wires with alligator clips
- conducting wires with suitable attachments for a power pack

METHOD

1. Copy the results table into your notebook, adding a title and rows as needed.
2. Copy the circuit diagram and label all the symbols used.
3. Set up the circuit as shown in the diagram.
4. Set the DC power supply to 6 V.
5. Attach both contacts to one of the materials.
6. Switch on power to the circuit and record your observations of the light bulb. Also record the ammeter reading.
7. Repeat steps 5 and 6 for each material.

DISCUSSION

1. What did you observe when the test material was a good conductor?
2. How did you know which of the materials was an insulator?
3. Based on your table, what do the materials that were good conductors have in common?
4. What do the good insulating materials have in common?

CONCLUSION

Copy and complete:

'The results show that: (*respond to the aim*)'.

ELECTRICITY IS A HAZARD. TAKE CAUTION.

RESULTS TABLE I11.4

Material	Did the light turn on?	Ammeter reading (mA)	Conductor or insulator?

INVESTIGATION 12.1

Energy efficiency

AIM

To investigate how transformations of energy are not completely efficient

MATERIALS

- weight (e.g. 50 g fishing sinker)
- string
- pencil
- desk
- scissors
- tape
- ruler
- large piece of white paper

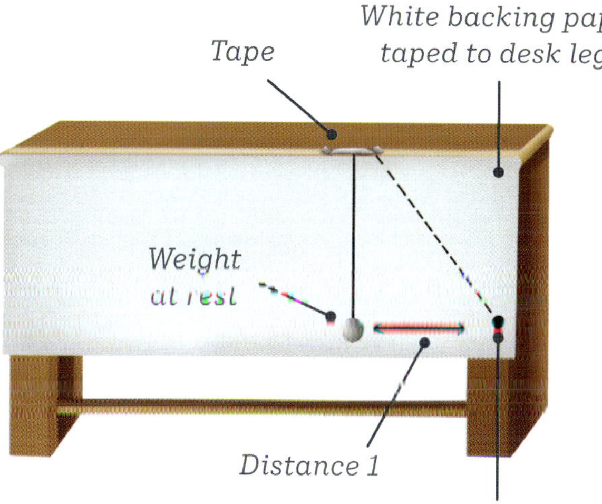

RESULTS

TABLE I12.1

Trial	Start distance (cm)	Finish distance (cm)

METHOD

1. Copy the results table into your notebook, adding a title and rows as needed.
2. Set up the materials as shown and check that the weight can swing freely.
3. Mark the position of the string when the weight is hanging at rest.
4. Pull the weight back a distance from the rest position and mark it on the background paper.
5. Release the weight and let it swing back and forth 10 times.
6. Mark, on the background paper, the place that the weight swings back to on the 10th swing.
7. Record the start and finish distances of the swing in your table.
8. Repeat steps 1–7, releasing the weight from various distances from the rest position.

QUESTIONS

1. What do you notice when you compare the start distance and finish distance values for each trial?
2. What type of energy does the weight have when it reaches the lowest point of its swing?
3. What type of energy does the weight have before it is released?
4. How can you explain the decrease in these values?
5. Draw a diagram that illustrates the transformation of energy in the weight.

INVESTIGATION 12.2

Insulation and heat transfer

AIM

To investigate the relationship between insulating material and the transfer of heat

MATERIALS

- selection of insulating materials, e.g. polystyrene, wool, aluminium
- 250 mL beakers (one for each material)
- thermometers (one for each beaker)
- watchglasses (one for each beaker)
- hot water
- scissors
- tape

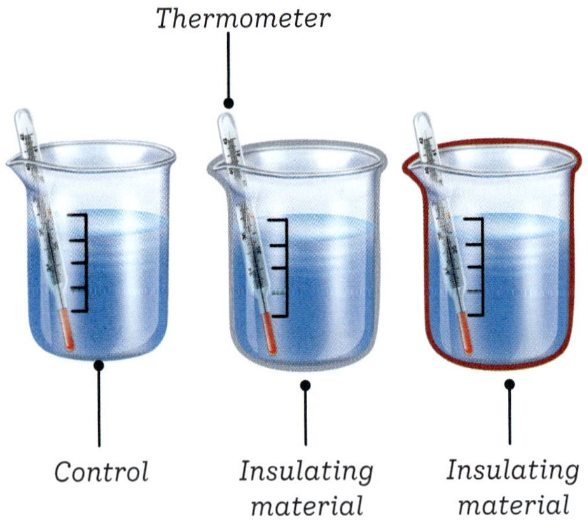

METHOD

1. Copy the results table into your notebook, adding a title and rows as needed.
2. Wrap the insulating material around each beaker and secure with tape.
3. Create a lid for each beaker with the insulating material.
4. Add equal volumes of hot water to the beakers.
5. Place a thermometer in each beaker and secure the watchglass lid with tape.
6. Allow enough time for the temperature reading on the thermometer to stabilise. Record this initial temperature in your table.
7. Record the water temperature for each beaker every 5 minutes for half an hour.

DISCUSSION

1. Which material was the best insulator? Give evidence to support your answer.
2. Which material was the worst insulator? Give evidence to support your answer.
3. Choose the results from one material for your table to graph.
4. Suggest a design for a thermos that aims to keep tea hot for as long as possible. What kind of materials could be used?

CONCLUSION

Copy and complete:
'The results show that: (*respond to the aim*)'.

RESULTS **TABLE I12.2**

Time (min)	Water temperature (°C)			
	Control	Material A	Material B	Material C
Initial				
5				
10				

INVESTIGATION 12.3

Atmospheric convection and the inversion layer

TEACHER DEMONSTRATION

AIM

To model how thermal pollution affects circulation in the atmosphere

MATERIALS

- 2 identical small, wide-mouthed jars
- stirring rod
- scissors
- large, shallow tray
- hot water from a tap
- very cold water
- red food colouring
- blue food colouring
- cardboard or waxed paper

METHOD

PART 1: INVESTIGATING CONVECTION

1 Your teacher will half fill one of the jars with iced water. They will add 1 drop of blue food colouring. Observe how it mixes. Your teacher will mix the food colouring in completely if needed.
2 Your teacher will slowly add more iced water to the jar until there is a slight bulge of water over the top of the rim of the jar. Then they will set this jar in the shallow tray.
3 Your teacher will carefully half-fill the other jar with hot water and add 1 drop of red food colouring. Observe how it mixes. Your teacher will mix the food colouring in completely if needed.
4 Observe as your teacher fills this jar with hot water all the way to the top and places the card or waxed paper over it. They will tap the card gently so that the card forms a seal with the rim of the jar.
5 Your teacher will take the jar with hot water in one hand, with the other hand on the card. They will quickly, and without hesitating, pick it up and flip the jar upside down.
6 Your teacher will place the card and the jar of hot water on the jar of cold water so that the rims line up exactly.
7 Your teacher will carefully remove the card without disturbing the jars. Record your observations.

PART 2: INVESTIGATING INVERSION

1 Your teacher will empty both jars and rinse them.
2 Watch as they repeat the steps in part 1, but this time put the jar with the red coloured hot water in the tray and put the card on top of the jar with the blue coloured cold water. They will turn the blue jar upside down and put it on top of the jar with red coloured hot water. Record your observations.

DISCUSSION

1 What happened in part 1 of the investigation? What happened in part 2, and how was it different?
2 Which jar represented the atmosphere with thermal pollution, and which represented the atmosphere with smog?
3 Why does the water mix so quickly when the glass of hot water is on the bottom?
4 If this situation happened in a large city, what would the air be like for the people living there?

CONCLUSION

Copy and complete:

'The results show that: (*respond to the aim*)'.

INVESTIGATION 13.1

Observing and classifying

AIM

To investigate different ways of classifying living things

MATERIALS

- notebook
- camera
- paper

METHOD

1 Visit an area near your school or home where there are animals and plants. This could be a natural space, such as a beach or area of bushland, or a built space, such as a park.
2 Record at least 10 different plants that you find in the area. Photograph each one if possible.
3 Record at least 10 different animals that you find in the area. Photograph each one if possible. If the animals belong to other people, such as pets being walked, ask permission before taking the photo.
4 In class, print out each of your photos and label it with the name of the organism. Alternatively, write the name of each animal or plant you recorded on an individual slip of paper.
5 Sort all of the organisms you found into 3–5 different groups. The organisms in each group should have similar features.

QUESTIONS

1 What features did you use to sort the organisms?
2 Were there any organisms that could have fitted into two or more groups? If so, how could you have changed the groups so that each organism only fitted into one group?
3 Compare your sorting groups to those of other students. How did they sort the organisms differently?

INVESTIGATION 13.3

Supermarket classification key

AIM

To investigate the design and use of classification keys

MATERIALS

- notebook

METHOD

1 Working in pairs, design a classification system for items you would find in a supermarket. Choose a particular set of items – such as fruits and vegetables, dairy, hygiene products, snacks – rather than try to accommodate every item. Your key should have three or four levels of classification.
2 Visit a local supermarket (or look online) and apply your key to at least 20 different items. Record where each item falls in your key.

QUESTIONS

1 Explain how you designed your key.
2 Did you find any items that should have fitted the key but didn't? If so, suggest a way to adjust the key so that the item would have fitted.
3 Compare your key to those of other students. How did they design theirs differently?

INVESTIGATION 13.4

Investigating features of marine animals

AIM

To investigate the physical and skeletal features of different marine animals

MATERIALS

- 1 whole fish
- 1 large, whole prawn
- 1 squid
- hand lens
- dissecting board
- dissecting kit
- disposable gloves

METHOD

1. Place the fish, squid and prawn on separate dissection boards.
2. In groups of three or four, observe the external features of the fish and record your observations.
3. Open the fish's mouth and look inside, using the hand lens to magnify the details. While looking in the mouth, use a dissection probe to open the gill covers on the head. Record your observations.
4. Carefully cut open the fish lengthwise so that you can see its skeleton. Record your observations.
5. Use a probe to open the prawn's mouth, and look inside through the hand lens. Try to identify whether the prawn has gills and, if so, where they are located. Record your observations.
6. Feel the outside of the prawn, then peel it and cut it in half. Record your observations.
7. Use a probe to open the squid's mouth and look inside through the hand lens. Try to identify whether the squid has gills and, if so, where they are located. Record your observations.
8. Feel the outside of the squid, then peel it and cut it in half. Record your observations.

QUESTIONS

1. Which of the three marine animals are vertebrates (with skeletons) and which are invertebrates (without skeletons)?
2. Draw a series of diagrams showing the external and internal features of each of the specimens.
3. What differences are there between the gill structures of the three animals?
4. How would you design a key to classify marine animals based on their skeletons and gill structures?
5. What internal physical features does each specimen contain that are unique or similar to each other?

WEAR EYE AND HAND PROTECTION. TAKE CAUTION WITH CUTTING IMPLEMENTS. DISPOSE OF ALL MATERIALS AS DIRECTED BY YOUR TEACHER. WASH YOUR HANDS AFTERWARDS.

INVESTIGATION 13.6

Growing mould

AIM

To investigate the conditions in which mould and fungus grow

MATERIALS

- 4 pieces of bread
- assorted kitchen scraps (e.g. fruit peels, carrot shavings, coffee grounds)
- disposable gloves
- water
- plastic sandwich bags
- paper plates
- hand lens

METHOD

1. Splash the pieces of bread with water so that they are damp. Place each piece in an individual plastic bag, but don't seal the bags.
2. Place two bags in an area where they will receive direct sunlight. Place the other two somewhere that has about the same temperature but that doesn't receive direct sunlight.
3. Place the food scraps on two paper plates. Put one plate next to bread specimens sunlight, and the other plate next to the bread in the darker spot.
4. Leave the bread and scraps alone for 2–4 days, then observe any spots of mould growing. Record your observations in your notebook, and sketch any stalks that are growing.
5. Wait a few more days until there is a large patch of mould on the bread and scraps. Observe the large patch using the hand lens and record your observations.

DISCUSSION

1. Which set of bread had the most growth on it, the one in the light or the one in the dark? Why do you think this is?
2. Did some kitchen scraps have more mould than others? Which types of scraps grew more mould?
3. What kinds of structures and growths did you observe through the hand lens?

CONCLUSION

Copy and complete:
'The results show that: (*respond to the aim*)'.

TAKE CAUTION WHEN WORKING WITH MOULD TO AVOID SPREAD OR CROSS CONTAMINATION. DISPOSE OF IT AS DIRECTED BY YOUR TEACHER. WASH YOUR HANDS AFTER THIS INVESTIGATION.

NEVER CONSUME ANYTHING IN THE SCIENCE LABORATORY.

INVESTIGATION 14.1

Examining cells under a microscope

AIM

To investigate the cells in a slide of onion tissue under a microscope

MATERIALS

- onion
- light microscope
- microscope slides and coverslips
- chopping board and knife
- tweezers
- dropper bottle of dilute iodine solution
- paper towel
- blunt dissecting needle or sharp pencil

METHOD

1 Watch your teacher demonstrate how to create a microscope slide.
2 Place 1 drop of iodine solution in the centre of the microscope slide.

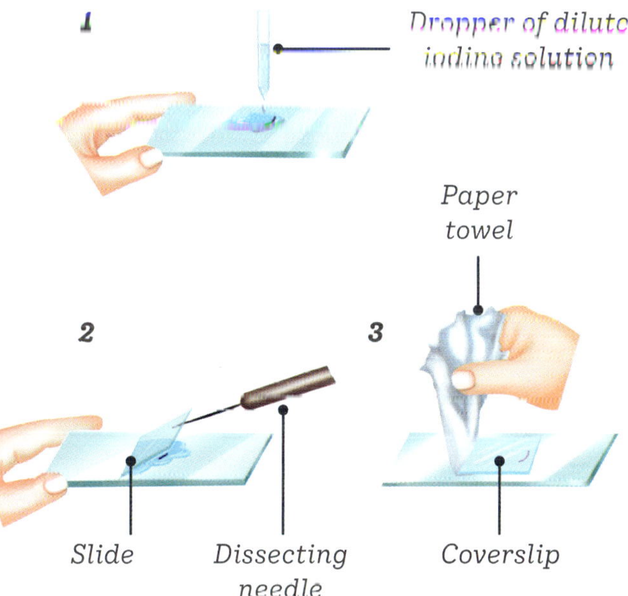

3 Using the knife and chopping board, cut a small piece from an onion ring.
4 Using your fingernails or tweezers, carefully peel the membrane away from the inner curve of the onion piece.
5 Place this membrane on the drop of iodine on your slide, trying not to let it curl up. Place a second drop of iodine solution on top of the onion membrane.
6 Use the blunt dissecting needle (or a sharp pencil) to carefully and slowly lower a coverslip over the onion membrane as shown.
7 Use a corner of a paper towel to gently soak up any extra liquid that squeezes out from under the coverslip.
8 Examine your slide under the microscope. Record your observations.

QUESTIONS

1 What features of the onion cells could be seen?
2 Suggest why other organelles could not be seen.
3 What was the role of the iodine in this investigation?
4 Name two differences between a plant and animal cell.

TAKE CAUTION USING THE KNIFE. IF YOU CUT YOURSELF, TELL YOUR TEACHER IMMEDIATELY AND SEEK FIRST AID.

INVESTIGATION 14.3A

Cellular respiration in yeast

AIM

To investigate the temperature that is most suitable for yeast respiration

MATERIALS

- 3 × 500 mL water bottles (empty)
- yeast
- sugar
- hot and cold water
- ice
- filter funnel
- thermometer
- 3 balloons
- matches
- teaspoon

METHOD

1. Copy the results table into your notebook, adding a title.
2. Add 2 teaspoons of yeast and 5 teaspoons of sugar to each empty bottle using a filter funnel.
3. Prepare water of approximately 10°C, 35°C and 60°C using ice, tap and hot water from a kettle. Add 100 mL of 10°C water to one bottle, 35°C water to the second bottle and 60°C water to the third.
4. Shake each bottle to mix the contents, and mark the level of the mixture in each bottle.
5. While you hold each bottle, have a classmate stretch a balloon over the mouth of each.
6. Measure the distance the yeast mixture rises every 5 minutes. Record your observations in your table.
7. Remove the balloon that has expanded to the largest size, and hold it so that the gas does not escape.
8. Release the gas to a lighted match and record what happens.

DISCUSSION

1. Which temperature created the most yeast activity? Use your results to justify your answer.
2. Which temperature created the least yeast activity? Use your results to justify your answer.
3. What was used as a reactant in respiration, and what product was formed?
4. Explain the effect of the gas on the lit match.

CONCLUSION

Copy and complete:
'The results show that: (*respond to the aim*)'.

TAKE CAUTION USING MATCHES. NEVER LIGHT A MATCH UNLESS INSTRUCTED.

RESULTS TABLE I14.3A

Temperature	Yeast activity (cm) every 5 min					Observations
10°C						
35°C						
60°C						

INVESTIGATION 14.3B

Energy from food

AIM

To investigate which foods release the most energy when burnt

MATERIALS

- piece of bread
- selection of other foods (e.g. pasta, cheese, potato, apple, broccoli, carrot, meat, fish)
- test tube
- water
- Bunsen burner
- heatproof mat
- tripod
- gauze mat
- retort stand with bosshead and clamp
- thermometer
- wooden handle with sharp metallic end (probe), or tongs
- measuring cylinder

METHOD

1 Copy the results table into your notebook, adding a title and rows as needed.
2 Cut all the food to the same size.
3 Add 5 mL of water to a test tube and secure the test tube upright using the retort stand, bosshead and clamp.
4 Put a thermometer inside the test tube and measure the initial temperature.
5 Hold the piece of bread with the probe or tongs.
6 Set up the Bunsen burner on the heatproof mat, with the tripod and gauze mat. Light the Bunsen burner and heat the bread over the flame until it ignites.
7 Put the burning bread under the test tube of water and note the highest temperature the water reaches. Record this in your table, along with the total change in temperature.
8 Repeat steps 3–7 with other foods.

DISCUSSION

1 Which food produced the highest temperature and energy change?
2 Explain why this food might have generated more heat when burned.
3 Identify three errors that may have resulted in inaccurate results in this experiment.
4 Suggest improvements to the method that would address each of these errors.

CONCLUSION

Copy and complete:
'The results show that: (*respond to the aim*)'.

AN OPEN FLAME IS A HAZARD. TAKE CAUTION. IF YOU BURN YOURSELF, TELL YOUR TEACHER IMMEDIATELY AND PLACE THE BURNT AREA UNDER COLD RUNNING WATER FOR 20 MINUTES.

RESULTS TABLE I14.3B

Food item	Initial temperature (°C)	Highest temperature (°C)	Change in temperature (°C)
Piece of bread			

INVESTIGATION 14.5A

Observing unicellular organisms

AIM

To investigate unicellular organisms in pond water using a microscope

MATERIALS

- pond water
- light microscope
- microscope slides
- coverslip
- dropper
- paper towel

METHOD

1 On a clean microscope slide, place 1 or 2 drops (depending on the size of the drop of pond water).
2 Gently add a coverslip and wipe away any extra water.
3 Put the prepared slide on the microscope and clip it down using the stage clips.
4 Observe your slide under low magnification. Record your observations.
5 Increase the magnification to high and observe the sample again. Record your observations.
6 Draw what you saw under low and high power.

QUESTIONS

1 Why was a microscope necessary to view these organisms?
2 How many organisms did you see under low magnification?
3 What happened when you increased the magnification of the field of view to high?
4 Describe the movement of the organisms. Were they moving fast or slow?
5 Tally how many of each organism you saw. Draw a column graph representing the number of organisms in a drop of pond water.

INVESTIGATION 14.5B

Observing specialised cells in multicellular organisms

AIM

To investigate specialised cells found in multicellular organisms

MATERIALS

- microscope or bioviewer
- prepared microscope or bioviewer slides of plant and animal cells

METHOD

1 Your teacher will prepare and distribute slides of plant and animal cells.
2 Place each prepared slide on the stage of the microscope and clip it in place.
2 Observe each slide under low power and then under high power.
2 Record your observations. Draw your observations and label any cell features that you see.

QUESTIONS

1 Were all the cells you drew similar in structure? How were the cells different?
2 Suggest the functions of the cells you studied.
3 Looking at the structure of the cells, which of these cells are plant cells and which of them are animal cells?

INVESTIGATION 14.7

Water transport in plants

AIM

To investigate how water is transported in plants

MATERIALS

- 3 celery sticks
- food colouring
- 3 × 200 mL beakers
- knife
- magnifying glass
- water
- gloves
- marker pen

METHOD

1 Label the beakers A, B and C. Add 100 mL of water to beakers A and B and leave beaker C empty. Add enough food colouring to beaker A to get an intense, dark colour.
2 Cut 2 cm off the bottom of each stick of celery.
3 Place one celery stick in each beaker, then leave them to stand overnight.
4 The next day, observe the leaves in the three celery sticks and record your observations.
5 Wearing gloves, remove the celery that was in the coloured water and slice it crosswise and lengthwise. Observe the pathway of the food colouring.
6 Draw what you see in your notebook.

DISCUSSION

1 What was the purpose of beakers B and C?
2 Explain what happened to the celery in beaker A in terms of the plant's water transport system.
3 What result did you obtain in beaker C? Why do you think this happened?
4 Describe how water is transported in plants, using observations or diagrams from your investigation to support your answer.

CONCLUSION

Copy and complete:
'The results show that: (*respond to the aim*)'.

TAKE CAUTION USING THE KNIFE. IF YOU CUT YOURSELF, TELL YOUR TEACHER IMMEDIATELY AND SEEK FIRST AID.

INVESTIGATION 15.1

Photosynthesis and respiration

AIM

To investigate the process of photosynthesis in plants

MATERIALS

- gloves
- water
- water weeds
- bromothymol blue solution
- colour scale for bromothymol blue pH
- measuring cylinder
- conical flask
- straws
- 3 test tubes with stoppers
- test-tube rack
- aluminium foil
- sun lamp

METHOD

1. Pour 75 mL of bromothymol blue into the flask. Record the colour of the solution.
2. Using a straw, slowly blow into the flask until the solution just turns yellow.
3. Check the colour of your solution against the colour scale. Record the colour and the pH value.
4. Pour an equal amount of the solution into three test tubes. Label them A, B and C.
5. Put a 5 cm water weed in tube A and stopper the tube. Do the same with tube B, then wrap it in aluminium foil so that no light can enter. Stopper tube C without adding anything to it.
6. Place the tubes upright in the rack, and place the rack about 20 cm from the sun lamp. Turn on the lamp, making sure the light reaches all three tubes equally, and leave them for 24 hours.
7. After removing the water weeds and foil, compare the solutions in all three tubes to the colour scale. Record your observations for each tube.

DISCUSSION

1. Why was tube C left empty except for the solution?
2. What happened in each test tube?

CONCLUSION

Copy and complete:
'The results show that: (*respond to the aim*)'.

TAKE CAUTION WHEN USING BROMOTHYMOL BLUE. USE HAND AND EYE PROTECTION. NEVER CONSUME ANYTHING IN THE SCIENCE LABORATORY.

INVESTIGATION 15.4

Dissecting a flower

AIM

To investigate the physical features of flowering plants

MATERIALS

- 1 large, whole flower
- hand lens
- dissecting board
- dissecting kit
- disposable gloves

METHOD

1 In groups of 3 or 4, observe the external features of the flower and record your observations.
2 Locate both parts of the stamen on your plant (the anther and filament). Record your observations.
3 Locate all three parts of the pistil on your plant (stigma, style and ovary). Cut open the ovary and count the number of eggs inside. Record your observations.

QUESTIONS

1 Why is it important for the anthers to be towards the top of the flower?
2 How many eggs were inside the plant's ovary? How could you tell they were eggs?
3 Explain how pollen would normally move from the male to female organs of the flower.

WEAR EYE AND HAND PROTECTION. TAKE CAUTION WITH CUTTING IMPLEMENTS. DISPOSE OF ALL MATERIALS AS DIRECTED BY YOUR TEACHER. WASH YOUR HANDS AFTERWARDS.

INVESTIGATION 15.5

Dissecting a sheep's heart

AIM

To investigate the physical features of a sheep's heart

MATERIALS

- 1 whole sheep's heart
- hand lens
- dissecting board
- dissecting kit
- disposable gloves
- paper towel

METHODS

1 In groups of 3 or 4, observe the external features of the heart and record your observations.
2 Locate the right side of the heart – it will feel softer than the left side. Cut it open using a scalpel to expose the right atrium and ventricle. Record your observations.
3 Cut open the left side of the heart in the same way. Record your observations and draw a diagram at each stage of the dissection.
4 Locate the tendons holding the heart valves. Use the probe to open the valves on both sides. Record your observations.
5 Find the large artery that leaves the left ventricle. Push the probe up the artery and record your observations.

QUESTIONS

1 Suggest why there would be a difference in the thickness of the left and right ventricle walls.
2 What function do the tendons have in the heart?
3 Describe the large artery coming out of the left ventricle. What function does this artery have?

WEAR EYE AND HAND PROTECTION. TAKE CAUTION WITH CUTTING IMPLEMENTS. DISPOSE OF ALL MATERIALS AS DIRECTED BY YOUR TEACHER. WASH YOUR HANDS AFTERWARDS.

INVESTIGATION 16.1

The contagion game

AIM

To investigate how diseases can spread through populations

MATERIALS

- small scraps of paper
- box to store scraps
- notecards or index cards
- pencils or pens
- masking tape

METHOD

1 Write your name on a scrap of paper and put it in the box. While you do this, your teacher will use the masking tape to divide the class into three areas, labelled 'susceptible', 'infected' and 'recovered'.
2 Write the numbers 1 to 5 down the side of your index card.
3 Walk up to another student, say hello and shake their hand. Both of you then write the name of the student you shook hands with on your index card, in the 1 position.
4 Repeat step 3 with four more students, writing their names in the 2–5 positions on your card. When your card is full, stand and wait in the 'susceptible' area.
5 When everyone is in the 'susceptible' area, your teacher will draw a name from the box. That student was infected with 'handshake disease'! Handshake disease is temporary but contagious, and always affects the next three people you shake hands with.
6 The infected student moves to the 'infected' area and reads out the first name on their card. That student also moves into the 'infected' area.
7 Each student in the 'infected' area reads the next name on their card. Those students move into the 'infected' area.
8 Continue reading names out as students move into the 'infected' area. When a student has read out the first three names on their card, they are cured and move into the 'recovered' area.
9 Continue playing until there is no one in the 'susceptible' area or until everyone in the 'infected' area has read out the first three names on their card.

QUESTIONS

1 Did anyone in the 'infected' area read out the name of a student who was already in that area? If so, what does this represent?
2 Did any students stay in the 'susceptible' area and never move into the 'infected' area? If so, what does this represent?
3 How does this game model the way diseases move through populations?

INVESTIGATION 16.2

Treating muddy water

AIM

To investigate how muddy water can be treated to remove contamination

MATERIALS

- gloves
- water containing clay or mud
- aluminium potassium sulfate (alum)
- limewater
- chlorine
- teaspoon
- stirring rod
- two beakers
- plastic flowerpot (must have a hole in the bottom)
- tripod
- sand
- gravel

METHOD

1. Copy the results table into your notebook, adding a title.
2. Pour 150 mL of muddy water into the beaker. Record its appearance and odour.
3. Add half a teaspoon of alum and 10 drops of limewater to the beaker of muddy water and stir.
4. Allow the water to stand so that floc forms and then settles at the bottom of the beaker. Record the appearance and odour of the water.
5. Place the flowerpot on the tripod, with the second beaker underneath. Pour a 2–3 cm layer of gravel into the bottom of the flowerpot, and a 2–3 cm layer of sand on top.
6. Pour the muddy water into the flowerpot and let it filter through the layers, through the hole and settle in the second beaker. Record its appearance and odour.
7. Add several drops of chlorine to the water and stir. Record the water's appearance and odour.

RESULTS TABLE I16.2

Treatment stage	Appearance of water	Odour of water
Untreated		
First		
Second		
Third		

QUESTIONS

1. At which stage of treatment was the water cleanest?
2. How did the smell of the water change during treatment?
3. How could these treatment stages be applied to sewage or other waste materials?

CHLORINE AND LIMEWATER ARE IRRITANTS. TAKE CAUTION. USE EYE AND HAND PROTECTION. NEVER CONSUME ANYTHING IN THE SCIENCE LABORATORY.

INVESTIGATION 17.1

Modelling a pond ecosystem

AIM

To investigate a pond ecosystem using a model

MATERIALS

- clear glass jar with lid
- sand
- pond water containing water fleas, water snails and microscopic organisms
- *Myriophyllum* (water weed)

METHOD

1. Add sand to your jar to form a 2–3 cm layer.
2. Slowly fill your glass jar with pond water until it is almost full. Make sure you have some water fleas and a snail in your jar.
3. Add 2 or 3 pieces of the plant to your jar.
4. Place the lid on your jar and seal it. Weigh the jar and record the initial mass in your notebook.
5. Place your jar near a window, but out of direct sunlight.
6. Every few days for 2 weeks, make observations about your ecosystem and re-weigh the jar to record the mass. Record your observations.
7. Draw a diagram of your model ecosystem.

QUESTIONS

1. What changes did you observe in your ecosystem over the 2 weeks?
2. Where do the different organisms gain their energy to survive in the jar?
3. Why was there a change in mass? Suggest a reason for this.
4. Draw one food chain for this ecosystem.

INVESTIGATION 17.2

Observing ecosystems

AIM

To investigate activity in a local ecosystem

MATERIALS

- pen and paper
- gardening gloves
- camera (optional)

METHOD

1 Find a small ecosystem (e.g. tree, pond, rose bush) in or near your school.
2 Record all the different plants in the ecosystem.
3 Look for small organisms living on the plants. Look on the underside of leaves, in the branches and grasses. Record the names of any organisms you find. Photograph your findings if possible.
4 Using gardening gloves, carefully turn over fallen leaves, and look under rocks and soil. Record any organisms that you observe. Photograph your findings if possible.
5 Sit quietly and record the different organisms you can see and hear in this ecosystem. Some organisms may move in and out of the ecosystem or may be nocturnal. Photograph your findings if possible.

QUESTIONS

1 Identify one producer and one consumer you found.
2 What is another name for a secondary consumer? Did you find any secondary consumers?
3 Did you see any evidence that decomposers were present in your ecosystem? Explain.
4 Draw two food chains using the information you gathered.

BEWARE OF DANGEROUS ORGANISMS SUCH AS SPIDERS AND STINGING PLANTS IN YOUR ECOSYSTEM. MAKE SURE YOU WEAR GARDENING GLOVES.

INVESTIGATION 17.3

Yeast as a decomposer

AIM

To investigate the action of yeast as a decomposer

MATERIALS

- pieces of banana, cut to 1 cm thickness
- sachet of active yeast
- 4 Petri dishes with lids
- water
- tweezers
- marker pen
- hand lens

METHOD

1 Using tweezers, place a piece of banana in each Petri dish. Add 1 mL of water to each dish.
2 Place the lids on two of the Petri dishes. Label one of these as 'no yeast, cold' and the other as 'no yeast, warm'.
3 Open the yeast sachet and sprinkle equal amounts of yeast evenly over the banana in the remaining two Petri dishes.
4 Place the lids on these two Petri dishes. Label one as 'yeast, cold' and the other as 'yeast, warm'.
5 Place the two Petri dishes labelled 'warm' in a warm place chosen by your teacher, and the two labelled 'cold' in a cold place chosen by your teacher.
6 Make and record a prediction about what will happen to the contents of each Petri dish over the next five lessons.
7 Record observations for each Petri dish over the next five lessons. Use a hand lens to look at the Petri dishes containing yeast.

QUESTIONS

1 Does decomposition happen faster in warm or cold conditions?
2 Where is the yeast getting oxygen and food from?
3 Some people have compost bins in their back yards to return nutrients into their vegetable gardens. How do compost bins prepare the nutrients? Why is this a benefit to a vegetable garden?
4 Often compost bins are black, have holes on the sides and contain earthworms. How do each of these features help decomposers?
5 Why was it important to use tweezers and keep the lid sealed during the investigation?

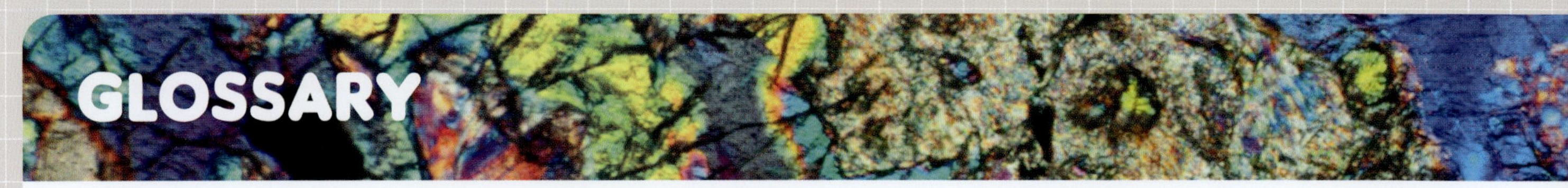

acceleration an increase in speed in the direction of movement

accurate very close to the true, exact value

adhesion the 'sticking' of molecules to other substances

aerobic respiration how living organisms produce energy using oxygen

agriculture the science or practice of farming

algae organisms living in water that make food by photosynthesis

alloy a mixture of a metal and one or more other elements

angiosperm a plant that produces seeds in fruit or flowers

annular ring-shaped

antibiotic a substance that kills or slows the growth of bacteria

aqueous containing water

aquifer a layer of rock that contains water

archaeologist someone who studies history by examining sites and objects

astronomer a scientist who studies space and the objects within it

atmosphere the mixture of gases surrounding Earth

atom the smallest unit of an element

attractive force a force that pulls objects towards each other

axis a real or imaginary line through the centre of an object

bacteria microscopic, single-celled organisms that can live in a range of environments

binary fission a process where very simple cells divide into identical halves

biodegradable able to be broken down, for example by bacteria

biodiversity the variety of organisms in an ecosystem

biofuel a substance produced from living things that can be burnt to create energy

biopsy tissue removed from a living body for testing

biosphere the parts of Earth where living things are found

blood vessel tube such as a vein or artery that carries blood in the body

body system a group of organs working together

boiling point the temperature when something changes from a liquid to a gas

bore a pipe or well dug into an aquifer to access water

brittle not able to be bent; will break if stressed

Bunsen burner a piece of equipment used in science that produces a single open gas flame

bryophyte a plant that doesn't contain vascular tissue

cancer the uncontrolled growth of cells in some part of the body, which then spread to other body parts

carnivore an organism that eats only meat

cartilage connective tissue that holds bones together

cast an object created when sediment or minerals fill a mould

cathode ray tube a glass tube with a beam of electrons passing through it

cell the smallest functional unit of an organism

cell membrane a thin layer around a cell that controls the substances coming in and out

cell wall a stiff layer around a plant cell that supports it

centrifuge a machine that spins fluid samples very quickly so that they separate into parts of different densities

charge the amount of electrical potential energy

chemical bond a force that holds atoms together as molecules and compounds

chemical change a change in properties with a new substance formed

chemical formula uses chemical symbols to show the number of atoms of each element in the compound

chemical reaction a process in which one or more substances are changed to new substances

chemical sedimentary rock sedimentary rock formed from layers of mineral crystals that have crystallised from water

chemical symbol a symbol of one or two letters used to represent an element

chemist a scientist who studies elements and compounds

chemotherapy a treatment of diseases such as cancer using chemical substances

chlorophyll the green pigment in chloroplasts that enables photosynthesis

chloroplast an organelle in a plant cell that allows plants to make food

chromosome a thread-like molecule of genetic information in the nucleus of a cell

circulate move freely or continuously through an area

classification the process of sorting things into groups or classes

clastic sedimentary rock sedimentary rock made of sediments cemented together

coefficient of friction a value which indicates how easily an object moves when interacting with another material

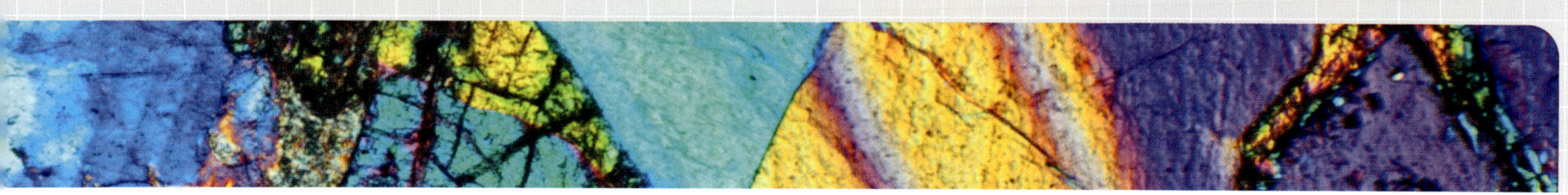

cohesion the 'sticking' of molecules to each other

collaboration working cooperatively together

component a part of an electric circuit in an electrical device

compound a substance containing atoms of two or more elements bonded together in fixed proportions

compress squash into a smaller space

condensation changing from a gas to a liquid

condenser a glass tube cooled by water that cools a gas to become a liquid

conduction transfer of heat from one place to another in solids

conductor a material that is good at transferring heat or electricity

conglomerate rock sedimentary rock made of large, rounded pebbles and fragments cemented together

consumer an organism that gains energy by consuming other organisms

contact metamorphism the process of change that happens to a rock over small areas, often near volcanoes

contract get smaller, shrink

controlled variables all the things that need to stay the same during an investigation

controversial creating public disagreement and debate

convection transfer of heat from one place to another in gases and liquids

core Earth's central layer, which has a liquid outer core and a solid inner core

corrosion breaking down or destruction of a substance, especially a metal, through a chemical reaction

crude oil oil that has not yet been separated into useable petroleum products

crust Earth's thin outer layer

crystal a solid substance made up of very ordered microscopic parts

crystallisation separation of a solution by evaporating the solvent, leaving behind solute crystals

cultivation preparing and using land for crops or gardening

cultural significance importance to a particular culture

current the overall movement of charge in one direction

cytoplasm a jelly-like fluid in which the other parts of a cell sit

decanting carefully pouring the liquid from a mixture, leaving the sediment behind

decomposer an organism that gains energy by breaking down dead material

decomposition the process of rotting and decay

deforestation the removal of trees to make land suitable for other uses

deform change shape

density the amount of matter in a certain volume

dependent variable the thing that will be measured and is altered by the independent variable

deposition a process in which sediment is left in a new place

dichotomous divided into two parts

differentiate change to have a particular function

differentiated cell a cell that has specialised functions

digestion the physical and chemical processes that break down food in the body

diode an electrical component that allows electric current to flow in only one direction

disinfection destroying bacteria, often using special light or chlorine

dissolve when a substance is mixed into another substance (usually a liquid), and the molecules stay mixed together without separating

distillation using heating to separate liquids with different boiling points

eclipse the blocking of the Sun's light from Earth

ecosystem a system of living things and their environment

efficient reduces or avoids waste

electric circuit a path along which electrons flow

electrolysis passing electricity through a substance to break it up

electromagnet a device that uses electricity to make a magnetic field

electron a negatively charged sub-atomic particle found around the outside of the nucleus of an atom

electrostatic charge the unmoving electric charge on the surface of an object

electrostatic force a non-contact force between any objects with an electric charge

element a pure substance made up of only one type of atom

embryo an early stage of life where the parts of an unborn organism are still being developed

energy the ability to do physical things such as move or change

enzyme a chemical that speeds up a reaction (e.g. digestion of food)

epidermis the outer layer of cells

erosion a process in which sediments are moved from one place to another

ethical relating to principles about what people think is 'wrong' and 'right'

eukaryote an organism with a nucleus and special structures inside its cell(s)

evaporation changing from a liquid to a gas

evidence facts and observations that can be used to support or oppose a theory

excretion the elimination of cellular waste from the body

expand get bigger, increase in size

experiment an investigation carried out under controlled conditions, to test a hypothesis

exploration processes undertaken to find rocks that contain minerals

extrusive igneous rock igneous rock formed at Earth's surface

equinox the two times each year when night and day are about the same length

fair test an investigation in which only one factor is changed and all other variables are kept the same

fertilise join a sperm with an ovum

field an area of space where objects are affected by a non-contact force

field line a line used to show the direction of a force within a field

fieldwork an investigation conducted in the natural environment, not a laboratory

filtrate the liquid that passes through a filter

foetus a later stage of development of an unborn baby

food chain a path of energy through an ecosystem

food web a system of interlocking food chains

force a push, pull or twist on an object when it interacts with another object

fossil preserved remains or traces of once-living things

fossil fuel a natural fuel formed over millions of years from the remains of living things

fossilisation the process of the formation of a fossil

fractional distillation a method that separates liquids because of their different boiling points

friction a force opposing the motion of surfaces in contact

froth flotation a method that uses special chemicals to separate minerals from their ores

fulcrum the point on which a lever turns when moving an object

galaxy a large system of stars

geological history how Earth has changed over time

germinate grow and put out shoots

glucose a type of sugar that is the energy source for cells

gravitational force the force that attracts physical objects with mass towards each other

gravity a force that attracts objects towards each other

grounding providing a pathway to conduct electricity down to the ground

groundwater water within rocks in the ground

gymnosperm a plant that produces seeds in cones

habitat the place where an animal or plant naturally lives

hazard something that can harm living things, objects or the environment

heat a type of energy

heating flame the blue (very hot) flame of a Bunsen burner (approx.1500°C), used for heating substances

herbivore an organism that eats only plants

hormone a chemical substance produced by the body that controls the activity of certain cells or organs

humane compassionate and sympathetic

hydrophilic water-attracting

hydrophobic water-repelling

hydrosphere all the water on Earth's surface

hygiene ways of doing things that support health and prevent disease, often through cleanliness

hypothesis a scientific statement that can be tested

igneous rock rock made by the cooling of molten rock

immiscible not able to be mixed

immune resistant to a particular illness or disease

impact the effect of a force

inclined tilted up at an angle from horizontal

independent variable the thing that is purposely changed during an investigation

inference an educated guess or judgement based on observations

insoluble something that does not dissolve

insulation material that acts as a barrier to heat flow

insulator a material that is poor at transferring heat or electricity

intensity how much heat is released from a fire front

interact act on each other

intrusive igneous rock igneous rock formed under Earth's surface

invertebrate organism without a backbone or spinal cord

irrigation supplying water to land or crops

joule a unit of energy for kinetic and potential energy

key a system for identifying characteristics

landform a natural feature of Earth's surface

lattice a three-dimensional shape like a cube made up of lots of atoms

lava molten (melted) rock at Earth's surface

law of conservation of energy a law stating that energy cannot be created or destroyed

lever a bar acted upon at different points by two forces

like poles magnetic poles that are the same (both north or both south); they repel each other

limitation a factor that limits or restricts

lithosphere Earth's rigid outer zone (crust and upper mantle), made up of tectonic plates

load any part in a circuit that receives power

lubrication a substance that makes a surface slippery or smooth

made resource a resource that is manufactured from natural resources

magma molten (melted) rock under Earth's surface

magnetic force a non-contact force that affects any object made of certain metals, such as iron

magnetic pole one of the two ends of a magnet

magnitude the size or power of an object, energy or force

malignant cells dividing uncontrollably and perhaps spreading to other parts of the body

malleable able to be bent and shaped

mantle Earth's middle layer, made up of an upper mantle and a lower mantle

mass the amount of matter in an object

matter particles that make up all physical substances; they have mass and take up space

meiosis complex cell division, where new cells are not identical to the original cell

melting point the temperature when something changes from a solid to a liquid

metabolism the chemical reactions that happen in the body

metamorphic rock rock formed by another rock that has been changed by heat and pressure

metamorphism the process of change that happens to a rock because of heat, pressure or both

microorganism an organism only visible under a microscope

microscope an instrument used to look at objects too small to see with the naked eye

mineral a naturally occurring inorganic (non-living) substance

mineral deposit rocks that contain a particular mineral

mineral ore a rock that contains useful metals

mitochondria the organelles where respiration happens

mitosis simple cell division, where new cells are identical to the original cell

mixture substances mixed together that can be physically separated

model a simplified way of explaining something complex and real based on evidence

molecule a group of atoms chemically bonded together

monoculture the practice of cultivating a single crop in a given area

motion the change in position of an object over time

mould a hollow impression formed by an imprint of an organism, or when the original bone or shell has dissolved

nanotechnology the field of science especially dealing with individual atoms and molecules

naturalist someone who studies nature and its history

natural resource a resource that is valuable in its natural form

net force the sum of all forces acting on an object

neutron a particle without charge in the nucleus of an atom

Nobel Prizes world-famous awards given each year for academic, cultural and scientific advances

non-biodegradable does not break down in the environment

non-renewable resource a resource that can run out, or one that takes longer than a human life span to be restored

nuclear medicine the branch of medicine to do with the use of radioactive substances in research, diagnosis and treatment

nucleus the control centre of a cell; DNA is found inside the nucleus

observation something you see and know to be true

ochre a mixture of iron oxide, clay and dirt, used as a natural paint

oil slick a layer of oil on the surface of water

omnivore an organism that eats both plants and meat

orbit the curved path a smaller object takes around another object

ore body a mineral deposit that is profitable to mine

organ a group of tissues with a specific function

organelle a cell structure that has a membrane around it (usually)

organic sedimentary rock sedimentary rock formed from the remains of plants or animals

organism an individual animal, plant or other living thing

palaeontologist a scientist who studies fossils

paper chromatography a technique used to separate coloured substances using a strip of paper and a solvent

parasite an organism that lives in or on another organism, causing it harm

particles a very small amount of matter

particle model a model used to describe the properties of solids, liquids and gases

passive design building design that considers climate and location to make the best use of natural heating and cooling

pathogen a microorganism that can cause disease

penumbra the outer part of the Moon's shadow on Earth

periodic table a table of all known elements and their chemical symbols

pesticide chemicals used on farms to protect crops by killing pests

phenomenon an observed event

photosynthesis the chemical reaction, powered by sunlight, that plants use to change carbon dioxide and water into sugars and oxygen

physical change a change in appearance with no new substance formed

physical property a characteristic that can be seen or measured (e.g. colour, shape, hardness)

pistil the female reproductive organs of a flower (anther and filament)

plankton microscopic, animal-like organisms that float or drift in water

polarise develop parts with positive and negative charges

polarity the direction of a force, such as a magnetic force

pollen the fine, powdery substance in the flowers of plants, which contains male sex cells

pollination the movement of pollen from the male part of the plant (anther) to the female part (stigma)

polymer a substance consisting of large molecules made up of many small repeating units

population all the living things of one species in a particular area

precipitation water falling from clouds as a solid (e.g. hail) or liquid (e.g. rain)

prediction a statement about the future based on observation and evidence

primary data first-hand data, from your own investigation

producer an organism that produces energy at the start of a food chain

product a substance formed during a chemical reaction

prokaryote an organism without a nucleus or structures inside its cell(s)

property characteristic or attribute of a substance

properties what a substance looks like and how it behaves

protist a microorganism, usually unicellular, that is not a fungus or bacteria – such as algae

proton a positively charged particle in the nucleus of an atom

protozoan a unicellular eukaryote that moves and feeds on organic matter

pulley a wheel with a grooved rim for carrying a cable

qualitative written descriptions and observations

quantitative numerical information and data

reactant a substance that takes part in a chemical reaction

recycle change something into something else that is useful

regional metamorphism the process of change that happens to a rock over large areas

rehabilitation processes that return the environment to close to how it was before mining

relative age which rocks are younger and which rocks are older when compared with one another

reliable provides consistent results when repeated

renewable resource a resource that cannot run out, or can be restored in a human life span

repulsive force a force that pushes objects away from each other

research gather data and information in an organised way to inform a hypothesis or investigation

residue the solid that does not pass through a filter

resistance a measure of how difficult it is for electricity to flow through an object

resource a source of something that is useful

respiration a chemical reaction that converts glucose to energy

reversible can be taken back to its previous state

revolve move in a circular path around another object

risk the chance that a hazard will cause harm

rotate spin on an axis

runoff water that runs off the surface of the land into waterways such as rivers

safety flame the orange (cooler) flame of a Bunsen burner (approx.300°C), used between heating substances sanitation facilities and services to ensure safe treatment of human waste (urine and faeces)

scientific model an idea used to explain and predict occurrences that we can't directly interact with

secondary data second-hand data, from someone else

sediment the solid that settles to the bottom of a liquid

sedimentary rock rock formed by sediments that have been pressed together

seismic wave a wave of energy caused by an earthquake or explosion

selective breeding breeding organisms with desirable traits

sequence the order of something

sewage semi-liquid human waste

sewerage pipes that carry sewage

smog a form of severe air pollution

solar energy energy made by the Sun

solidify become a solid

solstice the two times each year when night and day are the most different in length

soluble something that dissolves

solute a substance that is dissolved by a solvent

solution a mixture made up of a solvent and a dissolved solute

solvent a substance that dissolves a solute

species a single, specific type of living organism

spore a tiny part of some plants that is used to reproduce

stamen the male reproductive organs of a flower (anther and filament)

static friction a friction force that keeps an object in place on a surface

stationary not moving

stem cell a cell that can produce a different type of cell and divide to make more cells

stomata pores in the surface of a leaf; the site of gas exchange in plants

substance matter that has a fixed chemical make-up

sustainable able to be maintained at a certain rate or level

symbiont an organism that lives with a host organism, and both organisms benefit

system a set of simple things that work together as a more complex whole

taxonomy the area of science to do with classifying organisms

tectonic plate a plate made up of the mantle and crust of the lithosphere

temperature the measurement of how hot a substance is

temperature inversion weather conditions that stop fresh air from circulating

tendon connective tissue that connects muscle to bone

theory an explanation that can be supported or disproved using evidence

thermal relating to heat

tilt a sloping position or lean

tissue a group of cells with a similar structure and function

tracheophyte a plant that contains vascular tissue

transmissible able to be passed from one person to another

transpiration how water in the soil travels through plants

transfer move from one place or object to another

transform change from one type to another

tumour a swelling caused by an abnormal growth of tissue

umbra the inner part of the Moon's shadow on Earth

unicellular made of one cell

unlike poles magnetic poles that are different (south-north or north-south); they attract each other

urbanisation the creation of urban areas such as cities

valid measures what is intended to be measured

vascular tissue plant tissue that transports fluid and nutrients

vertebrate organism with a backbone or spinal cord

virus a tiny infectious agent that multiplies in its host

volume the amount of space an object takes up

waterborne carried or existing in water

water vapour the gas state of water

weathering a process in which rocks are worn down into smaller particles

weight the force of a gravitational field on the mass of a body

zygote the first single cell of new life

INDEX

ACKNOWLEDGEMENTS

The author and publisher are grateful to the following for permission to reproduce copyright material:

PHOTOGRAPHS

AAP/AP, **35** (bottom), Dean Lewins, **132**; Alamy Stock Photo /Arco/TUNS, **202**, /Martin Bond, **119** (bottom), /Byvalet, **12**, /John Cancalosi, **325** (bottom), /Daniel Cole, **295**, /Susan E. Degginger, **319** (bottom right), /David R. Frazier Photolibrary, Inc., **8** (bottom)Budimir Jevtic, **87**, /Dorling Kindersley, **211** (top), /Geoz, **319** (bottom left), **325** (top right), /GL Archive, **189** (bottom), /Martin Harvey/Avalon/Photoshot License, **50**, /Ralph Lee Hopkins/National Geographic Image Collection, **4** (bottom), **16** (top), /Jacek laskowski, **82**, /Stefan Jannides /redbrickstock.com , **113**, /Kateryna Kon/Science Photo Library, **67**, **226** (bottom centre), **232** (centre), /Ekaterina Kriminskaya, **78** (left), **94** (bottom centre)Chen Leopold , **174** (right), /Colin Monteath/Hedgehog House/Minden Pictures, **94** (top centre right), Cigdem Simsek, **73** (top), /Gerry Pearce , **266** (centre right), **278** (top centre right), /Martin Shields, **154** (centre), **283**, /Molekuul/Science Photo Library, **28** (top), /NG Images , **57** (bottom), /NOAA, **51** (bottom), /Paulo Oliveira , **265** (bottom), /Panther Media gmbh, **83** (schist), /Sean Pavone, **56** (top), /Siim Sepp, **319** (top)Pictorial Press, **32** (JJ Thomas), /Pjrstudio, **325** (top left), /Pulsar Imagens, **272**, **278** (bottom centre right), /Scenics & Science, xiii–xi (banner), **80**, **217** (top), **280–1**, **380–1**, /Science History Images , **175** (bottom), **254**, /Sciencephotos, **181**, /Sciepro/Science Photo Library, **239** (bottom left), /Ragnar Th Sigurdsson/ARCTIC IMAGES, **72** (right), /Steve Bloom Images, **199** (bottom), /Robbreece/room the Agency , **3** (top), /Fabrizio Troiani/age fotostock, **78** (bottom), **94** (bottom left), **144** (left), /US Coast Guard Photo, **35** (top), /Vintagecorner, **19** (bottom), /Valery Voenny, **83** (slate, shale), **78** (top), **94** (bottom right, shale), **321**, /Taras Vyshnya, **90**, **324**, /Andrew Watson/John Warburton-Lee Photography, **31**, /Andreas Werth/imagebroker, **84** (left), /Björn Wylezich, **76** (right), **94** (jade), /Tong Yu, **163** (LED), /zev rad/bibleland, **30**, **31** (top), /ZUMA Press, Inc, **196** (centre); Image courtesy of ANSTO, **68**; Antarctic Photo Library /Keith Vanderlinde/NSF, **67** (bottom); Farm House by Archterra Architects/Photography by Douglas Mark Black, **193**; CERN /Maximilien Brice, **23** (bottom); City of Sydney/Turf Design Studio/Ethan Rohloff Photography, **131**; COLAS/Robert Brice, **111** (top); CSIRO, **69** (left), /Andrea Wild (CC BY 3.0), **275** (bottom); Fotolia.com/Georgios Kollidas, **32** (J Dalton); Getty Images /Auscape/Universal Images Group, **274**, /Galerie Bilderwelt /Hulton Archive, **57** (top), /Dr Jeremy Burgess/Science Photo Library, **243** (left), /by Ruhey/Moment, **174** (top), /Matthew Cavanaugh for The Washington Post, **258**, **262** (bottom left), /Cavan Images, **62**, **70** (centre), /China Photos, **240** (bottom), /Jeffrey Coolidge/Stone, **164** (top), /Steve Gschmeissner/Brand X Pictures/Science Photo Library, **271**, **278** (bottom centre left), /Harry How, **137** (bottom), /jack0m/DigitalVision Vectors, **29** (centre), **32** (NaCL), /Jeffry Hamilton/DigitalVision, **20** (top), **33** (top), /Miriam Maslo/Brand X Pictures/Science Photo Library, **253** (bottom), /Photos.com, **105** (bottom), /paranyu pithayarungsarit/Moment, **155** (bottom), /Science Photo Library, **96** (centre left), **98**, /Sciepro/Science Photo Library, **230** (bottom right), /John Sirlin/EyeEm, **167**, **172** (bottom centre), /STR/AFP, **97** (top), /Gillianne Tedder, **92**, **94** (centre left), /Thomas Trutschel/Photothek, **43** (bottom), /VCG, **189** (top), **196** (bottom), /Taylor Weidman/Bloomberg, **120**, /William West/AFP, **165**, **172** (centre), /Olivia ZZ/Moment, **9** (bottom); ImageSource, **249** (top right); iStockphoto/2199_de, **161**, /4x6, **174** (left), **176** (bottom left, bottom right), /abadonian, **93**, **203** (bottom left), /adventtr, **72** (top), **229** (top), /AGorohov, **219** (right), /Ales-A, **136** (bottom), /allFOOD, **54** (centre left), /aluxum, **28** (bottom), /alvarez, **136** (right), **203** (top left), **214** (top right), /Andrey_Kuzmin, **311**, /Antagain, **266** (centre left, left), **278** (top centre left, top left), /AnthiaCumming, **255** (top), **262** (top left), /AOosthuizen, **264** (centre left), /ArtbyBart, **52**, /artisticco, **128–9**, **134** (top), /avid_creative, **176** (top right), **186** (top right), /AzmanL, **198** (right), /BanksPhotos, **154** (bottom), /Barcin, **118**, /bdspn, **48**, /Bigpra, **130**, /Bill Oxford, **186** (centre left), /BlackJack3D, **iv** (bottom), **45**, **305**, /Bouillante, **269** (bottom), /brandtbolding, **216** (centre), /brittak, **234** (centre), /Michael Burrell, **317** (left), **43** (top), /busypix, **297**, /Byjeng, /CarolinaSmith, **96** (top), /carroteater, **211** (bottom), /CBCK-Christine, **21**, /cbpix, **20** (bottom), **269** (top), **278** (centre left), /ChaoticMind75, **2** (left), /Chelsea Sampson, **234** (right), /Chillim, **339**, /choja, **47** (bottom), **54** (centre), /choness, **29** (bottom), **32** (salt), /ChristinLola, **2** (right), /ClaraNila, **209** (bottom), /Claudiad, **262** (centre bottom left), /ClaudioVentrella, **332–3**, **389**, /colematt, **239** (top left), **250** (centre), /Dafinchi, **76** (left), **94** (pyrite), /Daft_Lion_Studio, **184** (bottom), /Daniiielc, **112** (top), **122** (centre right), /danyckar, **18** (top left), /Darryl Peroni, **122** (centre left), /Steve Debenport, **285** (right), /defun, **207**, /Devonyu, **294** (bottom), **348** (top left), /DieterMeyrl, **239** (top right), /Django, **288**, /DonNichols, **264** (centre right), **278** (centre), **312**, **348** (bottom left), /Wenjie Dong, **119** (top), **122** (bottom), /dottedhippo, iv (top), v, **72** (left), /dra_schwartz, **213** (top), /eestingnef, **29** (top left), /ElementalImaging, **16** (centre), /fcafotodigital, **5** (top), /Fertnig, **252** (top), /Filip_Krstic, **198** (centre), /Floortje, **371**, /Fly_dragonfly, **47** (top), /FotoDuets, **60** (top), /fotosr, **277**, /[illegible], **71** (bottom), /Fylatto, **110** (bottom), /[illegible], **3** (bottom), /[illegible], **348**, /[illegible], **130**, /[illegible], **200** (top right), **207** (top), **214** (centre top), /g-miner, **18** (bottom left), /Grahamphoto23, **110** (bottom left), /gremlin, **136** (top), /hdagli, **114**, /hdere, **365** (right), /Henrik_l, **268**, /Highwaystarz-Photography, **180** (bottom), /hometowncd, **69** (right), /HRAUN, **273**, /iagodina, **203** (centre right), /idizimage, **iv–vii** (banner), /Ilya_Starikov, **294** (top right), /imamember, **96** (bottom right), **100** , /industryview, **37** (bottom), /Inhauscreative, **168**, /ISvyatkovsky, **3** (bottom), /Ithinksky, **203** (bottom right), **267** (top), /IvancoVlad, **159**, **172** (top left), /Ja_Het, **316**, /jacus, **61** (bottom), **70** (centre right), /janrysavy, **154** (left), **328**, /jarun011, **255** (bottom), **262** (top right), /jessicaphoto, **104** (bottom), /jgroup, **105** (top), /joggiebotma, **164** (right), /JohnWNixon, **216** (right), /joingate, **285** (left), /JoKMedia, **294** (bottom right), /Jonah_Photos, **264** (bottom right), /jonya, **335**, /JuSun, **216**, (top left), **226** (bottom left), /jxfzsy, **83** (gneiss), /kaanates, **203** (bottom centre), **214** (bottom centre left), /kazakovmaksim, **243** (right), /kckate16, **186** (centre), /Kenshi991, **241**, /Kerrick, **278** (centre right), /kimberrywood, **240** (top), /Kirillm, **77** (right), **94** (emerald), /Kitsada Wetchasart, **186** (bottom right), **191**, /kokkai, **257**, **262** (centre right), /koto_feja, **172** (bottom right), /KristyKolloy, **7**, /Kuzmik_A, **34** (left), **38**, /kyoshino, **294** (left), **348** (bottom centre), /leezsnow, **117** (top), **122** (top), /levkr, **226** (right), /lila-es, **293** (centre), /lostinbids, **124** (centre left),

/lovleah, **267** (bottom), /Luisrftc, **293** (right), /Magnilion, **188** (left), /maki_shmaki, **174** (bottom), /malerapaso, **110** (top), **146**, **154** (top), /Aldo Manganaro, **56** (right), /MarcelStrelow, **223**, /Loren Mariani, **111** (bottom), /MarkMirror, **264** (top), /Medesulda, **256**, /mayakova, **306**, /MicrovOne, **10** (top), **16** (bottom), /MikeLaptev, **2** (centre), /Milkos, **252** (bottom), /minemero, **46**, /MJ_Prototype, **59** (right), **70** (top right), /mladn**61**, **176** (top left), **186** (top left), /mmz, **152** (bottom right), /MollyBrownNZ, **188** (right), /monkeybusinessimages, **218** (right), /Moof, **36**, /msk.nina, **163** (wire), /Mark Murphy, **290** (scalpel), /nano, **214** (bottom left), /Naypong, **252** (right), /NeilLockhart, **348** (bottom right), /Nerthuz, **252** (left), /NNehring, **216** (bottom left), **221** (top), /npix, **18** (bottom centre), **32** (top left), **33** (bottom), /Okea, **37** (top), **54** (top left), **343**, /olga_sweet, **34** (right), /ooyoo, **116**, **122** (bottom right), /ordachelr, **18** (bottom right), /ourlifelooklikeballoon, **290** (mortar and pestle, scale), /Chansom Pantip, **60** (bottom), **70** (centre left), /PeopleImages, **18** (top centre), /PeterHermesFurian, **219** (left), /PeterJamesSampson, **206**, /petesphotography, **125** (top), /Pgiam, **354**, /PhillipMinnis, **124** (bottom left), /photovs, **110** (centre right), /PicturePartners, **64**, **365** (top), /pixologicstudio, **244**, /pjmorley, **96** (centre right), **99**, **108** (bottom), /Popartic, **190**, **196** (top), /Todd Powell, **110** (centre left), /pterwort, **248**, /quentinjlang, **115**, /Rainer von Brandis, **110** (bottom right), /RealCreation, **235** (bottom), /Remains, **262** (centre top left), /rep0rter, **77** (left), **94** (garnet), **172** (centre right), /Tetiana Rostopira, **112** (bottom), /Rost-9D, **270** (left), /rusm, **226** (top left), /Chris Ryan, **140** (bottom left, bottom right), /sandsun, **186** (centre right), /sasha85ru, **172** (bottom left), /SebastianGauert, **163**, /SerrNovik, **171**, /shayes17, **75**, /sherjaca, **124** (centre right), /sidewaysdesign, **53**, /simonkr, **261** (bottom), /skipro101, **275** (top), /Slonov, **245** (top), **250** (bottom), /Smoczyslaw, **293** (left), /Spiderstock, **350**, /spukkato, **44**, /ssucsy, **4** (top), /stockcam, **136** (left), **152** (top left), /Suzifoo, **51** (top), **54** (top right), /t_kimura, **39**, **54** (bottom left), **163** (lamp), /Tatiana Pankova, **290** (thermometer), /TheBusman, **29** (top right), /Thomas–Soellner, **9** (top), /ThomasVogel, **294** (centre), /tiler84, **61** (top), /Wesley Tolhurst, **266** (right), **278** (top right), /TomekD76, **56** (left), **58** (top left, top right), /tomispin, **196** (right), /TonlStl, **284** (bottom left), /Tuomas_Lehtinen, **163** (battery), /Turnervisual, **266** (centre), **278** (top centre), /ugurhan, **217** (bottom), /UroshPetrovic, **378**, /valterdias, **356–7**, /vavlt, **230**, **232** (bottom right), /VIDOK, **373**, /Viktor_Kitaykin, **210**, **214** (bottom right), /volschenkh, **289**, /VvoeVale, **83** (phyllite), **321**, /WachiraS, **170**, /Wavebreakmedia, **148**, /Wittayayut, **157**, /Noppasin Wongchum, **209** (top), /wundervisuals, **188** (bottom), **196** (left), /xxmmxx, **369**, /yakochun, **291** (bottom), /yodiyim, **239** (bottom centre left), /yodiyim, **261** (top), **262** (bottom right), /zetter, **117** (bottom); V.E. Vasquez, D.A. Ebert, & D.J. Long. **2015**. Etmopterus benchleyi n. sp., a new lanternshark (Squaliformes: Etmopteridae) from the central eastern Pacific Ocean. Journal of the Ocean Science Foundation 17: 43–55, **201;** Patrick J. Krug, **199** (top); NASA, **106–108;** NASA image by Robert Simmon and Marit Jentoft-Nilsen, based on MODIS data, **126;** NASA/JPL-Caltech/MSSS and PSI, **91;** Nature Picture Library/MYN/Niall Benvie, **198**, **204–5**, **208** (bottom), **214** (bottom centre right); Newspix/Amos Aikman, **133**, **134** (bottom); October Films Ltd, **235** (top); Reuters/Albert Gea, **253** (top); Science Photo Library, **5** (bottom), /ISM, **218** (left), /Lawrence Lawry, **155** (top), /Sheila Terry, **26** (bottom), **65**, **70** (bottom), /Geoff Tompkinson, **49**, **54** (bottom centre), /Wim Van Egmond, **270** (right), **248** (bottom left), /ZEPHYR, **150**; Science Source, **23** (top), /GIPhotoStock, **8** (top), /Lawrence Migdale, **19** (top), /Jerome Pickett-Heaps, **225**; Shutterstock.com/3dmitruk, **138**, **152** (top centre), /6493866629, **144** (right), /Akimov Igor, **282**, /AlexLMX, **180** (top), **186** (bottom left), /Anatomy Insider, **245** (bottom), /ANGHI, **208** (top), /Ann.and.Pen, **200**, /Art2ur, **58–59**, **317** (right), /Benoist, **137** (top), /Yasemin Olgunoz Berber, **175** (top), /BlueRingMedia, **242**, **250** (top), /Olga Bolbot, **249** (bottom, top left), **251**, /Chesky, **192**, /fen deneyim, **162** (left, right), /Dja65, **10** (bottom), /donatas1205, **vi** (left), /Dirk Ercken, **124** (bottom right), /Fablok, **316**, /flight of imagination, **163** (closed switch, open switch), /Shawn Hempel, **259** (bottom), /Humdan, **24** (top), **32** (bottom), /Rosa Jay, **vi** (right), /Daniel Jedzura, **42**, **54** (bottom right), /journeyman777, **56** (bottom), /Kateryna Kon, **212**, /Kamenetskiy Konstantin, **140** (top left, top right) , **152** (top right), /Winston Link, **291** (top), /Lipskiy, **163** (cell), /luca pbl, **276**, **278** (bottom right), /Macrovector, **229** (bottom), /Magic mine, **239** (bottom centre right), /Gabriele Maltinti, **141**, /marktucan, **84** (right), /Marques, **326**, /metamorworks, **188** (top), /New Africa, **234** (left), /NosorogUA, **259** (top), /NsdPower, **348** (top right), /Nsit, **290** (glasses), /OBprod, **344**, /Oleksandrum, **96** (bottom left), /Orhan Cam, **265** (top), /Pat_Hastings, **13**, **296**, /Pattikky, **203** (centre left), **214** (centre bottom), /Peefay, **224**, **232** (centre left), /ribeiroantonio, **89** (top), /Rich Pixel Photography, **81**, /RomanVX, **89** (bottom), /RTimages, **15**, /Vladimir Sazonov, **6**, **70** (top left), /sciencepics, **228**, **232** (bottom left), /Georgy Shafeev, **66**, /Shamzami, **89** (right), **94** (centre right), /ShutterStockStudio, **14**, /Soleil Nordic, **290** (forceps, stand, clamps), /Spasiblo, **145** (top left, top right), **152** (bottom left), /stockshoppe, **246–7**, /Rattiya Thongdumhyu, **220**, **232** (top left), /ValentinaKru, **104** (top), /Wave_Movies, **221** (bottom), **232** (top right), **386–7**, /WindVector, **260**; stock.adobe.com/Juulijs, **32** (E Rutherford); Stocksy United/Evgeniya Savina, **i**, **iii**; Up the Resolution, **365** (left); USAP/Josh Landis, NSF, **125** (bottom); Wellcome Collection/Fernán Federici (CC BY 4.0), **231**; Wildstone Arts/L. Sue Baugh/Lynn Martinelli © 2003, **73.**

OTHER MATERIAL

NSW Science K–10 Syllabus © Copyright 2012 NSW Education Standards Authority, **xiii–1**; **4**, **6**, **8**, **10**, **12**, **14**, **20**, **22**, **24**, **26**, **28**, **30**, **36**, **38**, **40**, **42**, **44**, **46**, **48**, **50**, **52**, **58**, **60**, **62**, **64**, **66**, **68**, **74**, **76**, **78**, **80**, **82**, **84**, **86**, **88**, **90**, **92**, **98**, **100**, **102**, **104**, **106**, **112**, **114**, **116**, **118**, **120**, **126**, **128**, **130**, **132**, **138**, **140**, **142**, **144**, **146**, **148**, **150**, **156**, **158**, **160**, **162**, **164**, **166**, **168**, **170**, **176**, **178**, **180**, **182**, **184**, **190**, **192**, **194**, **200**, **202**, **204**, **206**, **208**, **210**, **212**, **218**, **220**, **222**, **224**, **226**, **228**, **230**, **236**, **238**, **240**, **242**, **244**, **246**, **248**, **254**, **256**, **258**, **260**, **266**, **268**, **270**, **272**, **274**, **276**